FREE ELECTRON LASERS

Edited by **Sándor Varró**

Free Electron Lasers

Edited by Sándor Varró

Published by InTech

Janeza Trdine 9, 51000 Rijeka, Croatia

Publishing Process Manager Oliver Kurelic

Technical Editor Teodora Smiljanic

Cover Designer InTech Design Team

Additional hard copies can be obtained from orders@intechopen.com

Free Electron Lasers, Edited by Sándor Varró
p. cm.
ISBN 978-953-51-0279-3

For EU product safety concerns :
IN TECH d . o . o . , Prolaz Marije Krucifikse Kozulić 3 , 51000 Rijeka , Croatia ,
info@intechopen . com or visit our website at intechopen . com .

Meet the editor

Sándor Varró is a scientific adviser at the Wigner Research Centre for Physics of the Hungarian Academy of Sciences, in Budapest. In the field of nonlinear laser-matter interactions and in theoretical quantum optics he has made seminal contributions to many important realms of physics ranging from electrons in strong fields, high-harmonic generation and attosecond physics, via phase-space distributions and Wigner functions, quantum statistics of single surface plasmons, to a novel description of quantum noise and black-body radiation. His results on the quantum phase properties of squeezed states are widely recognized. Prof. Varró has published more than a hundred scientific papers in high-ranked international journals, and has authored many book chapters. As a guest scientist he has worked at several research sites in Europe and USA. He had a stay of two years as an Alexander von Humboldt Fellow at the Max-Planck-Institut für Quantenoptik, Garching, Germany. He received the Schmid Award of the Eötvös Lóránd Physical Society and the Jánossy Award in Hungary. Prof. Varró is a member of the Committee of Laser Physics of the Hungarian Academy of Sciences, and of the European Physical Society.

Contents

Preface

The idea of using periodic magnetic fields for properly bending the trajectories of relativistic electrons, and making them generate coherent tunable radiation, appeared as early as the middle of the last century. Further study of this subject, concerning both fundamentals and practice, received a new impetus at the beginning of the seventies. Since then, a huge amount of knowledge has been accumulated in this rapidly developing field of research, creating a new class of lasers, the *free electron lasers*.

This volume consists of a collection of current studies on free electron lasers. *Free Electron Lasers* covers the analysis of fundamentals and questions of design of these devices functioning from the infrared to ultraviolet or xuv wavelength regimes. In addition to the comparison with conventional lasers, analyses of some related topics concerning near-field and cavity electrodynamics, compact and table-top arrangements and strong radiation–induced exotic states of matter are presented. Digital control and diagnostics of such devices and radiation safety issues are also discussed.

Free Electron Lasers provides a selection of research results, from basics to some interesting applications and new ideas of current interest.

Sándor Varró
Wigner Research Centre for Physics, Hungarian Academy of Sciences,
WIGNER RCP, SZFI, Budapest,
Hungary

1

Introduction to the Physics of Free Electron Laser and Comparison with Conventional Laser Sources

G. Dattoli[1], M. Del Franco[1], M. Labat[1], P. L. Ottaviani[2] and S. Pagnutti[3]
[1]*ENEA, Sezione FISMAT, Centro Ricerche Frascati, Rome*
[2]*INFN Sezione di Bologna*
[3]*ENEA, Sezione FISMET, Centro Ricerche Bologna*
Italy

1. Introduction

The Free Electron Laser (FEL) can be considered a laser, even though the underlying emission process does not occur in an atomic or a molecular system, with population inversion, but in a relativistic electron beam, passing through the magnetic field of an undulator. Several parameters (such as gain, saturation intensity, etc.) are common to both devices and this allows a unified description. The origin of Laser traces back to the beginning of the last century, when Planck derived the spectral distribution of the radiation from a "blackbody" source, a device absorbing all the incident radiation and emitting with a spectrum which only depends on its temperature. A perfectly reflective cavity, with a little hole, can be considered a blackbody source; the radiated energy results from the standing wave or resonant modes of this cavity. The Planck distribution law, yielding the equilibrium between a radiation at a given frequency ω and matter at a given temperature *T*, is the following:

$$I(\omega)d\omega = \frac{\hbar\omega^3}{\pi^2 c^2\left[\exp\left(\frac{\hbar\omega}{kT}\right)-1\right]}d\omega \tag{1}$$

$$\hbar = \frac{\mathrm{h}}{2\pi},\ \omega = 2\pi\nu,\ \nu = frequency\ .$$

$I(\omega)d\omega$ is the energy per unit surface, per unit time, emitted in the frequency interval $\omega, \omega + d\omega$. Its fundamental nature can be stressed by noting that it crucially depends on three constants *h, k and c* the Planck, the Boltzmann constants and the light velocity, respectively. Its derivation required the assumption that the changes in energy are not continuous, but discrete. The Plank theory prompted Bohr to make a further use of the discretization (or better quantization) concept to explain the atom stability. He assumed that the electrons, in an atomic system, are constrained on a stationary orbit, whose energetic level is fixed and can only exchange quanta of radiation with the external environment. This point of view allowed Einstein to formulate the theory of spontaneous and stimulated emission, which paved the way to the laser concept. An atomic system with two states of energy E_1 and E_2 is shown in Fig. 1, 1 being the lower energy state.

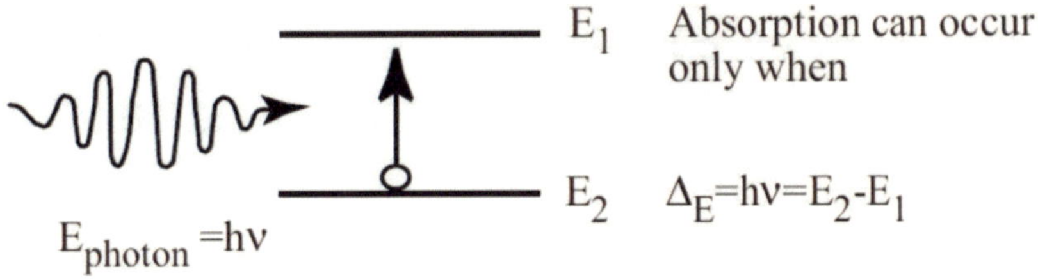

Fig. 1. Absorption process of a photon and consequent state excitation

The excitation of the system to the higher state (see Fig. 1) can result, with a probability $B_{1,2}$, from the absorption of a photon with an energy of:

$$\varepsilon = \hbar\omega, \; \omega = \frac{E_2 - E_1}{\hbar} \tag{2}$$

The principle of minimum energy requires that the system, in the excited state, decays to the lower state spontaneously, by emitting a photon of the same energy (see Fig. 2). The crucial step, made by Einstein, was the assumption that there is another process of decay, referred as stimulated, for which the transition to the lower state (see Fig. 3) occurs, with probability $B_{2,1}$, in the presence of a photon of energy equal to the energy difference between the two states as defined in Eq. (2).

We consider now a system of N atoms, among which N_1 are in the lower state and N_2 in the higher state ($N_1 + N_2 = N$), in interaction with an electromagnetic field of intensity $I(\omega)$.

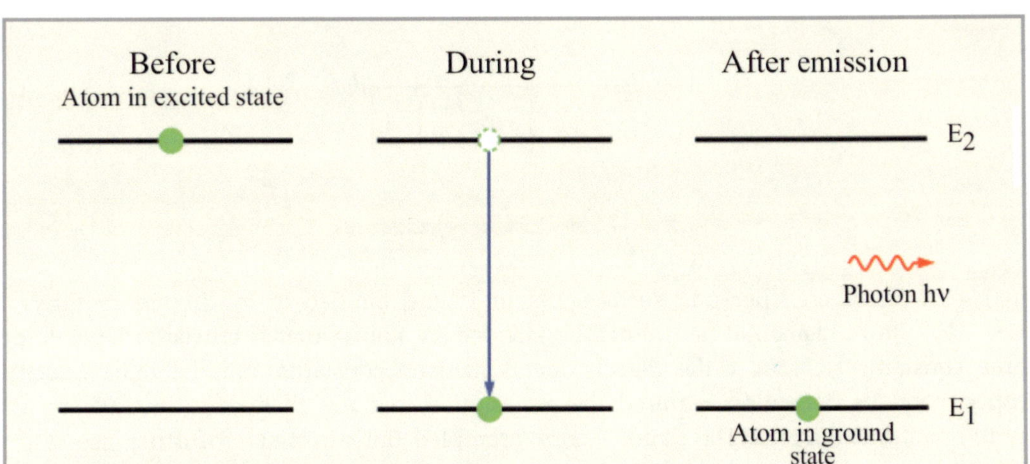

Fig. 2. Spontaneous emission process

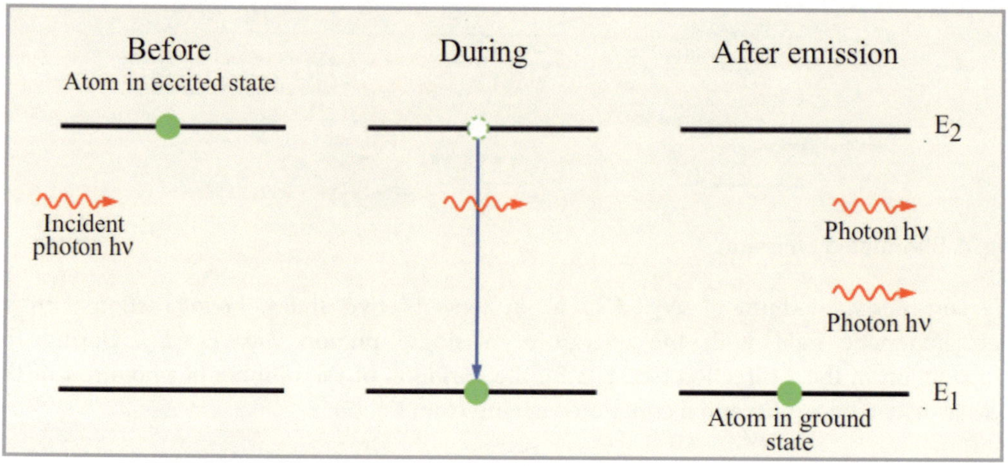

Fig. 3. Stimulated emission process

The equilibrium condition of the atomic system with the radiation is therefore given by:

$$N_1 B_{1,2} I(\omega) = N_2 B_{2,1} I(\omega) + N_2 A_{2,1} \tag{3}$$

where $A_{2,1}$ represents the spontaneous emission contribution, independent of $I(\omega)$. In the conditions of thermodynamic equilibrium at temperature *T*, the number of atoms per unit volume in a given state of energy E_i is described by the Boltzmann distribution law: $N_i \propto e^{-iE_i/KT}$. From this relation, one can derive the ratio $\frac{N_1}{N_2}$ in terms of energy difference and absolute temperature, namely $\frac{N_1}{N_2} = \exp\left(\frac{\hbar\omega}{kT}\right)$, which combining with Eq. (3), yields:

$$I(\omega) = \frac{A_{2,1}}{B_{1,2} exp(\frac{\hbar\omega}{kT}) - B_{2,1}} \tag{4}$$

Assuming that the absorption and stimulated emission processes are perfectly symmetric, i.e. that $B_{m,n} = B_{n,m}$, and using the Wien law, the spontaneous emission can be written as:

$$A_{n,m} = \frac{\hbar\omega^3}{\pi^2 c^2} B_{m,n} \tag{5}$$

Eq. (4) provides therefore the Planck distribution law. The importance of this result stems from the fact the Planck distribution has been derived on the basis of an equilibrium condition and on the new concepts of spontaneous and stimulated emission. The main difference between the two processes is that, in the case of the spontaneous emission, the atoms decay randomly in time and space, generating photons uncorrelated in phase and direction; in the other, the atoms decay at the same time, generating "coherent" photons, with phases and directions corresponding to those of the incident photon. Given a certain number of atoms, all (or most of them) in the higher level, one can get a substantive amplification process from one single photon (see Fig. 4).

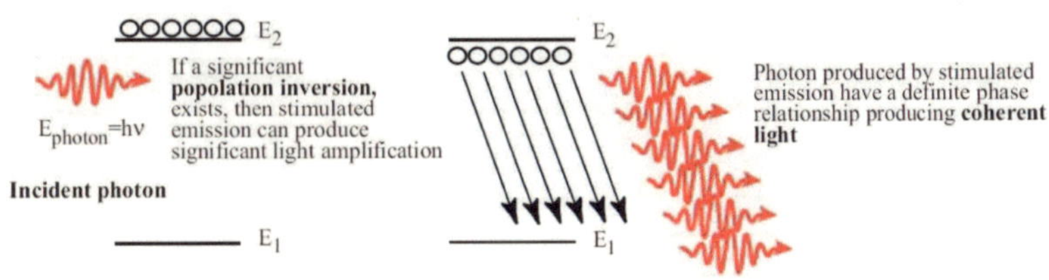

Fig. 4. Stimulated emission

We consider a medium of $N_1 + N_2 = N$ systems of two states, in interaction with an electromagnetic field with the energy of a single photon $\hbar\omega = E_2 - E_1$. During the propagation in the z direction (see Fig. 5), the variation of the number of photons n of the field is driven by a differential equation deriving from Eq. (3):

$$\frac{dn}{dz} = (N_2 - N_1)b_{1,2}n + a_{2,1}N_2 \tag{6}$$

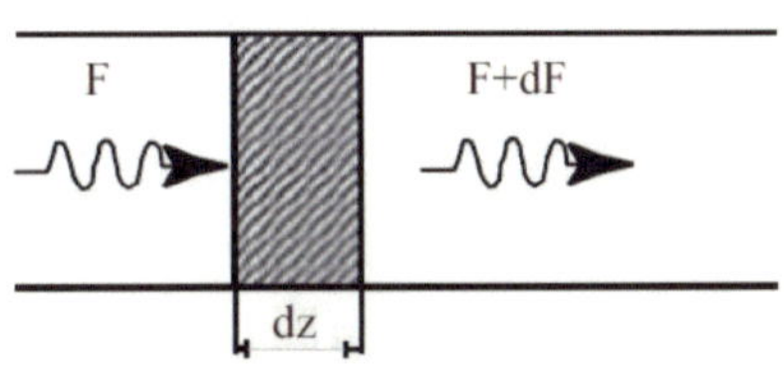

Fig. 5. Increment of the radiation

The coefficients of emission and absorption have been redefined to take into account the physical dimensions. The solution of Eq. (6) is:

$$n(z) = n_0 exp(\alpha z) + \beta\frac{exp(\alpha z) - 1}{\alpha}, \alpha = (N_2 - N_1)b_{2,1}, \beta = a_{1,2}N_2 \tag{7}$$

where n_0 is the initial number of photons of the field. Eq. (7) consists of two parts, and one (the spontaneous emission term) is independent of the initial number of photons. We consider the case in which the spontaneous emission contribution can be neglected. If $N_2 < N_1$, i.e. if there are less atoms in the excited than in the ground state, then $\alpha < 0$ and Eq. (7) becomes: $n(z) = n_0 exp(-|\alpha| z)$, corresponding to the Beer-Lambert law of absorption and the coefficient α can be interpreted as a coefficient of linear absorption. When $N_2 = N_1$, the number of photons remains constant. The amplification is possible if $N_2 > N_1$, i.e. when a population inversion has occurred (namely the system has been brought to the "excited" state through some mechanism). In this case, the coefficient α is understood as the small signal gain coefficient. The amplification can be seen as a chain reaction: a photon causes the decay of the excited state, creating a "clone", and, following the same process, both lead to the "cloning" of other two photons, and so on. Adding the spontaneous emission, the chain can be started without the need of an external field.

These few introductory remarks provide the elements of the light amplification, we will see how it can be implemented to realize a laser oscillator.

2. Laser oscillator

The radiation amplification, based on the stimulated emission mechanism, requires an active medium in which population inversion occurs and an environment which can provide a feedback to maintain the system in operation as an oscillator (see Fig. 6).

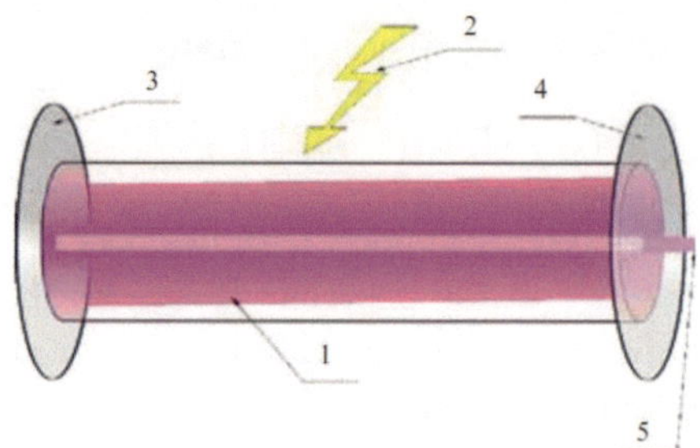

Fig. 6. Components of an oscillator laser: 1) Optical active medium, 2) Energy delivered to the optical medium, 3) Mirror, 4) Semi-reflective mirror, 5) Output laser beam

Without entering into the details of the population inversion mechanism, we can notice that it cannot be obtained simply by heating the active medium. Indeed, if we consider a two states system, the electrons obey the statistics of Fermi-Dirac and we have:

$$\frac{N_2}{N_1} = \frac{1}{\exp\left(\frac{\hbar\omega}{kT}\right)+1} \tag{8}$$

The population inversion process is not the result of a heating. At a temperature close to 0° K, the higher state is depopulated while at high temperature ($T \to +\infty$) $\frac{N_2}{N_1} = \frac{1}{2}$, i.e. the two states are equally populated. An extensively used mechanism for population inversion is the optical pumping: the active medium is slightly more complex than a two states system, which remains a convenient abstraction, but hardly actually feasible.

Since the spontaneous emission is essentially isotropic (see Fig. 7), a first direction selection can be operated by embedding the medium in an optical cavity, consisting, in its most simple configuration of two facing mirrors.

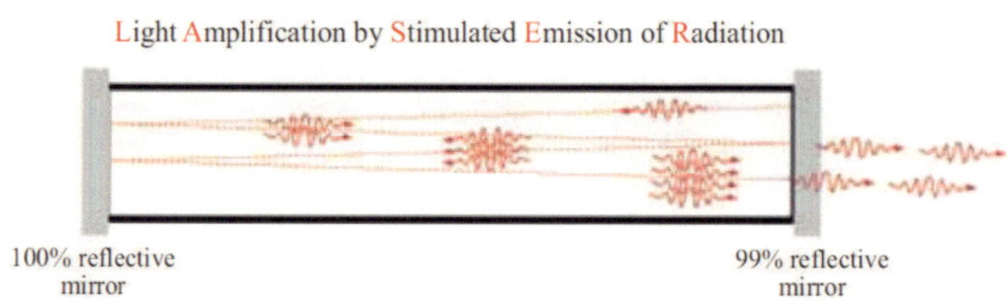

Fig. 7. Selection of the transverse mode in the optical cavity

The emitted photons result selected according to their direction, increasing the spatial coherence of the radiation at each pass in the cavity. Being assumed the existence of a gain, one expects the radiation intensity to increase and that, at each reflection on the semi-reflective mirror, a fraction of this radiation is extracted off the cavity. Since only a part of the radiation is reflected, the evolution of the system will strongly depend on the losses of the cavity and we can expect a saturation mechanism that will drive the system to an equilibrium state. As for other physical systems, the saturation mechanism is nonlinear and controlled by the radiation intensity itself. We can now write the rate equations, i.e. the evolution equations of the laser intensity inside the cavity:

$$I_{n+1} - I_n = [(1-\eta)G(I_n) - 1]I_n \tag{9}$$

n represents the number of round trips inside the optical cavity and η the cavity losses (we do not yet distinguish the active and the passive [1] losses, so that η are the total losses). In order to take into account the effects of saturation, we assume that the gain $G(I)$ is a decreasing function of the intensity, namely

$$G(I) = \frac{g}{1+\frac{I}{I_s}} \tag{10}$$

I_s, introduced in a phenomenological way, is a quantity of paramount importance representing the saturation intensity. For $I = I_s$, the gain corresponds to one half of the small signal gain g, which we no longer refer to as α to take into account the fact that the active medium is not a simple two sates system. Using the following approximation

$$I_{n+1} - I_n \cong T_R \frac{dI}{d\tau} \tag{11}$$

Equation (11) can be rewritten in terms of a differential equation:

$$\frac{dI}{d\tau} = [(1-\eta)G(I) - \eta]I. \tag{12}$$

τ is a dimensionless time related to the actual time t and to the length of the optical cavity L_c according to: $\tau = \frac{t}{T_R}$, $T_R = \frac{2L_c}{c}$, with T_R the duration of a round trip in the cavity. Transforming Eq. (12) into:

$$\frac{dX}{g[(1-\eta)\frac{1}{1+X} - r]X} = d\tau \tag{13}$$

$$X = \frac{I}{I_s}, r = \frac{\eta}{g}$$

it becomes obvious that for $X \ll 1$, i.e. far from saturation, the intensity evolves exponentially:

---

[1]The active losses correspond to the transmission of the radiation to the external environment, while the passive losses correspond to the absorption of the radiation by the medium and the cavity mirrors.

$$X = X_0 exp([(1-\eta)g - \eta]\tau) \quad (14)$$

It is also obvious that the equilibrium ($\frac{dX}{d\tau} = 0$) is reached when the intensity in the cavity is such that $G(I) = \frac{\eta}{1-\eta}$ and therefore

$$I_E = [\frac{1-\eta}{\eta} g - 1] I_s \quad (15)$$

Since in most cases $\eta << 1$, the extracted power at equilibrium (assuming that the losses are only active losses) is given by:

$$I_{out} = \eta I_E \cong g I_s \quad (16)$$

This relation is an additional evidence of the importance of the saturation intensity, which along with the gain defines the power which can be extracted from the cavity. The evolution of the laser power inside the cavity as a function of the number of round trips is reported in Fig. 8.

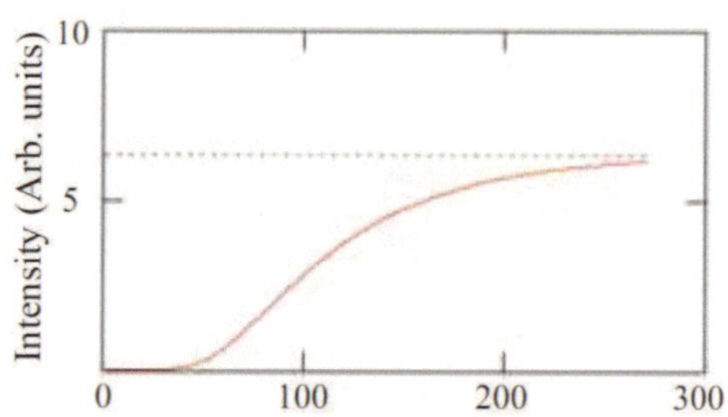

Fig. 8. Laser intensity (arb. unit) vs round trip number for g=15% and h=2%

The intensity first increases exponentially and then slows because of the saturation mechanism. When the gain equals the losses, the system reaches a stationary state and the power can be extracted off the cavity.

Despite not mentioning it before, the population inversion requires a sufficient level of power, which we refer to as pump, that will be further partially transformed into laser power. The ratio between the extracted laser power and the pump power represents the efficiency of the system.

We have seen that the optical cavity is an essential element of an oscillator, since it confines the electromagnetic radiation, selects the transverse and longitudinal modes, thus creating the spatial and temporal coherence of the radiation.

The radiation field inside the optical cavity, whose components have a small angle with respect to the cavity axis, is successively reflected by the mirrors. This enables to select the components travelling in a direction parallel to the cavity axis. The interfering of the waves in the cavity leads to the formation of stationary waves at given frequencies, depending on the distance between the mirrors. In the case of planar and parallel mirrors, separated by a distance L, the stationary condition for the wave frequencies is: $\nu_n = n\frac{c}{2L}$, n being integer. The various frequencies obtained, varying n, are referred as longitudinal modes of the cavity. For each longitudinal mode, various transverse sections are possible, which modify the transverse intensity distribution in the plans perpendicular to the optical axis. Those are referred as transverse modes. From a conceptual point of view, an optical cavity consists of two parallel mirrors (see Fig. 9).

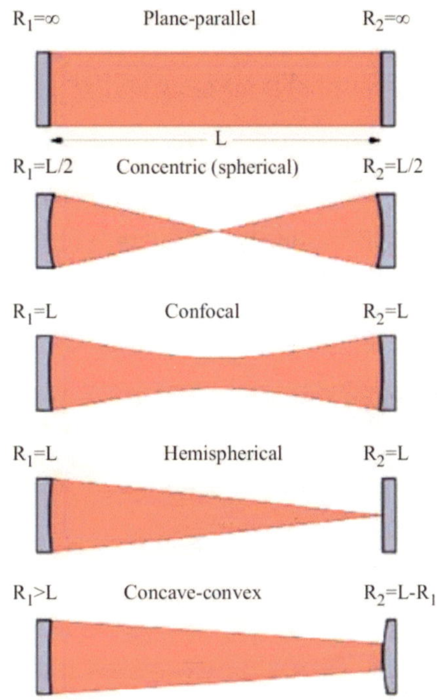

Fig. 9. Examples of optical cavities

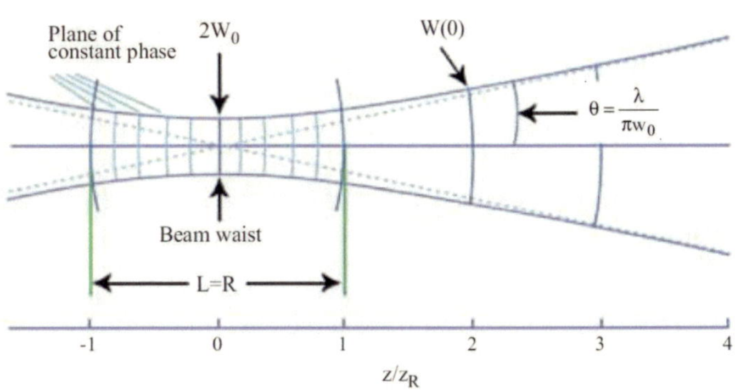

Fig. 10. Gauss-Hermite mode and main reference quantities

The mirrors configuration sets the field shape inside the cavity, as illustrated in Fig. 9. Two flat mirrors, i.e. with an infinite radius of curvature, lead to a constant transverse mode distribution. When the mirrors are confocal, the transverse mode has a parabolic longitudinal profile, which we will discuss more in details later. Additional configurations are shown in Fig. 9. In the case of a confocal cavity, the transverse modes are the Gauss-Hermite modes and, as previously mentioned, the longitudinal profile of the mode is parabolic (see Fig. 10).

The fundamental mode corresponds to a Gaussian distribution defined as:

$$I(r,z) = I_0\left(\frac{w_0}{w(z)}\right)^2 e^{-\frac{2r^2}{w(z)^2}}, \qquad (17)$$

where r is the distance to the axis, w is the rms spot size of the transverse mode:

$$w(z) = w_0[1+(\frac{z}{z_R})^2]\ ,\ w_0 = \sqrt{\frac{\lambda L}{2}}\ ,\ z_R = \frac{\pi w_0^2}{\lambda} \qquad (18)$$

w_0 is the beam waist, i.e. the minimum optical beam transverse dimension, z_R the Rayleigh length, corresponding to the distance for which $w(z) = \sqrt{2}w_0$. For the fundamental mode, $z_R = L/2$. The divergence of the mode is related to the former parameters according to:

$$\theta = \frac{\lambda}{\pi w_0^2} \qquad (19)$$

θ is referred as the angle of diffraction in the far field approximation.

The longitudinal modes are equally separated in frequency by Δf, which depends on the cavity length according to:

$$\Delta f = \frac{2c}{L} \qquad (20)$$

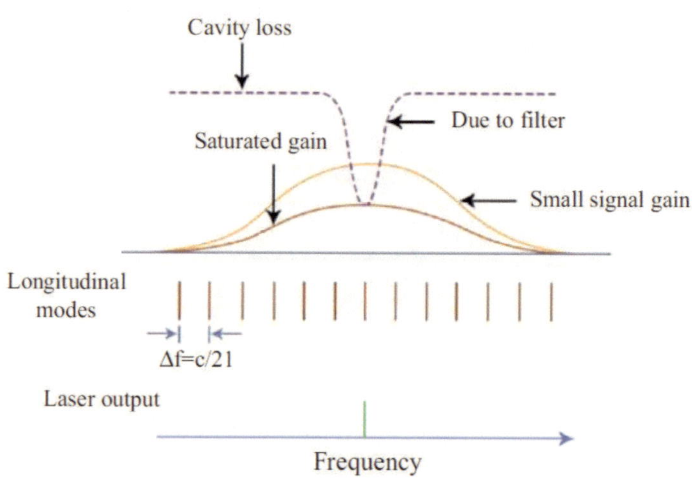

Fig. 11. Gain and loss spectra, longitudinal mode locations, and laser output for single mode laser operation

The gain bandwidth of the active medium performs the mode selection and therefore allows the amplification of the modes within this bandwidth only. Those modes oscillate independently with random phase and beating. A longitudinal mode-locking enables to fix the longitudinal modes phases so that they all oscillate with one given phase [2]. It seems obvious that, in theory, varying the losses and therefore the gain profile would allow to suppress the competition in between modes and choose for instance the one with higher gain, as illustrated in Fig. 11.

But it's also obvious that this is obtained to the detriment of the output laser power. We will now discuss alternative and less drastic methods to suppress the destructive mode competition using the previously mentioned mode-locking.

In a mode-locked laser, the output radiation does not fluctuate in a chaotic manner. It consists of a periodic train of pulses (see Fig. 12), with specific duration and temporal separation.

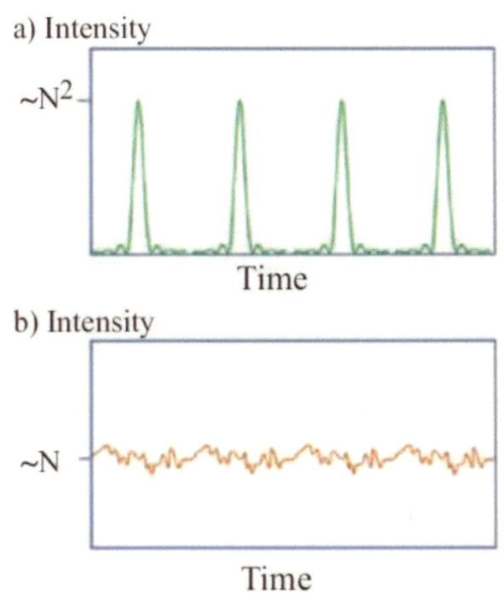

Fig. 12. Gain and loss spectra, longitudinal mode locations, and laser output for single mode laser operation(a) Mode-locked laser output with constant mode phase. (b) Laser output with randomly phased modes

We consider an electromagnetic field in an optical cavity consisting of N_m longitudinal modes locked in phase with a Δf frequency gap[2]:

$$A(z,t) = E_0 \sum_{m=-\frac{N_m-1}{2}}^{m=\frac{N_m-1}{2}} e^{2\pi i m \Delta f (t - z/c)} \tag{21}$$

For simplicity, we assumed that all the modes have the same amplitude E_0. The sum can be performed using the identity $\sum_{m=0}^{m=q-1} a^m = \frac{a^q - 1}{a-1}$, which leads to:

[2]One can notice that the frequency gap is not always the one given in Eq. (30). It can be larger if some modes are suppresses using for instance an etalon.

$$A(z,t) = E_0 \frac{sin[\pi N_m \Delta f (t - z/c)]}{sin[\pi \Delta f (t - z/c)]} \qquad (22)$$

The intensity of the laser is defined as the squared moduleus of the former expression, which gives, at z =0:

$$I(t) \propto E_0^2 [\frac{sin[\pi N_m \Delta f .t]}{sin[\pi \Delta f .t]}]^2 \qquad (23)$$

The result of Eq. (23) is illustrated in Fig. 12. The radiation consists of a train of peaks, separated by a distance $T = \frac{1}{\Delta f}$ with a width $\frac{1}{N_m \Delta f}$. The intensity of the peaks is proportional to the square of the number of involved modes $I_p \propto N_m^2 E_0^2$, while the average intensity is proportional to N_m, i.e. $I_M \propto N_m E_0^2$. Finally, it is important to notice that in the case of random phases, the output laser intensity corresponds to the average intensity of the mode-locked laser and that the fluctuations have a correlation time equal to the duration of one single pulse of the mode-locked laser. Since the former remarks are of notable interest for the laser applications and for further discussions, we provide with additional precisions on the radiation pulse train presented in Fig. 13.

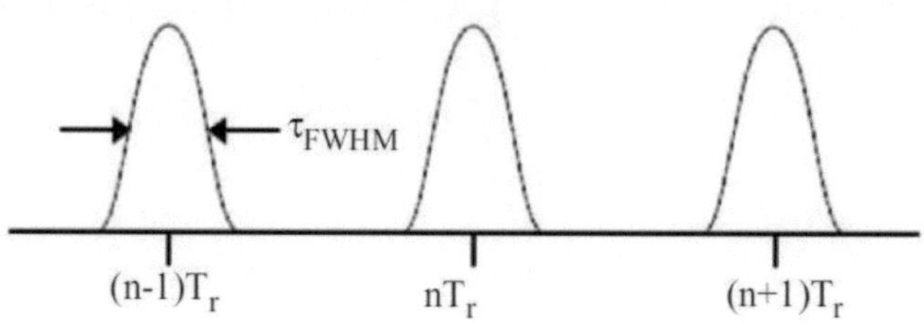

Fig. 13. Periodic pulse train

The train is characterized by an energy per pulse E_p, a delay between pulses T_p, a pulse duration δ_T (full width half maximum), an average power P_m and a peak power P_p [3]. All these quantities are linked as it follows: $P_p = \frac{E_p}{\delta_\tau}$, $P_m = P_p \frac{\delta_\tau}{T_R}$, the ratio $\frac{\delta_\tau}{T_R}$ corresponds to the duty cycle. The mode-locking techniques allow to generate ultra-short pulses.

Laser devices are usually said to provide light with temporal and spatial coherence.

The temporal coherence is a measure of the degree of mono-chromaticity of an electromagnetic wave. As a consequence, we can write:

$$\tau_c \Delta f \cong 1 \qquad (24)$$

[3]We recall that the electric field is $E_p = \sqrt{2 Z_0 I_p}$, with $I_p = \frac{P_p}{A_{eff}}$ the peak intensity and A_{eff} the effective area of the optical mode, and Z_0 is the vacuum impedance.

In the case of a mode-locked laser, the coherence time is essentially given by the distance in between the radiation packets. The coherence length l_c can be calculated using Eq. (24):

$$l_c(=c\tau_c) \cong \frac{\lambda^2}{\Delta\lambda} \tag{25}$$

The spatial coherence of a wave is the measurement of the temporal auto-correlation between two different points of a same transverse plane of the wave (see Fig. 14).

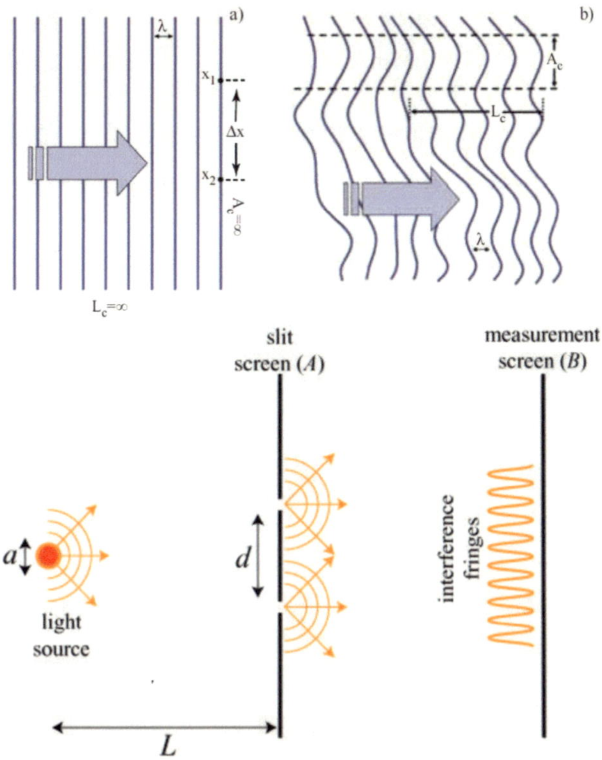

Fig. 14. Examples of wave-fronts with a) infinite and b) finite coherence length c) Coherence area

The spatially coherent part, belonging to the so called coherent area, of a wave passing through a slit produces a diffraction figure. The concept of coherence area can be quantified using the Van Cittert Zernike theorem, which states that the coherence area of a light source is given by:

$$S_c \cong \frac{L^2\lambda^2}{\pi a^2}$$

with L the distance to the source and d its diameter.

The former discussion allowed to give a general idea of the issues relative to conventional lasers. We will now discuss analogous issues relevant to FEL devices.

3. The free electron laser

In this and in the forthcoming sections we will discuss laser devices operating with an active medium, consisting of free relativistic electrons [3,5]. A few notions and a glossary of relativistic kinematics are therefore necessary. The total energy E of a particle of mass m moving with a velocity υ is given by:

$$E = m_0 \gamma c^2, \gamma = \frac{1}{\sqrt{1-\beta^2}}, \beta = \frac{v}{c}. \tag{26}$$

β is the reduced velocity and γ the relativistic factor. The total energy corresponds to the sum of the kinetic and mass energy, the mass energy being given by $m_0 c^2$. The factor γ is a measure of the particle energy, and γ-1 of the kinetic energy. In addition: $\beta = \sqrt{1-\frac{1}{\gamma^2}}$.

In the high energy case $\gamma >> 1$, the reduced velocity becomes $\beta \cong 1 - \frac{1}{2\gamma^2}$

and the particle is said "ultra-relativistic".

We now consider the main characteristics of the electrons: a charge $|e| \cong 1.6\times10^{-19}$ C and a mass $m_e \cong 0.51$ MeV (1 MeV=10^6 eV, 1 eV=1.6×10^{-19} J). The adjective relativistic is used when the kinetic energy of the electrons is of the order of a few MeV. Finally, a charged particle beam is associated to a power equal to the product of its energy with its current:

$$P(MW) = I(A)E(MeV). \tag{27}$$

A 20 MeV beam with a current of 2 A has a power of 40 MW. Such power is delivered to the beam by the accelerator. From now on, we will refer to LINAC accelerator, i.e. linear accelerators, where the acceleration is performed in radiofrequency cavities. The electron beam power corresponds in the case of free electron lasers, to the pump power in the case of conventional lasers.

Free electrons passing through a magnetic field produces flashes of Bremsstrahlung radiation. Synchrotron light sources rely on this process. We consider the case of a magnetic field delivered by an undulator:

$$\vec{B} = B_0(0, sin(\frac{2\pi z}{\lambda_u}), 0) \tag{28}$$

Such field is oscillating in the vertical direction with a peak value B_0 and a periodicity λ_u. It can be obtained with two series of magnets with alternative *N-S* orientation, as illustrated in Fig. 15.

The Lorentz force, due to the undulator field, introduces a transverse component in the electron motion, initially exclusively longitudinal. The electron motion in the magnet is governed by (Gaussian units):

$$\frac{d\vec{p}}{dt} = -\frac{e}{c}\vec{v} \times \vec{B}. \quad (29)$$

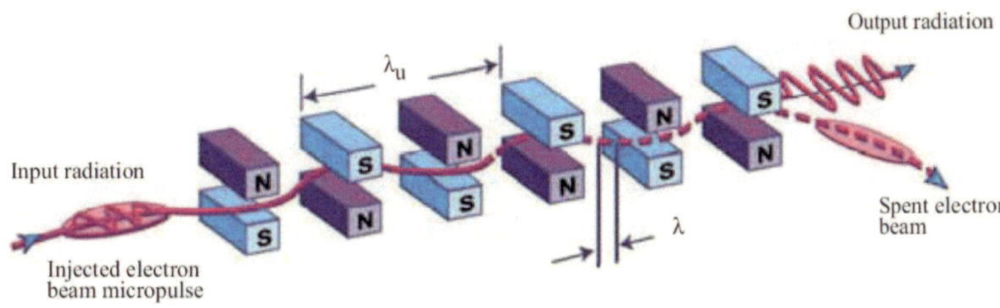

Fig. 15. Electron motion in an undulator

Averaging over one undulator period, one can derive the transverse and longitudinal velocities:

$$v_x \cong \frac{cK}{\sqrt{2}\gamma},\ v_z \cong c[1 - \frac{1}{2\gamma*^2}],\ \gamma* = \frac{\gamma}{\sqrt{1+\frac{K^2}{2}}}, K = \frac{eB_0\lambda_u}{2\pi m_0 c^2}$$

The K parameter is introduced to take into account the effect of the transverse motion on the longitudinal velocity. The electrons moving in the undulator emits radiation and this process can be viewed as a kind of spontaneous emission, The frequency selection mechanism can be understood by referring to Fig. 16.

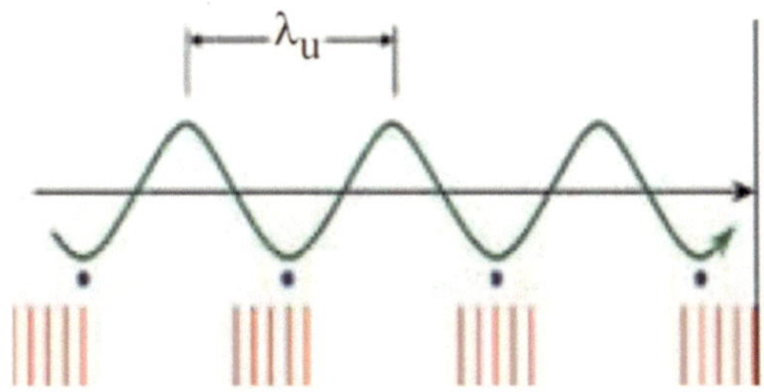

Fig. 16. Phase advance of the radiation with respect to the electrons

After one undulator period, the emitted radiation, travelling at the light velocity, has slept ahead of the electron beam by a quantity

$$\delta \sim (c - v_z)\frac{\lambda_u}{c} = \frac{\lambda_u}{2\gamma^2}(1 + K^2/2) \quad (30)$$

Since δ represents a phase advance of the electromagnetic wave, constructive interference with the radiation emitted at the next undulator period, requires the condition:

$$\delta = n\lambda, \quad (31)$$

n being an integer. From Equation (30) and Eq. (31) finally comes the definition of the undulator radiation wavelength:

$$\lambda_n = \frac{\lambda_u}{2n\gamma^2}(1+\frac{K^2}{2}) \tag{32}$$

n=1 represents the fundamental wavelength, while n>1 represent the higher harmonics. In a first step, we just consider the radiation on the fundamental and we try to characterize it, giving an evaluation of its relative bandwidth. The radiation pulse at the end of the undulator has a structure of a step function with a length $N\delta$, corresponding to a pulse duration $\Delta\tau$ of: $\Delta\tau \cong \frac{N\delta}{c}$, the corresponding spectral distribution is given by the Fourier transform of this pulse (see Fig. 17), and the spectral width $\Delta\omega$ results defined according to the Parseval relation $\Delta\tau\Delta\omega \cong \pi$.

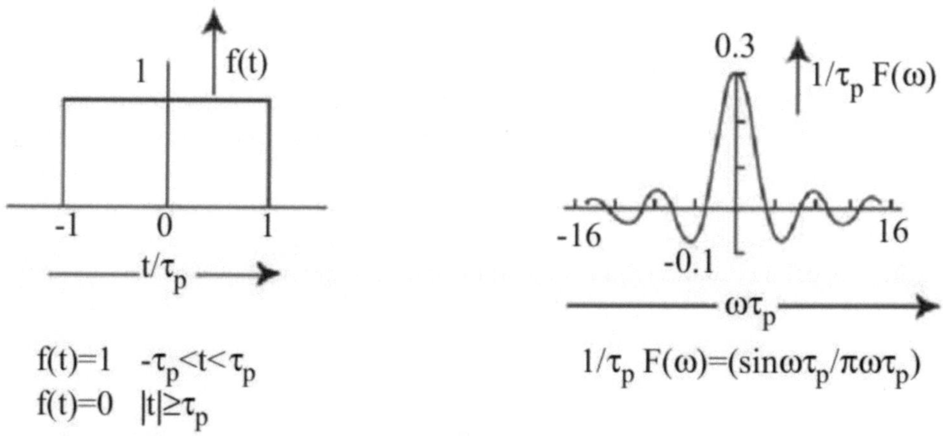

Fig. 17. Step function and corresponding Fourier transform

Thus getting for the relative bandwidth

$$\frac{\Delta\omega}{\omega} = \frac{1}{2N} \tag{33}$$

The spectral intensity profile of the radiation emitted by a single undulator is therefore just given by

$$f(\nu) \propto [\frac{sin(\frac{\nu}{2})}{\frac{\nu}{2}}]^2 \quad \nu = 2\pi N\frac{\omega-\omega_0}{\omega_0} \tag{34}$$

$f(\nu)$ is illustrated in Fig. 18 and can also be written as:

$$f(\nu) = 2Re[\int_0^1 (1-t)e^{-i\nu t}dt] \tag{35}$$

ν, the so-called detuning, is a very useful parameter, which is used to define the spectral profile with respect to the central reference wavelength. Since ν is a dimensionless parameter, the width of the profile at half of the maximum intensity is close to 2π.

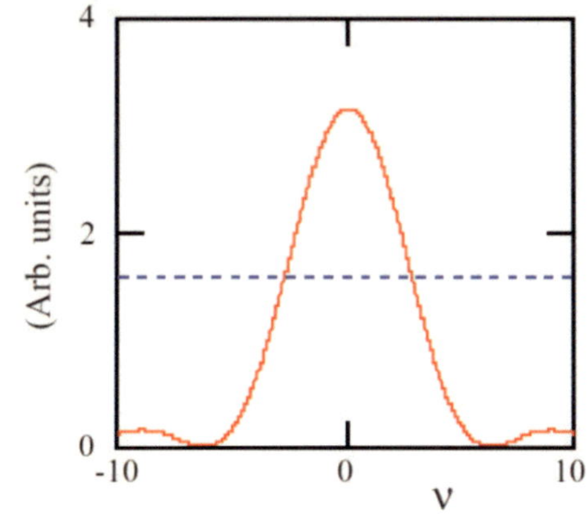

Fig. 18. Function $f(\nu)$

An oscillator FEL is similar to a conventional laser oscillator. Indeed, we find again a spontaneous emission process from a free electron beam in an optical cavity, the storing of the radiation inside the cavity, and the amplification of this radiation throughout a mechanism of stimulated emission. Fig. 19 also shows that during the interaction between the electrons and the radiation, the electron beam is being modulated in energy which results in a spatial modulation (bunching) at the radiation wavelength.

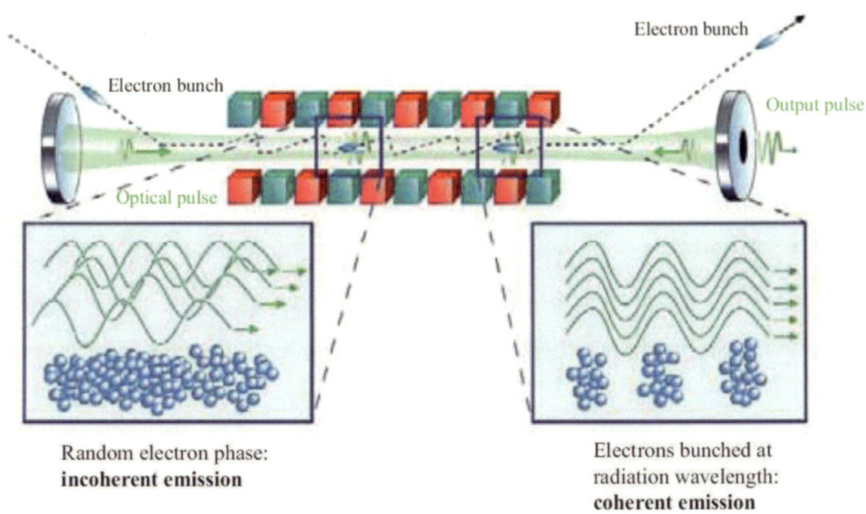

Fig. 19. Oscillator FEL

This modulation enables the production of coherent radiation. The mechanism of interaction in between the electrons and the radiation of a FEL can be considered as the combination of two competing effects:

Energy loss from the electron and then amplification of the incident photon, absorption of the stimulating photons with consequent yielding of energy to the electrons.

Such mechanism can be understood quite easily using a reference system in which the electrons are nonrelativistic. In such system, the electrons see two electromagnetic fields: the laser field travelling in the same direction, and the undulator field, which proceeds in the opposite direction ("Virtual quanta" model of Weizsacker and Williams). In Figure 20, we reported the processes of stimulated emission and absorption, similar to the processes of stimulated emission in conventional lasers.

1. the stimulated absorption can be interpreted as a backward scattering of the photons by the laser field. The electrons gain energy, which corresponds to a negative gain;
2. the stimulated emission can be interpreted as a forward scattering of the photons by the undulator field, with a frequency close to the laser frequency. The electrons loose energy, corresponding to a positive gain.

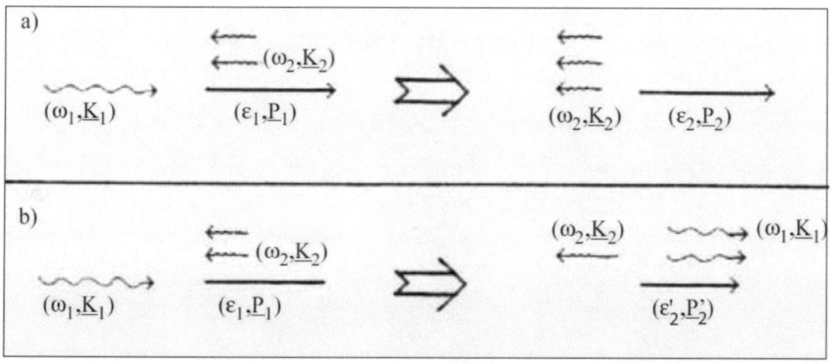

Fig. 20. Stimulated emission and absorption processes, similar to the stimulated emission process in the case of conventional lasers. $\omega_1=c/k_1/=$ laser frequency; $\omega_2=c/k_2/=$ undulator radiation frequency; $\varepsilon_1,p_1=$ initial energy and moment of the electrons; $\varepsilon_2,p_2=$ final energy and moment of the electrons. a) Backward stimulated scattering (negative gain). b) Forward stimulated scattering (positive gain

The spontaneous emission process is interpreted using Fig. 20 as the scattering of "undulator photons", without the presence of co-propagating electrons.

Within such a context the gain process can be viewed as the balance between stimulated absorption and emission processes, namely

$$G(\nu) \propto f(\nu_+) - f(\nu_-) \tag{36}$$

$f(\nu_{+/-})$ refers to the cases illustrated in the Figure, to emission and absorption.

Since $\nu_+ - \nu_- \propto \delta E_{rec}$, where δE_{rec} represents the energy loss of one single electron via recoiling, the expansion of Eq.(36) at the lowest order leads to (this assumption holds for low gain devices as we will discuss later):

$$G(\nu) \propto -\frac{\delta f(\nu)}{\delta \nu} \tag{37}$$

Taking into account that the radiation is emitted by an electron beam at a given current, the gain can be eventually expressed as follows

$$G(\nu) = -\pi g_0 \frac{\delta f(\nu)}{\delta \nu} = 2\pi g_0 \int_0^1 t(1-t)sin(\nu t)dt \;,\; g_0 = 4\pi \frac{|J|}{I_0}(\frac{N}{\gamma})^3(\lambda_u K^* f_b)^2$$

$$I_0 = \frac{4\pi}{Z_0 c}\frac{m_e c^3}{e} = 1.7045.10^4 A \;,\; f_b = J_0(\frac{K^{*2}}{2(1+K^{*2})}) - J_1(\frac{K^{*2}}{2(1+K^{*2})}) \quad K^* = \frac{K}{\sqrt{2}} \tag{38}$$

g_0 is the small signal gain coefficient, I_0 the Alfven current. The gain curve is reported in Fig. 21 (one can note that the maximum of the gain does not correspond to the maximum of the spontaneous emission intensity, but located at $\nu_0 \cong 2.6$). J is the current density of the electron beam and f_b the Bessel factor which takes into account the non-perfectly sinusoidal trajectory of the electrons in the linear polarized undulator.

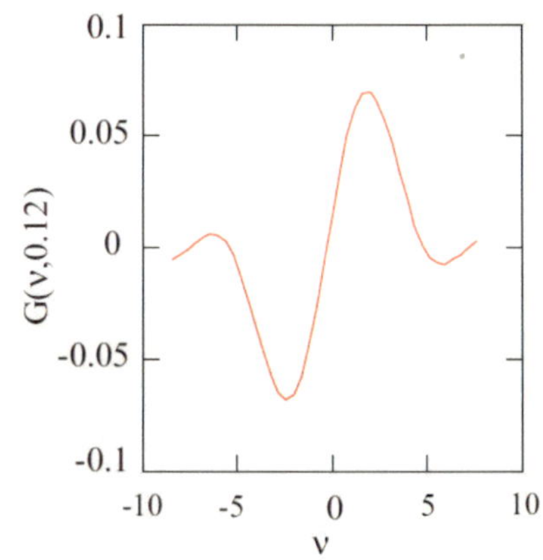

Fig. 21. FEL gain curve

It is worth mentioning that the gain curve is asymmetric. It consists of one part of positive gain and one part of absorption where the electrons, instead of giving energy to the radiation, absorb some energy from the radiation.

Now that the existence of a gain is clarified, we will determine the saturation intensity. The FEL process consists in a power transfer from the electron beam to the laser. The gain curve is asymmetric but the FEL process is maintained until the Kinematic conditions are such that a positive gain is guaranteed and that a sort of saturation is reached. The energy loss can be deduced from the width of the positive gain region, related to the detuning parameter (see Fig. 33 and Eq. (36)) by:

$$\Delta \nu \cong 4\pi N \frac{\Delta \gamma}{\gamma} \cong 2\pi \tag{39}$$

This equation yields the energy variation of the electron beam induced by the FEL interaction:

$$\frac{\Delta \gamma}{\gamma} \cong \frac{1}{2N} \tag{40}$$

We finally get the expression of the power, transferred from the radiation to the electrons (see Eq. (27)):

$$P_L \cong \frac{P_E}{2N} \tag{41}$$

According to what we discussed for conventional lasers and assuming that a similar dynamics is valid in the case of FEL, the saturation intensity can be expressed as:

$$I_S \cong \frac{I_E}{2Ng_0},\ I_E = \frac{|J|E}{e}$$

or in more practical units:

$$I_S[\frac{MW}{cm^2}] = 6.9312.10^2 \frac{1}{2}(\frac{\gamma}{N})^4(\lambda_u[cm]K^* f_b)^{-2} \tag{42}$$

In conclusion, we demonstrated that an oscillator FEL behaves like a conventional oscillator laser and therefore that most of the relevant theoretical description can be applied to the FEL.

4. FEL Oscillator model

In the former paragraph, we have shown a full parallel (in terms of gain and saturation intensity) between FEL and conventional lasers. Pursuing the analogy we can write the saturation of the FEL gain as (we impose an upper limit on the small signal gain coefficient to fix the limit of validity of this treatment)

$$\begin{gathered} G(X) = \frac{G_M}{F(X)}, X = \frac{I}{I_s} \\ G_M = 0.85 g_0,\ \text{for}\ g_0 \le 0.3 \end{gathered} \tag{43}$$

G_M represents the maximum gain calculated at $\nu_0 \cong 2.6$ and:

$$F(X) = 1 + \alpha X + \beta X^2, \alpha + \beta = 1, \alpha = 2(\sqrt{2} - 1), \beta = 3 - 2\sqrt{2} \tag{44}$$

One can notice that in the case of the FEL, the saturation results slightly different from the conventional laser (presence of a quadratic term), but this is just a technical detail which does not modify the physics of the process. The use of rate equations as defined in Eq .(9) enables to obtain the evolution of the laser power in the cavity as

$$I_{r+1} - I_r = [(1-\eta)G(X_r) - \eta]I_r, \tag{45}$$

As already seen, the signal increases initially in the exponential mode, then the increase slows down and finally stops (or nearly) when the gain equals the losses. The equilibrium power in the cavity is obtained from the condition $I_{r+1} = I_r$, which implies:

$$G(X_e) = \frac{\eta}{1-\eta} \tag{46}$$

Together with Eq. (43) and (44), this leads to the expression of the equilibrium intensity in the cavity:

$$I_e = (\sqrt{2}+1)(\sqrt{\frac{1-\eta}{\eta}G_M} - 1)I_S \tag{47}$$

The solution of Eq. (45), (46) and (47) can be written as:

$$I_{r+1} = I_0 \frac{[(1-\eta)(G_M+1)]^r}{1+\frac{I_0}{I_e}\left\{[(1-\eta)(G_M+1)]^r - 1\right\}} \tag{48}$$

I_0 is the initial radiation intensity, due for instance to the spontaneous emission. An example of evolution of the laser signal in the cavity is presented in Fig. 22.

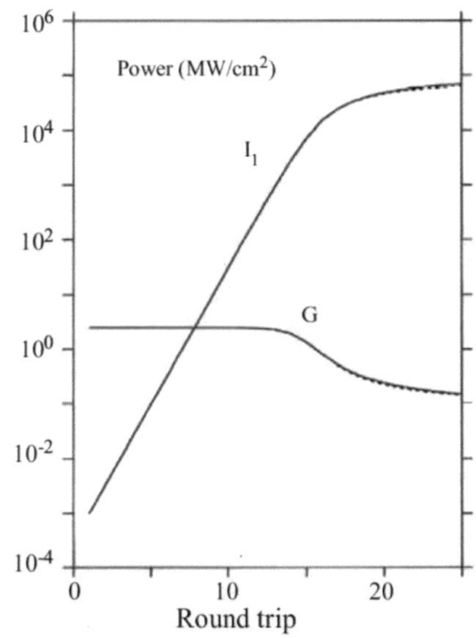

Fig. 22. Example of FEL signal evolution in the cavity up to saturation. Comparison between simulation (continuous line) and analytic calculation (dotted line). The lower curve represents the gain reduction due to the increase of the intracavity intensity.

The evolution is typically sigmoidal, as in the case of the conventional lasers.

We discuss the mechanism of mode-locking in the case of oscillator FEL. From the gain curve, whose width is given by:

$$(\Delta f)_{FEL} = \frac{c}{2N\lambda} \tag{49}$$

can be retrieved the number of coupled modes:

$$N_m = \frac{(\Delta f)_{FEL}}{(\Delta f)_e} = \frac{L}{N\lambda} \tag{50}$$

It is worth noticing that such value is inversely proportional to the slippage $N\lambda$. The physical meaning of this relation will be discussed later. In the absence of a specific mechanism of coupling of the longitudinal modes, the oscillator FEL modes also oscillate independently producing a fluctuating output radiation. But in the case of oscillator FEL, there is a natural coupling mechanism, resulting from the electron packets themselves, delivered by the accelerator with a pulse train structure of finite duration.

In Figure 23 is illustrated the structure of an electron beam delivered by a Radio-Frequency accelerator, which consists of a series of macro pulses themselves composed by a train of micro pulses which we assume to be Gaussian ($f(z)=\frac{1}{\sqrt{2\pi}\sigma_z}e^{-\frac{z^2}{2\sigma_z^2}}$.)

The current of a micro pulse corresponds to the peak current, while the current of the macro pulse corresponds to the average current. The case is similar to the one discussed for pulses in conventional lasers.

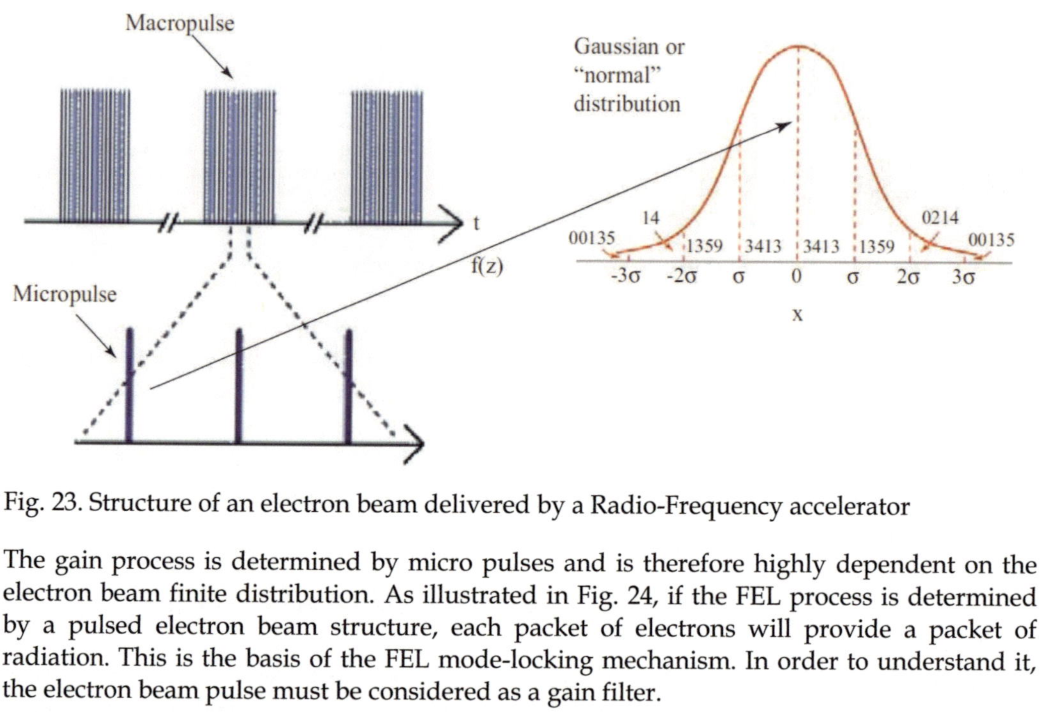

Fig. 23. Structure of an electron beam delivered by a Radio-Frequency accelerator

The gain process is determined by micro pulses and is therefore highly dependent on the electron beam finite distribution. As illustrated in Fig. 24, if the FEL process is determined by a pulsed electron beam structure, each packet of electrons will provide a packet of radiation. This is the basis of the FEL mode-locking mechanism. In order to understand it, the electron beam pulse must be considered as a gain filter.

This can be done considering the Fourier transform of the electron beam in the frequency domain of the FEL

$$\tilde{f}(\nu,\mu_c)=\frac{1}{\sqrt{2\pi}}\int_{-\infty}^{\infty}f(z)e^{-i(k-k_0)z}dz \tag{51}$$

$$=\frac{1}{\sqrt{2\pi}}\int_{-\infty}^{\infty}f(z)e^{-i\frac{\nu}{N\lambda}z}dz\,,\ \propto\frac{1}{\sqrt{2\pi}\mu_c}e^{-\frac{\nu^2}{2\mu_c^2}},\ \ \mu_c=\frac{\Delta}{\sigma_z};\Delta=N\lambda$$

Δ is referred as the slippage length. The Fourier transform of *f(z)* is Gaussian too and in the following, we will use the normalized form: $\tilde{f}(\nu,\mu_c)=\frac{1}{\sqrt{2\pi}\mu_c}e^{-\frac{\nu^2}{2\mu_c^2}}$

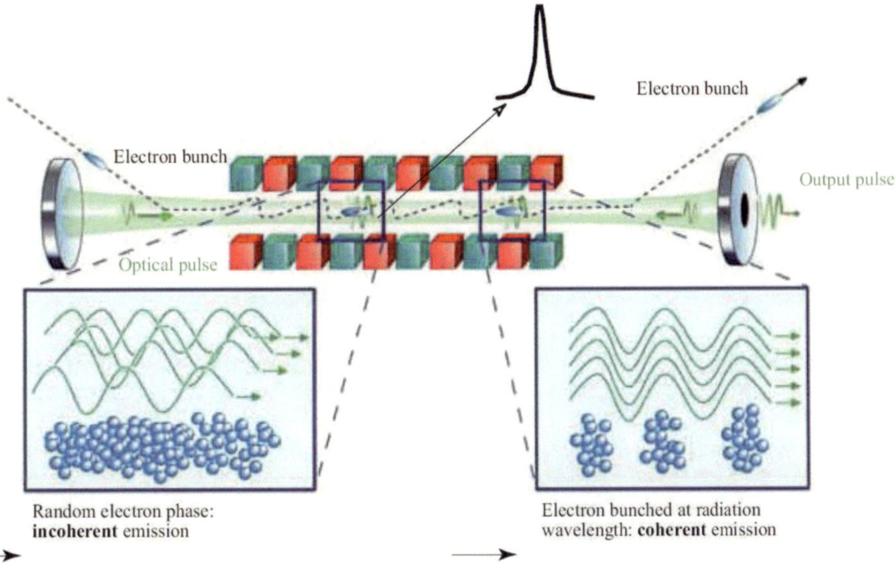

Fig. 24. Formation of a laser pulse from an electron beam pulse

The filter, representing the electron beam, is therefore regulated by the quantity μ_c, referred as longitudinal coupling coefficient. This quantity has an all-purpose physical meaning, which will be discussed in detail later. Here, we just highlight that when μ_c increases, the number of coupled modes increases, and that for a continuous beam ($\sigma_z \to \infty$) the operation becomes essentially single mode. The distribution derived from the former equation becomes a Dirac peak. The gain "filtered" by the electron beam can be written in terms of a convolution:

$$G_p(\nu,\mu_c)=\int_{-\infty}^{\infty} G(\nu+y)\tilde{f}(y,\mu_c)dy=2\pi g_0\int_0^1 t(1-t)sin(\nu t)e^{-(\mu_c t)^2/2}dt \quad (52)$$

It appears from this relation that for positive coupling coefficient values, the gain is decreased, simply because there are more modes oscillating in phase, and according to Eq. (52) that those coupled modes are within the rms band: $\sigma_\nu \cong \mu_c$.

5. FEL equations

The previous considerations are essentially qualitative, to make the forthcoming arguments quantitative we cast the FEL field evolution equation in the form

$$\frac{da}{d\tau}=i\pi g_0\int_0^{\tau} \tau' a(\tau-\tau')e^{-i\nu\tau'}d\tau' \quad a_n=|a_n|e^{i\phi_n},|a|^2=8\pi^2\frac{I}{I_s} \quad (53)$$

The previous equation can be exploited to specify the field growth before the saturation. It must be stressed that it contains something more, with respect to the considerations leading to the definition of the gain curve derived in the previous section, which were based on the assumption that the field remains constant during the interaction inside the undulator. Eq. (53) holds under more general assumptions, it includes the effects of high gain, i.e. the corrections associated to the fact that the field can vary significantly during the interaction.

The FEL gain is the relative variation of the input field in one interaction ($G = \frac{|a|^2 - |a_0|^2}{|a_0|^2}$).

The FEL gain curve can be derived from Eq. (88), using the low gain approximation satisfactory for small signal gain coefficient values lower than $30\,\%$. For higher values, the gain curve is no more anti-symmetric. The maximum gain is no more simply proportional to g_0 and further corrections should be included, namely

$$G_M = G(g_0) \cong 0.85 g_0 + 0.192 g_0^2 + 4.23 \times 10^{-3} g_0^3, g_0 \le 10 \tag{54}$$

Eq. (53) considers an ideal electron beam, i.e. one with all electrons at the same energy. In reality, the energy distribution of the electron beam can be assumed Gaussian: $f(\varepsilon) = \frac{1}{\sqrt{2\pi}\sigma_\varepsilon} e^{-\frac{\varepsilon^2}{2\sigma_\varepsilon^2}}$, $\varepsilon = \frac{\gamma - \gamma_0}{\gamma_0}$, with σ_ε being the electron beam relative energy spread. The energy distribution induces an analogous distribution in the frequency domain of the FEL gain:

$$\tilde{f}(\nu, \mu_\varepsilon) = \frac{1}{\sqrt{2\pi}(\pi\mu_\varepsilon)} e^{-\frac{\nu^2}{2(\pi\mu_\varepsilon)^2}}, \quad \mu_\varepsilon = 4N\sigma_\varepsilon \tag{55}$$

Following the same procedure of convolution, as in the case of the longitudinal modes distribution, Eq. (53) is modified as following:

$$\frac{\partial a}{\partial \tau} = i\pi g_0 \int_0^\tau \tau' a(\tau - \tau') e^{-i\nu\tau'} e^{-\frac{1}{2}(\pi\mu_\varepsilon \tau')^2} d\tau' \tag{56}$$

The reduction of the gain due to the energy distribution is similar to the reduction of the gain due to the inhomogeneous broadening in conventional lasers and the parameter μ_ε is one of the inhomogeneous broadening parameter of the FEL. The reduction of the gain due to such effect can be quantified according to $G \cong \frac{G_M}{1 + 1.7\mu_\varepsilon^2}, \mu_\varepsilon << 1$. Further sources of inhomogeneous broadening will not be considered here.

The inclusion of the electron packet shape and of the consequent optical pulse phenomenology can be done by taking into account the fact that the photon packet, having a higher velocity that the electrons, slips over the electron packet (see Fig. 38). Therefore, Equation (56) should be modified according to:

$$\frac{\partial a(z,\tau)}{\partial \tau} = i\pi g_0(z + \Delta\tau) \int_0^\tau \tau' a(z + \Delta\tau', \tau - \tau') e^{-i\nu\tau'} e^{-\frac{1}{2}(\pi\mu_\varepsilon \tau')^2} d\tau' \tag{57}$$

$g_0(z+\Delta\tau)$ is the small signal gain coefficient which takes into account the electrons distribution $f(z)$, $f(z)=\frac{1}{\sqrt{2\pi}\sigma_z}e^{-\frac{z^2}{2\sigma_z^2}}$. and the slippage. The former equation is sufficiently general to describe the dynamics of the pulse propagation with the effects of a high gain.

The coupling parameter μ_c has an additional physical meaning which is worth mentioning: it measures the overlapping of the electron beam with the photon beam. A very short electron beam remains superimposed to the photon beam for a very short time, which drives another reduction of the gain. The reduction of the gain due to the multi-mode coupling and spatial overlap can be quantified as following[4]: $G \cong \frac{G_M}{1+\frac{1}{3}\mu_c}$.

An important aspect of the FEL dynamics is the non-linear harmonic generation, resulting from the bunching mechanism, which itself is a consequence of the energy modulation mechanism (see Fig. 25). When the electrons are micro-bunched at a wavelength which corresponds to a submultiple of the fundamental wavelength, coherent emission is produced at higher order harmonics, throughout the mechanism of nonlinear harmonic generation.

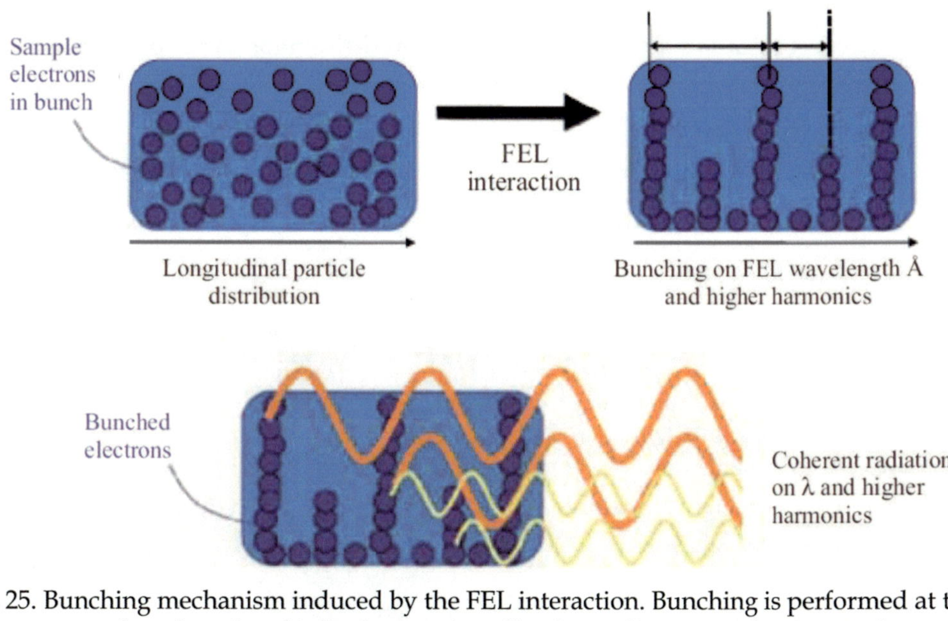

Fig. 25. Bunching mechanism induced by the FEL interaction. Bunching is performed at the resonant wavelength and on higher harmonics, allowing radiation production on the fundamental and higher harmonics

The intra cavity evolution of the fundamental and of the higher order harmonics is presented in Fig. 26.

[4]Both Eqs. (97) and (100) are valid in the case of small gain ($g_0 < 0.3$). For more general hypothesis, the expressions are slightly more complicated and will not be reported here for simplicity.

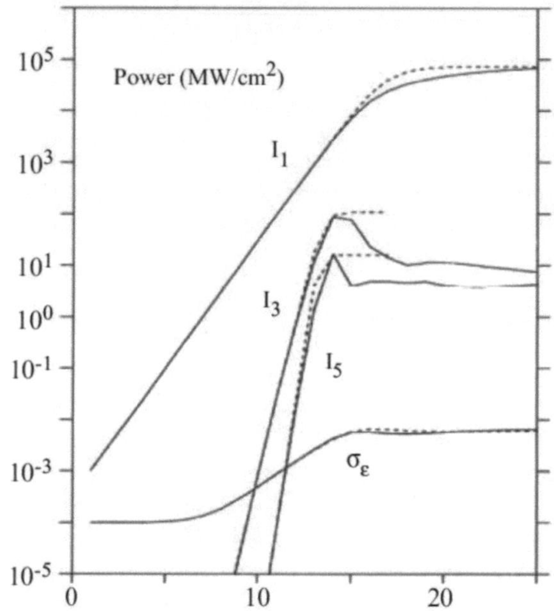

Fig. 26. Example of intra cavity evolution of the fundamental and of the high order harmonics (third and fifth). The Figure also reports the energy spread increase. Continuous lines: numerical calculation. Dashed line: analytical approximation

The dynamics of the process is understood as follows: the fundamental increases, until it reaches a sufficiently high power level to induce a bunching, that can lead to non-linear harmonic generation. The saturation mechanism comes from the combined effects of the energy loss and energy spread increase of the electron beam due to the same interaction. The maximum harmonic power is related to the maximum of the fundamental power according to[5]:

$$P_n^* = \frac{1}{2}\frac{\sqrt{n}}{n^3}\sqrt{\frac{n-1}{2}}\, g_{0,n} \frac{P_E}{4N} \tag{58}$$

where $g_{0,n}$ is the small signal gain coefficient of the harmonic n. Note that the coefficient $\frac{P_E}{4N}$ is the maximum power P_L achievable on the fundamental. The higher order harmonics, opposite to the fundamental, are not stored in the cavity. Therefore the powers presented in Fig. 26 are relative to the harmonic generation in one single round trip.

The non-linear harmonic generation mechanism is similar to the frequency-mixing mechanism in the case of non-linear optics. Eq. (58) can be understood as the non-linear response of the medium (the electron beam) to the laser electric field, so that it can then be written in terms of amplitude of the harmonics electric field:

---

[5]Note that in the linear polarized undulators, radiation is emitted on-axis only at odd order harmonics (3,5…)

$$E_n^* = \chi_n E_L \tag{59}$$

$$\chi_n = \sqrt{\frac{g_{0,n}}{2}\frac{\sqrt{n}}{n^3}}\sqrt{\frac{n-1}{2}}$$

Before concluding this paragraph, we will describe the profile of the pulses in a FEL oscillator and the method which can be implemented to shape the pulses simply using the cavity parameters.

7. Self amplified spontaneous FEL

One of the limiting factors of the FEL oscillator are the mirrors, which restrict the spectral range to wavelengths longer than a few hundreds of nm. Therefore, the soft X-ray region seems out of the range of operation of the FEL oscillator. It is possible to implement high gain mechanism which enables to reach saturation in one single pass in the undulator. One possible scheme is given in Fig. 27. A high peak current electron beam is injected in a long undulator, and amplifies the signal emitted in the first periods of the undulator. Such FEL configuration is referred as Self Amplified Spontaneous Emission (SASE) FEL. The main steps of the process are: Spontaneous emission, Energy modulation, Longitudinal bunching of the electrons over a distance of the order of the spontaneous radiation wavelength, Coherent emission, Saturation

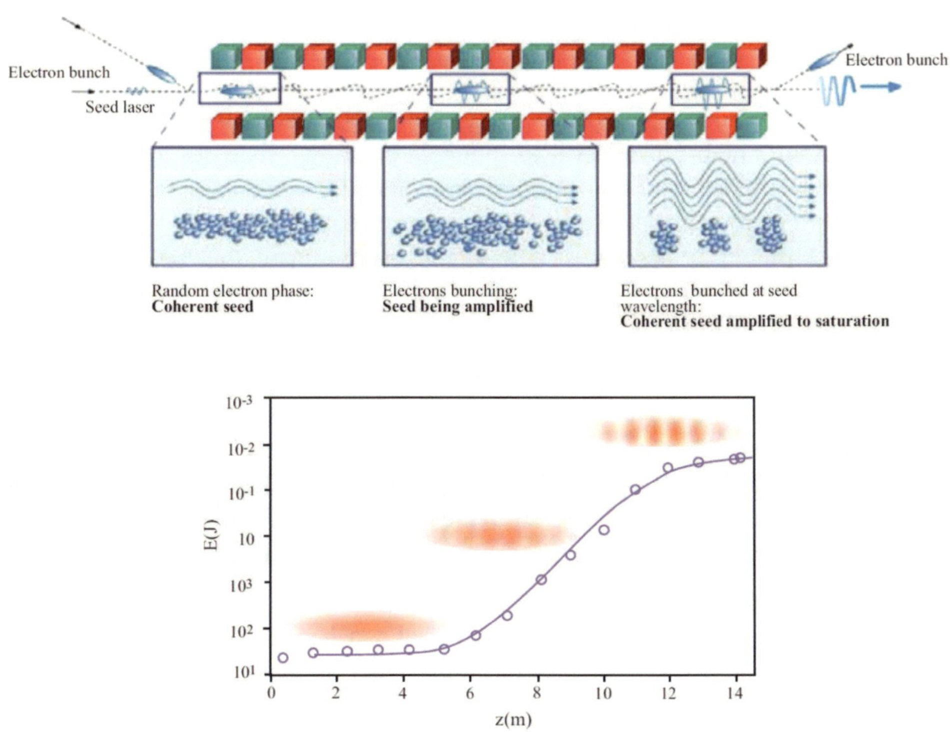

Fig. 27. SASE FEL process. Microscopic image of the bunching mechanism and laser signal increase along the undulator

In order to understand these steps, we go back to the integral form of the FEL equations (see Eq. (53)), which can be rewritten as follows:

$$e^{i\nu\tau}\frac{\partial a}{\partial \tau} = i\pi g_0 \int_0^{\tau} (\tau - \tau')a(\tau')e^{i\nu\tau'}d\tau' \tag{60}$$

$$e^{i\nu\tau}\frac{da}{d\tau} = i\pi g_0 \hat{D}_\tau^{-2}[a(\tau)e^{i\nu\tau}]$$

D_τ^{-2} indicates an integration (or a negative derivative) repeated twice.

We also used the following Cauchy equality for the integrations
($\hat{D}_x^{-n}f(x) = \frac{1}{(n-1)!}\int_0^x (x-\xi)^{n-1} f(\xi)d\xi$, $\hat{D}_x^{-n}f(x) = \int_0^x dx_1 ... \int_0^{x_n - 1} dx_n f(x_n)$).

Deriving both parts of Eq. (60) with respect to the temporal variables, the former integral equation can be transformed into an ordinary third order equation:

$$(\hat{D}_\tau^3 + 2i\nu\hat{D}_\tau^2 - \nu^2\hat{D}_\tau)a(\tau) = i\pi g_0 a(\tau)$$

$$\hat{D}_n^n = \frac{d^n}{d\tau^n} \tag{61}$$

$$a|_{\tau=0} = a_0, \hat{D}_\tau a|_{\tau=0} = 0, \hat{D}_\tau^2 a|_{\tau=0} = 0$$

The solution of this equation is obtained with a standard method and can be written as following:

$$a(\tau) = \frac{a_0}{3(\nu+p+q)} e^{-\frac{2}{3}i\nu\tau}((-\nu+p+q)e^{-\frac{i}{3}(p+q)\tau} +$$

$$+2(2\nu+p+q)e^{\frac{i}{6}(p+q)\tau}(cosh(\frac{\sqrt{3}}{6}(p-q)\tau)$$

$$+i\frac{\sqrt{3}\nu}{p-q}sinh(\frac{\sqrt{3}}{6}(p-q)\tau) \tag{62}$$

$$p = \left[\frac{1}{2}(r+\sqrt{d})\right]^{\frac{1}{3}}, q = \left[\frac{1}{2}(r-\sqrt{d})\right]^{\frac{1}{3}}$$

$$r = 27\pi g_0 - 2\nu^3, d = 27\pi g_0\left[27\pi g_0 - 4\nu^3\right]$$

It is not so easy to handle from the analytical point of view, nor easy to interpret from the physical point of view. A more transparent form can be obtained with ν =0 in Eq. (61). This is justified since we are considering solutions with a high gain, and that since in this case, maximum gain is reached for small values of the detuning parameter. In the ν =0 case, Eq. (61) becomes much more simple:

$$\hat{D}_\tau^3 a(\tau) = i\pi g_0 a(\tau) \tag{63}$$

The general solution is a linear combination of the three roots which is given by:

$$a(\tau) = \sum_{j=0}^{2} a_j R_j(\tau) = \sum_{j=0}^{2} a_j e^{\delta_j \tau} \tag{64}$$

δ_j are the three complex roots of $(\pi g_0)^{\frac{1}{3}}$. The solution is then finally given by:

$$a(\tau) \propto e^{(\pi g_0)^{\frac{1}{3}} \frac{\sqrt{3}}{2} \tau} \tag{65}$$

In addition, using $\tau = \frac{z}{N\lambda_u}$, the former relation becomes:

$$a(z) \propto e^{\frac{z}{2L_G}} \quad L_G = \frac{\lambda_u}{4\pi\sqrt{3}\rho} \quad \rho = \frac{1}{4\pi}(\frac{\pi g_0}{N^3})^{\frac{1}{3}} \tag{66}$$

L_G is referred as the gain length, while ρ is referred as the Pierce parameter and constitutes one of the fundamental parameters of the SASE FEL. Finally, including the three roots in the expression, the evolution of the power in the small signal gain regime is given by:

$$|a(\tau)| = A(z)|a_0|^2, \; A(z) = \frac{1}{9}(3 + 2cosh(\frac{z}{L_{g,1}}) + 4cos(\frac{\sqrt{3}}{2}\frac{z}{L_{g,1}})cosh(\frac{z}{2L_{g,1}}))$$

This relation enables to describe the initial zone of non-exponential growth.

The former equation only describes the intensity increase of the laser, not its saturation. The physical mechanism which determines the saturation is not different from the one which has been previously discussed in the case of the oscillators. To simplify the treatment, we assume that the evolution is just the one relative to the exponential growth, so that we can write the differential equation relative to the growth process as following:

$$\frac{d}{dz}P(z) = \frac{P(z)}{L_g}(1 - \frac{P(z)}{P_F}), \; P(0) = P_0$$

A quadratic non linearity has been added in order to take into account the effects of saturation. P_F indicates the final power in the saturation regime, that we will specify later. The solution of the former equation can be obtained easily, using the transformation: $\frac{d}{dz}T(z) = -\frac{1}{L_{g,1}}(T(z) - \frac{1}{P_{F,1}}), T(z) = \frac{1}{P_1(z)}$ and can be written:

$$P(z) = \frac{P_0}{9}\frac{e^{z/L_g}}{1 + \frac{P_0}{P_F}(e^{z/L_g} - 1)} \tag{67}$$

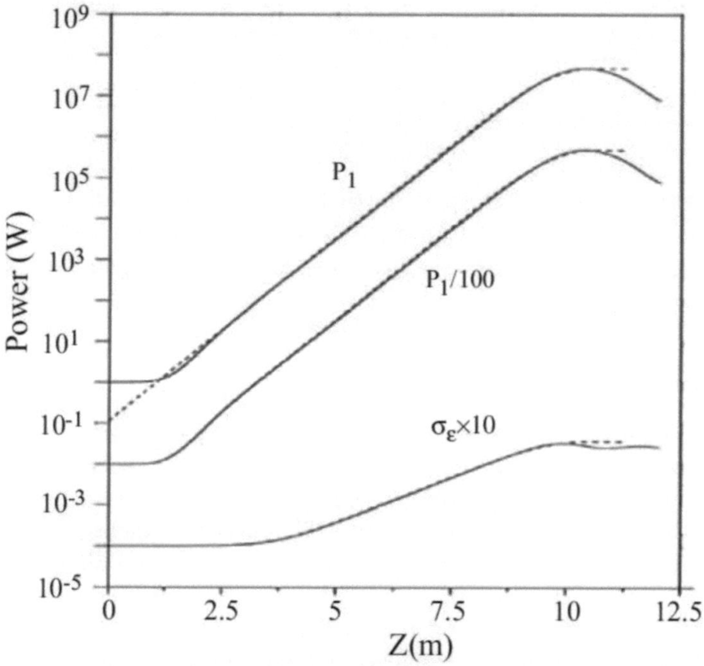

Fig. 28. Comparison between exact and approximated solutions, the upper curve represents the comparison with eq. (67) the mid curve provides the intensity growth (reduced by a factor 100 to avoid overlapping) predicted by eq. (68), the lower curve yields the evolution of the FEL induced energy spread.

In the more general case, the exponential is substituted by $A(z)$:

$$P(z) = \frac{P_0}{9} \frac{A(z)}{1 + \frac{P_0}{P_F}(A(z) - 1)} \tag{68}$$

The saturation power still remains to be defined. We already noted that when the gain coefficient increases, nonlinear elements tend to modify the maximum of the gain curve and its anti-symmetric shape.

$$G(\omega) = \frac{1}{\sqrt{2\pi}\rho} e^{-\frac{(\frac{\omega-\omega_0}{\omega_0})^2}{2\rho^2}}. \tag{69}$$

The relative width of the gain curve results proportional to the ρ parameter. According to what has been discussed in the former paragraphs, one expects the ratio between the FEL power and the electron beam power to be as well proportional to ρ.A more accurate analysis demonstrated that this is true, though with an additional numerical factor [6]:

$$P_F \cong \sqrt{2}\rho P_E \tag{70}$$

The saturation length is obtained by imposing $P(z) \cong P_F$ in Eq. (67), thus finding

$$Z_F = \ln\left(\frac{9P_F}{P_0}\right)L_g \tag{71}$$

Corresponding to the number of periods $N_F \cong \frac{1}{\rho}$

To get some numerical examples, we can note that if one limits the undulator length to a few tens of meters, given that the undulator period is of the order of a few centimetres and that $P_F \cong 10^8 P_0$, the value of the ρ results around 10^{-3}. A comparison between "analytical" and numerical predictions is given in Fig. 28.

8. SASE FEL and coherence

In the former paragraphs, we saw that the longitudinal coherence in FEL oscillators operated with short electron bunches, is guaranteed by the mode-locking mechanism. In the case of the FEL operated in the SASE regime, since there is no optical cavity, we can no longer talk about longitudinal modes, strictly speaking. Nevertheless, we can repeat the same argumentation on the filter properties of the electron beam to understand if something analogous to mode-locking can be defined for the cases of FEL operated in the SASE regime. As we already saw in the case of the oscillators, the effect of mode-locking is guaranteed in the interaction region where the electrons see a field with constant phase. This region is limited to one slippage length. We repeat the same procedure of Fourier transform which enabled to obtain Eq. (52). Using for the frequency domain the variable $\xi = \frac{\omega - \omega_0}{\omega_0}$ and changing the variable for the space domain according to $\tau = \frac{z}{c}$, Δ being the slippage length, we obtain the integral:

$$\tilde{f}(\xi) \propto \frac{1}{\sqrt{2\pi}}\int_{-\infty}^{\infty} f(\tau)e^{-i\xi\tau}d\tau \tag{72}$$

from which:

$$\tilde{f}(\xi) = \frac{1}{\sqrt{2\pi}\,\tilde{\mu}_c}e^{-\frac{\xi^2}{2\tilde{\mu}_c^2}} \quad \tilde{\mu}_c = \frac{\lambda}{\rho\sigma_z}$$

We used the approximation $\rho \cong \frac{1}{N}$. The former relation ensures that a coherence length l_c exists and that it is around λ / ρ (a more correct definition will be given later). If the electron beam length is about the coherence length, there would not be any problem of longitudinal coherence because all the modes inside the gain curve would be naturally coupled. Since in practical cases $\lambda / \rho \ll \sigma_z$ and since there is no clear definition of the longitudinal modes, we should define "macro modes", i.e. a sort of extension of the longitudinal modes. The number of "macro modes" is given by:

$$M_L \cong \frac{\sigma_z}{l_c} \tag{73}$$

Such macro modes introduced in the FEL Physics at the end of the 70's by Dattoli and Renieri are referred as *supermodes* (see [4] and references therein).

Each of these modes have a spatial distribution given by:

$$m(z) \cong \frac{1}{\sqrt{2\pi l_c}} e^{-\frac{z^2}{2l_c^2}}, \ l_c = \frac{\lambda}{4\pi\sqrt{3}\rho\sigma_z}$$

The relevant Fourier transform $\tilde{m}(\xi)$ provide us with the frequency distribution. The total spectrum is given by:

$$S(\xi) = \sum_{n=1}^{M_L} \tilde{m}_n(\xi) \tag{74}$$

The phase of each component is totally random. The spatial distribution of the optical pulse is given by the Fourier transform of the spectrum and the results is shown in Fig. 29.

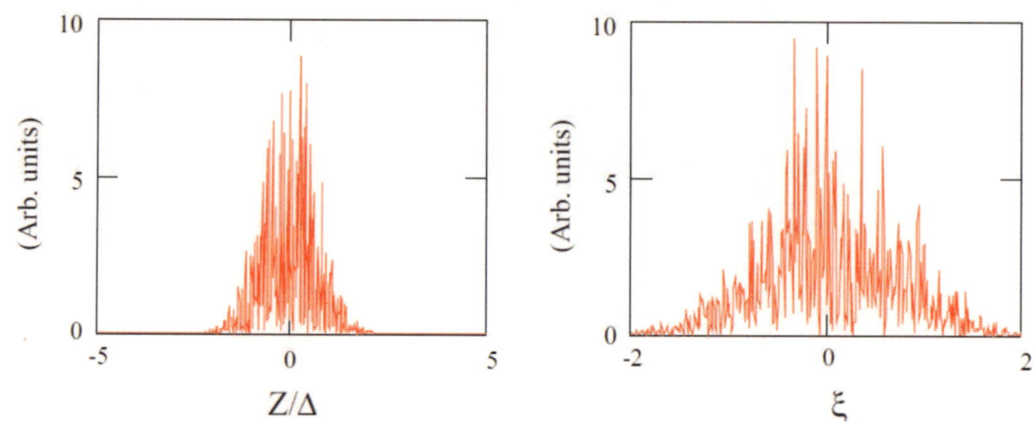

Fig. 29. Temporal and spectral distribution of the SASE FEL

Each randomly phased supermode makes one optical pulse which results in a series of spikes separated by a fixed distance: the coherence length (see Fig. 30).

The evolution of each supermode is nearly independent from the others. There is only the slippage which enables to create a coherence zone (of the same order of the coherence length) and produces a sort of smoothing of the chaos, while the field increases along the undulator (see Fig. 31).

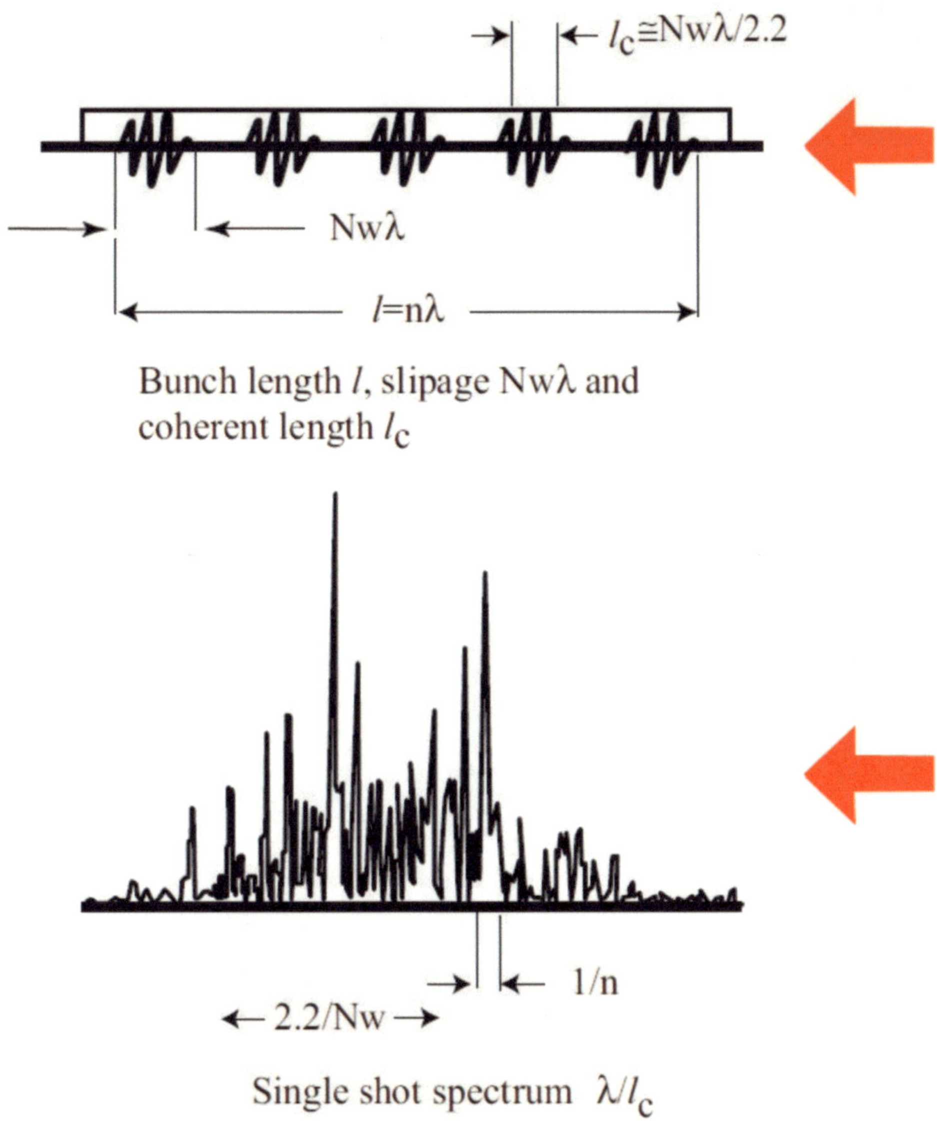

Fig. 30. Coherence length and spikes distribution

Dealing with a partially chaotic phenomenon, we should also define a probability distribution of the energy of these spikes [7]:

$$P(E) = \frac{M^M}{(M-1)!} x^{M-1} e^{-Mx}, \; x = \frac{E}{\langle E \rangle}$$

Such distributions are shown in Fig. 32. it is obvious that, when the number of supermodes increases, the distribution narrows and the average quadratic deviation is $\sigma_E \cong \frac{1}{\sqrt{M}}$

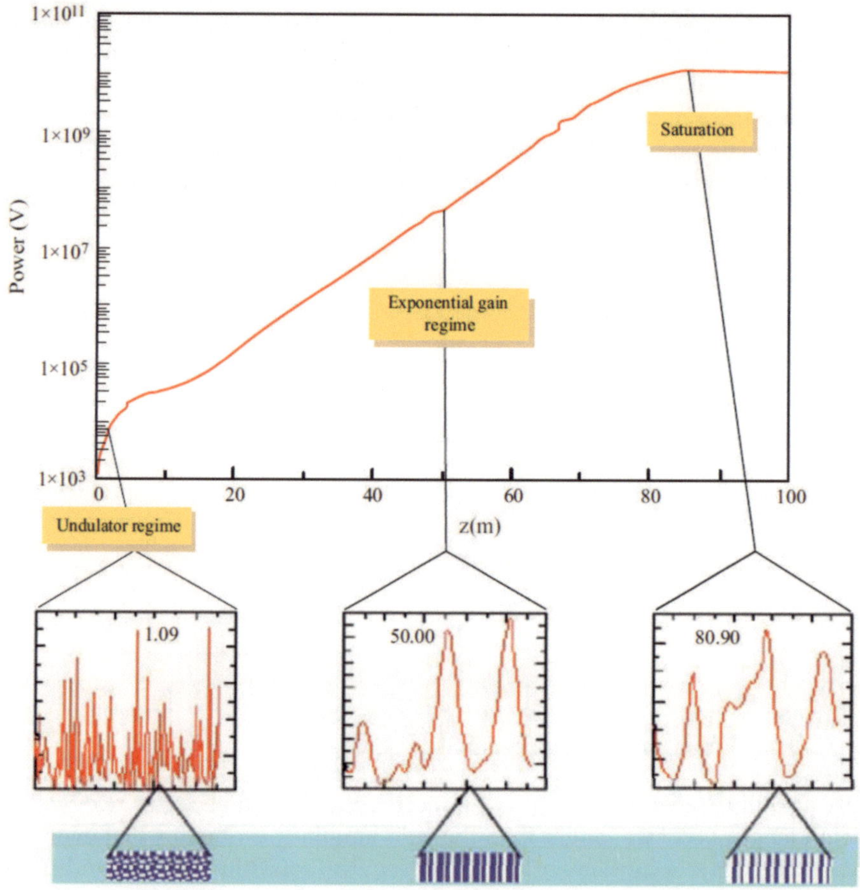

Fig. 31. Evolution of the power while signal increase and fluctuation smoothing

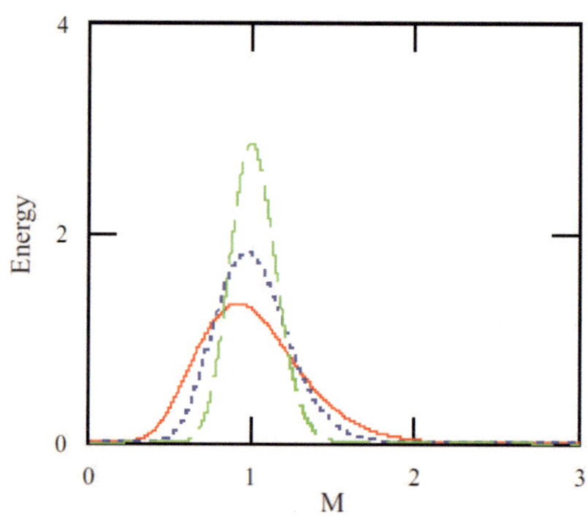

Fig. 32. Energy distribution of the FEL spikes as a function of $x = \frac{E}{\langle E \rangle}$, for various values of M. Continuous line: M=10, dotted line: M=20 and segmented line: M=50

Clarified the notion of longitudinal coherence, we move on to the definition of the transverse coherence. As previously, the problem of the definition of a transverse coherence results from the fact that there is no more optical cavity, which prevent us from defining strictly speaking transverse modes. We should note that in the high gain case, we will no longer be able to talk about free modes but guided modes, because of the strong distortion in the propagation caused by the interaction itself. An example of such behaviour is given in Fig. 33.

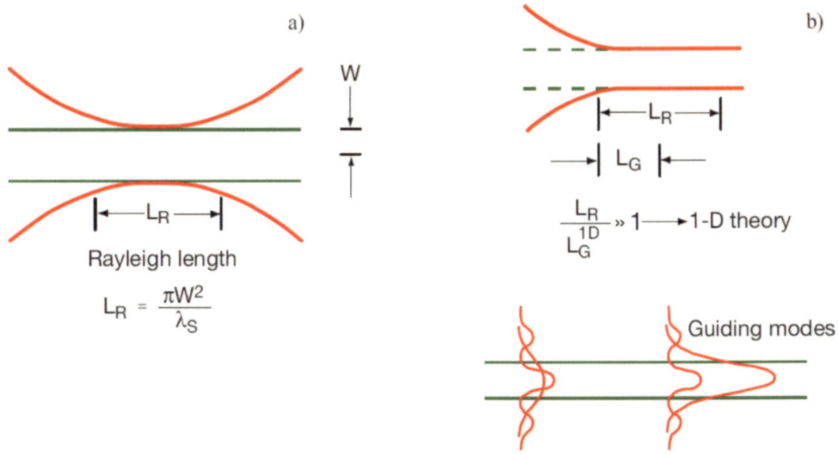

Fig. 33. a) Gaussian mode and b) guided mode

The Figure also illustrates the gain guiding mechanism, which becomes relevant as soon as the Rayleigh length becomes longer than the gain length. It is obvious that even in this case the ρ parameter plays a fundamental role. Indeed, the condition for the guided modes can be written as follows:

$$\rho >> \frac{\lambda_u \lambda}{4\pi^2 \sqrt{3} w_0^2} \tag{75}$$

Without optical cavity, the waist is essentially related to the transverse section of the electronic beam.

Before concluding this section, we note that the effects of inhomogeneous broadening and relative reduction of the gain in SASE can be treated using the same procedure given in the case of a FEL oscillator. In the case of energy spread, the parameter which quantifies the importance of the associated inhomogeneous broadening is

$$\tilde{\mu}_\varepsilon = \frac{2\sigma_\varepsilon}{\rho} \tag{76}$$

The macroscopic consequence is an increase of the saturation length given by:

$$\tilde{L}_g = \chi L_g \quad \chi = 1 + 0.185 \frac{\sqrt{3}}{2} \tilde{\mu}_\varepsilon^2 \tag{77}$$

The analysis of a SASE FEL system can then be done calculating quantities such as the saturation length and saturation power. One possible combination of the various parameters is given in Fig. 34, where we reported the saturation length, the power and the wavelength of a SASE FEL.

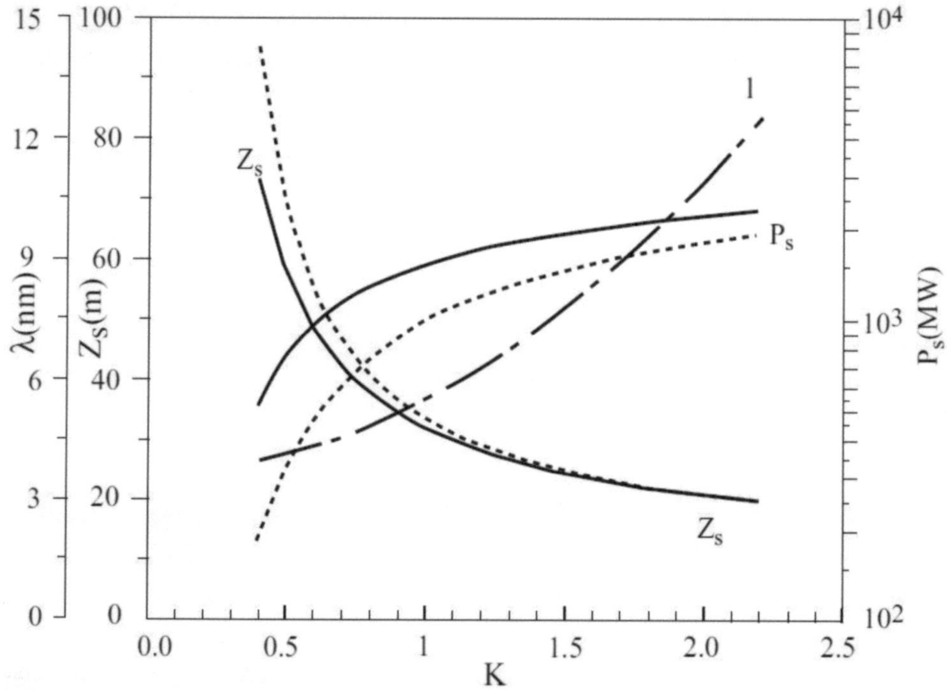

Fig. 34. Saturation length, power and wavelength as a function of the K parameter for a SASE FEL operating with a 2.8 cm period undulator and a 1 GeV energy, 5×10^{-4} energy spread and 1 kA current electron beam

9. FEL in the SASE regime and example of application

Several SASE FEL have been proposed as fourth generation light sources. These FEL will deliver radiation in the extreme ultraviolet and X-ray ranges, with a brightness at least 10 orders of magnitudes times larger than the one presently available on the synchrotron light sources. The evolution of the brightness in time is represented in Fig. 35. The so-called brightness is defined by *Number of photons per second per source unit area, per unit solid angle, per 0.1% bandwidth* $(photons / sec / mm^2 / mrad^2 / 0.1\%\Delta\omega)$.

The 10 orders of magnitude jump offered by the SASE FEL results from:

1. While conventional synchrotron light sources use the simple principle of spontaneous emission, SASE FEL are based on stimulated emission which guarantees a photon gain of $G = \frac{1}{9} e^{\frac{L_u}{\lambda_u} 4\pi\sqrt{3}\rho}$, with $L_u = N\lambda_u$ the undulator length. Assuming $N\rho \cong 1$, one gets that with respect to the conventional sources, the photon flux is higher by roughly eight orders of magnitude.

2. The conventional sources use electron beams accelerated in a storage ring, while SASE FEL use an electron beam accelerated in a LINAC (linear accelerator) of high energy. The transverse dimensions of the emitting source (the electron beam) is therefore 100 times smaller in the case of the SASE FEL that in the case of the synchrotron sources. According to what has been said in (1) and given that the brightness is inversely proportional to the transverse section of the beam, the magnification factor reaches ten orders of magnitude.

Another interesting aspect of the SASE FEL is the pulse duration of the radiation: as illustrated in Fig. 36, SASE FEL can produce ultra-short radiation pulses.

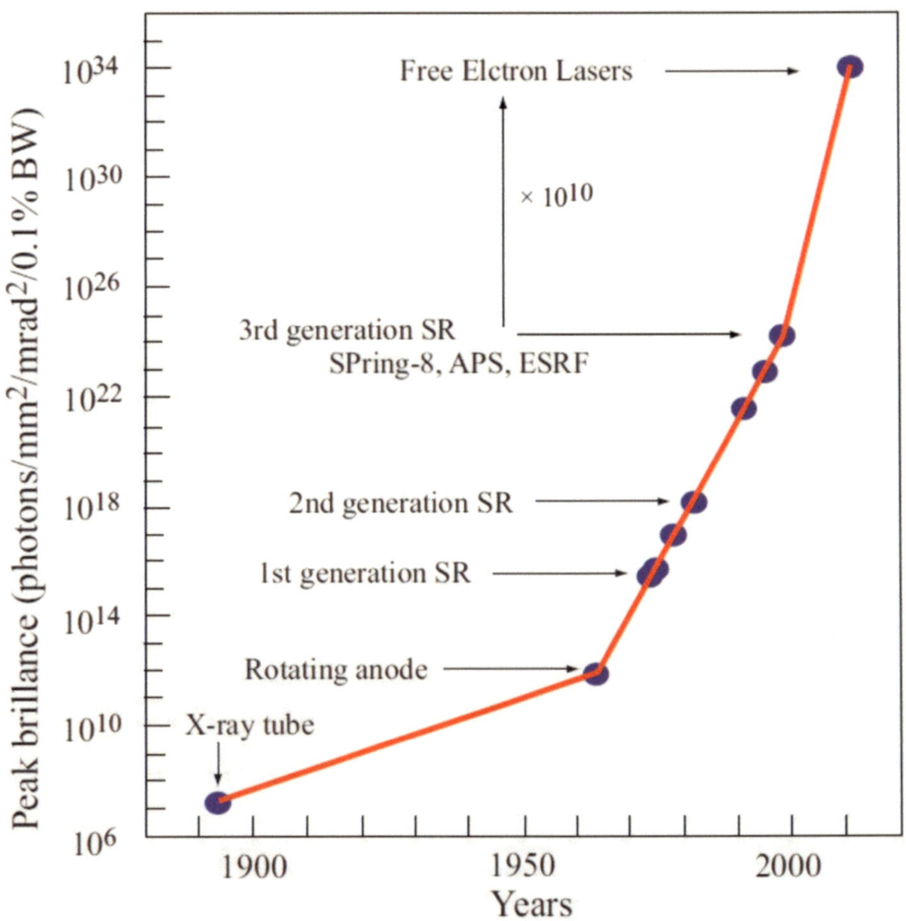

Fig. 35. Evolution of the brilliance of the X-ray conventional sources over the years and perspectives offered by FEL sources

The shortness of the X-ray pulse is related to the duration of the current pulse, which generated it. The electron beam length can be understood as follows. Assuming a maximum charge of 1 nC (10^{-9} C) and given that a SASE FEL needs currents of at least 1 kA to obtain a reasonable saturation length, it turns obvious that the duration of the electron beam should be smaller than a few ps.

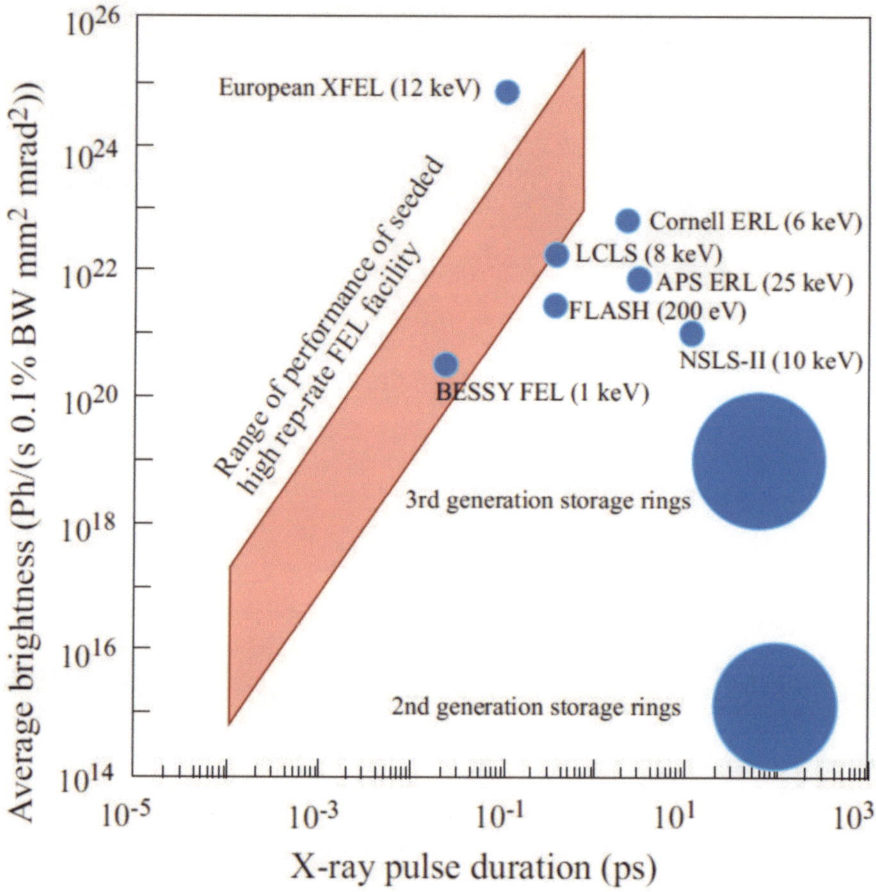

Fig. 36. Average brightness vs pulse duration of the SASE FEL sources

In conclusion: the SASE FEL can provide radiation in the X-ray range with duration of the order of the ps with extremely high brightness. Such characteristics make the SASE FEL sources of high interest for various types of applications.

10. References

[1] A. Yariv "Quantum Electronics", John Wiley and sons USA (1998)

[2] U. Kelly, *Ultrafast Laser Physics, Institute of Quantum Electronics, Swiss Federal Institute of Technology*. ETH (German: Eidgenössische Technische Hochschule) Höngerberg-HPT, CH-8093 Zürich, Switzerland.

[3] C.A. Brau, *Free-Electron Lasers*, Academic Press, Oxford (1990)

[4] G. Dattoli, A. Renieri, and A. Torre, *Lectures on Free Electron Laser Theory and Related Topics* (World Scientific, Singapore, 1993)

[5] F. Ciocci, G. Dattoli A. Torre and A. Renieri, *Insertion Devices for Synchrotron Radiation and Free Electron Laser* (World Scientific, Singapore 2000)

[6] G. Dattoli, P.L. Ottaviani, S. Pagnutti, *Booklet for FEL design: A collection of practical formulae*, ENEA Report RT/2007/40/FIM (2007)

[7] E.L. Saldin, E.A. Schneidmiller, M.V. Yurkov, *The Physics of Free Electron Lasers*, (Springer-Verlag Berlin 1999)

2

Design and Simulation Challenges of a Linac-Based Free Electron Laser in the Presence of Collective Effects

S. Di Mitri
Sincrotrone Trieste
Italy

1. Introduction

The generation of Free Electron Laser (FEL) radiation relies on the extraction of electromagnetic energy from kinetic energy of a relativistic electron beam by propagating it along the axis of a periodic lattice of alternating magnetic dipolar fields, known as undulator. This forces the beam to undulate transversally, thus causing the electrons to emit electromagnetic radiation. The fundamental wavelength emitted is proportional to λ_u/γ^2 , where λ_u is the undulator period, typically a few centimeters long, and γ is the relativistic Lorentz factor of the electrons, which typically reaches several thousand for X-ray emission. The main figures of merit of an FEL are extremely high brilliance, close to full transverse and longitudinal coherence, a bandwidth approaching the Fourier limit and a stable and well characterized temporal structure in the femtosecond time domain. We can identify two general ways to generate X-rays with an FEL. The Self Amplified Spontaneous Emission (SASE) [1–4] relies on the interaction of electrons and photons that are emitted by the electron beam itself. The electron bunching that generates the coherent emission of radiation starts to grow from the natural noise of the initial electron distribution. For this reason, the SASE output radiation is relatively poor in longitudinal coherence. In the High Gain Harmonic Generation (HGHG) scheme [5–11], instead, the initial energy modulation is driven by an external seed laser. It is then transformed into density bunching in a dispersive section inserted in the undulator chain. In this case, the output FEL properties reflect the high longitudinal coherence of the seed laser.

The FEL high brilliance, high intensity and shot-to-shot stability strongly depends on the electron beam source. As an example, an FEL requires a high peak current to increase the number of photons per pulse and reach power saturation at an early stage in the undulator. Magnetic bunch length compression is one way to increase the electron bunch current. It is carried out via ballistic contraction or elongation of the particles path length in a magnetic chicane. The linac located upstream of the magnetic chicane is run off-crest to establish a correlation between the particle longitudinal momentum with respect to the reference particle and the z-coordinate along the bunch, i.e. the bunch head has a lower energy than the tail. In the magnetic chicane, due to their lower (higher) rigidity, leading (trailing) particles travel on a shorter (longer) path than trailing (leading) particles. Since all particles of the ultra-relativistic beam travel in practice at the speed of light, the bunch edges approach the centroid position and the total bunch length is finally reduced. Unfortunately, magnetic compression is very often enhancing single particle and collective effects that may degrade the electron beam quality. Delivering a high quality electron beam and machine flexibility to serve a broad range of potential applications imposes severe requirements on the final electron beam parameters and the machine design. The primary goal of the machine design is that of preserving the 6-D electron beam emittance, ϵ, at the electron source level. Liouville's theorem [12] states that the phase space hypervolume enclosing a chosen group of particles is an invariant of the Hamiltonian system as the particles move in phase space, if the number of particles in the volume does not change with time. This volume is the emittance of the particle ensemble. In its more general sense, Liouville's theorem applies to Hamiltonian systems in which the forces can be derived from a potential that may be time dependent, but must not depend on the particles' momentum. Thus, the following collective phenomena limit the applicability of the Liouville's theorem to particles motion in an accelerator: collisions, space charge forces – intended as short-range inter-particle Coulomb interactions –, wake fields, electromagnetic radiation emission and absorption. In order to preserve the initial 6-D volume along the entire electron beam delivery system, all the afore-mentioned effects have to be analytically evaluated and simulated. In particular, we will focus on the particle motion in a single-pass, linac-driven FEL in the presence of the following short-range effects:

i) space charge (SC) forces [13–16];

ii) geometric longitudinal and transverse wake field in the accelerating structures [17–18];

iii) coherent synchrotron radiation (CSR) emission in dispersive systems [19–27];

For a clearer illustration of these topics, we will initially assume the particle motion being uncoupled in the transverse and in the longitudinal phase space. A good transverse coherence of the undulator radiation is ensured by the following limit [28] on the transverse normalized beam emittance: $\gamma\epsilon \leq \gamma\lambda/(4\pi)$, where $\lambda/(4\pi)$ is the minimum phase space area for a diffraction limited photon beam of central wavelength λ. Typically, $\gamma\epsilon \approx 1$ mm mrad for λ in the nm range. Note that the local emittance, referred to as "slice", can vary significantly along the bunch to give hot-spots where lasing can occur. In fact, in contrast to linear colliders where particle collisions effectively integrate over the entire bunch length, X-ray FELs usually concern only very short fractions of the electron bunch length. The integration length is given by the electron-to-photon longitudinal slippage over the length of the undulator, prior to FEL power saturation. The FEL slippage length is typically in the range 1–30 μm, a small fraction of the total bunch length. Thus, the electron bunch slice duration can reasonably be defined as a fraction of the FEL slippage length.

The longitudinal emittance or more precisely the energy spread for a given electron bunch duration, has to be small enough to permit the saturation of the FEL intensity within a reasonable undulator length. At saturation of a SASE FEL , $P \approx \rho P_e$ where P_e is the electron beam power and ρ, the so-called Pierce parameter [29], is the FEL gain bandwidth expressed in terms of normalized energy. It is seen to be a measure of the efficiency of the interaction, with typical values in the X-ray regime of $10^{-4} \leq \rho \leq 10^{-3}$. The relative energy spread of the electron beam at saturation is $\sigma_\delta \approx \rho$. Thus, if there is an initial electron energy spread approaching the maximum, which occurs at an FEL saturation of $\sigma_\delta \geq \rho$, then the FEL interaction is greatly reduced. For a seeded FEL such as in a HGHG scheme, the total energy spread $\sigma_{\delta,tot}$ is approximately given by the quadratic sum of the uncorrelated term $\sigma_{\delta,un}$, the energy modulation amplitude induced by the seeding laser $\Delta\delta$, and the residual energy chirp

$\sigma_{\delta,ch}$. The maximum acceptable deviation from the desired flatness of the longitudinal phase space is limited by $\sigma_{\delta,tot} \leq \rho$. At the same time, the FEL harmonic cascade is effective only for $\Delta\delta \geq N\sigma_{\delta,un}$ with N the ratio between the seed wavelength and the harmonic wavelength at which the final undulator is tuned. So, if $N = 10$ to produce, as an example, 20 nm and we want $\sigma_{\delta,tot} \leq 1 \cdot 10^{-3}$, we require a final energy chirp $\sigma_{\delta,ch} \leq 10^{-3}$ and a final slice energy spread (here assumed to be as uncorrelated for conservative calculations) $\sigma_{\delta,un} \leq 150$ keV at the beam energy of 1.5 GeV. Typical electron beam parameters of the fourth generation linac-based FELs (from infrared to X-rays spectral range) are listed in the following: 0.1–1 nC charge, 0.5–2 mm mrad normalized emittance, 0.5–3 kA peak current, 0.05–0.1 % relative energy spread and 1–50 GeV final electron energy.

2. Short-range space charge forces

The electron beam generation from a metallic photo-cathode [30,31] in the γ<200 regime is dominated by SC forces, which scale like γ^{-2} in the laboratory frame, and by image charge forces in the immediate cathode vicinity. The dynamics in these regions is usually simulated with quasi-static 3-D codes like GPT [32] or Astra [33] (the static electric field is calculated in the beam frame and both electric and transverse magnetic fields are included in the laboratory frame). They can model the electron gun azimuthal asymmetries (due to imperfections) and also large beam aspect ratios. Special beam current shaping at the cathode can be included: this is required to ameliorate the linearization of the longitudinal phase space in the presence of high beam charge (> 100 pC) and strong longitudinal geometric wake field, with a further positive impact on the final energy chirp [34]. These codes predict an uncorrelated energy spread out of the RF photo-injector in the range 1–3 keV rms [35–37]. The radiative force related to the variation of the bunch total electromagnetic energy during acceleration has also been recognized as a new source of local energy spread [38]. This new physics can only be studied with codes that correctly calculate the beam fields from the exact solutions of the Maxwell's equations that is the full retarded potentials. The slice energy spread is of crucial importance for the suppression of the so-called microbunching instability [39–41]. In fact, the velocity spread of the relativistic electrons acts as a low pass filter effect for density and energy modulations generated in the Gun. While the conversion of energy and density modulation amplitudes happens over the SC oscillation wavelength of $\sim$ 1 m, only wavelengths longer than 10's of μm survive out of the Gun [42].

Even when the electrons reach energies as high as $\sim$50 MeV, SC forces have to be considered in two cases. In the first case, the longitudinal electric field generated by clusters of charges or density modulation along the bunch can still be sufficiently high to induce an energy modulation as the beam travels along the accelerator. Such an energy modulation translates into density modulation when the beam passes through a dispersive region (this might happen in a magnetic bunch length compressor or in a dispersive transfer line), with a consequent degradation of the energy and the current flatness. This dynamics is assumed to be purely longitudinal and it is discussed in detail in Section 6. In the second case, SC forces might be enhanced because of the very high charge density achieved with the bunch length compression. Although 3-D tracking codes can be used to simulate the compression, an analytical estimation of the impact of these forces on the transverse dynamics is still possible. Following [15] we find that the rms transverse envelope equation for a bunched beam in a linac is:

$$\sigma'' + \frac{\gamma'}{\gamma}\sigma' + K\sigma = \frac{k_s}{\gamma^3\sigma} + \frac{\epsilon_{th}}{\gamma^2\sigma^3} \tag{1}$$

Here, the standard deviation of the beam transverse size σ is assumed to be a function of the axial position s along the linac, $\gamma' = d\gamma/ds$ is the accelerating gradient, $K = (eB_0)^2/(2mc\gamma)^2$ is the focusing gradient of a solenoid of central field B_0, $k_s = I/(2I_A)$, $I_A = 17kA$ being the Alfven current, and ϵ_{th} is the thermal emittance, which is mainly due to the photoemission process at the cathode surface: it is a Liouville invariant throughout acceleration. We now consider the following invariant envelope solution for the beam size:

$$\bar{\sigma} = \frac{1}{\gamma'}\sqrt{\frac{k_s}{\gamma(1/4+\Omega^2)}} \tag{2}$$

where $\Omega = \sqrt{K}\gamma/\gamma'$. The laminarity parameter, ρ_L, is defined as the ratio of the "space charge term" driven by k_s and the "emittance term" driven by ϵ_{th} in eq.1, computed with the substitution $\sigma = \bar{\sigma}$:

$$\rho_L = \left(\frac{k_s}{\epsilon_{th}\gamma\gamma'\sqrt{1/4+\Omega^2}}\right)^2 \tag{3}$$

If $\rho_L \gg 1$, the particles motion is dominated by SC forces with negligible contribution from the betatron motion. By computing the laminarity parameter as function of the beam parameters along the transport system, we can identify machine areas where $\rho_L \gg 1$, which should be investigated more carefully with 3-D codes. As an example, for an electron linac driven by a standing wave photoinjector with no external focusing, $\Omega^2 = 1/8$, the energy at which the transition occurs, $\rho_L = 1$, can be quite high:

$$\gamma_{tr} = \sqrt{\frac{2}{3}\frac{2k_s}{\epsilon_{th}\gamma'}} \tag{4}$$

often corresponding to several hundreds of MeV. Unfortunately, the transition from SC dominated ($\rho_L \gg 1$) to quasi-laminar ($\rho_L \ll 1$) motion cannot be described accurately by this model because, by definition, $\bar{\sigma}$ is a valid solution of eq.1 only for $\rho_L \gg 1$.

We consider the following example: a 400 pC bunch time-compressed to reach 0.5 kA and a 800 pC bunch compressed to reach 1 kA. In both scenarios we assume $\epsilon_{th} = 0.6$ mm mrad, $\gamma' = 39.1$ (corresponding to 20 MV/m in a S-band linac); one- and two-stage compression is adopted at the energy of, respectively, 300 and 600 MeV. The transition energy computed with eq.4 is 500 MeV for the low charge and 1 GeV for the high charge. We can conclude that the beam dynamics is not SC-dominated in the case of low charge/two-stage compression and only weakly dominated in the high charge/two-stage compression. The one-stage compression is performed at an energy well below the computed thresholds for both charges. Thus, a careful study of the 3-D beam dynamics in the presence of SC forces should been carried out for this option. Figure 1 shows the laminarity parameter, eq.3 with $\Omega = 0$, computed on the basis of a particle tracking, performed with the `elegant` code [43], from the injector end (100 MeV) to the linac end (1.5 GeV), in the configuration of one-stage compression. No external solenoid focusing is considered. The laminarity parameter and the peak current are computed as the average value over the bunch core, which runs over $\sim 80\%$ of the total bunch length.

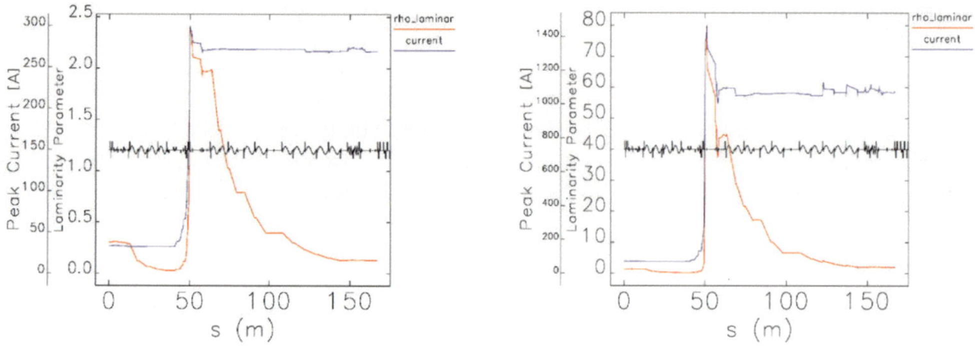

Fig. 1. Laminarity parameter and peak current computed from particle tracking. Left: 400 pC compressed by a factor of 6.5. Right: 800 pC compressed by a factor of 10. In both cases, $\rho_L \gg 1$ immediately downstream of the first magnetic compression, where the peak current in the bunch core rises to 1 kA. Then ρ_L falls down with $\sim \gamma^{-2}$ dependence.

3. Short-range geometric longitudinal wake field

3.1 Analytical model

For a longitudinal charge distribution λ_z, the energy loss of a test electron due to the electromagnetic wake of leading electrons is given by the geometric wake potential [17, 18]:

$$W(z) = -\int_z^{\infty} w(z-z')\lambda_z(z')dz' \tag{5}$$

where *w(z-z')* is the Green's function, also called "wake function", that emulates the effect of the wake fields as generated by a single particle. Because of the principle of causality, the wake is zero if the test electron is in front of the wake source. If the beam is much shorter than the characteristic wake field length s_0 [44] and if the structure length L is much longer than the catch-up distance $a^2/(2\sigma_z)$, where a is the cell iris radius and σ_z is the rms bunch length, then the wake field is said to be, respectively, in the *periodic structure* and in the *steady state* regime:

$$\frac{a^2}{2L} \ll \sigma_z \ll s_0 \tag{6}$$

The characteristic length $s_0 = \left(0.41a^{1.8}g^{1.6}/L_c^{2.4}\right)$ [44] is function of the cell iris radius a, of the cell inner width g and of the iris-to-iris distance L_c. In the very special case of periodic structure, steady state regime and very short electron bunches, the wake function assumes a simple form. For the longitudinal component we have in [V/C/m]:

$$w_L(0^+) = \frac{Z_0 c}{\pi a^2} \tag{7}$$

Here Z_0=377 Ω is the free vacuum impedance. Typical values are s_0=1.5 mm, s_1=0.5 mm, σ_z=40–100 μm and $a^2/(2L_c)$=2–20 μm. So, while the steady state approximation is always satisfied, the periodic structure approximation might be not. Nevertheless, it was found that by computing the short-range wake numerically and fitting it with a simple function, one can obtain a result that is valid over a large range of z (position along the bunch) and over a useful range of parameters [45]:

$$w(z) = \frac{Z_0 c}{\pi a^2} \cdot e^{\left(-\sqrt{z/s_0}\right)} \left[\frac{V}{C \cdot m}\right] \tag{8}$$

Depending on the specific geometry of the accelerating structure, eq. 8 can be modified with additional terms whose dependence on z is a polynomial. In such cases, the wake amplitude and the polynomial coefficients are determined by fitting procedures (see [46] as an example).

3.2 Energy loss

The longitudinal wake potential induces a total energy loss of the electron beam so that the relative energy change at the bunch length coordinate $\bar{z}$ is [17]:

$$\delta_w(\bar{z}) = -\frac{e^2 L}{\gamma m c^2} \int_0^\infty w(z) n(\bar{z} - z) dz \tag{9}$$

where $n(\bar{z} - z)$ is the longitudinal particle distribution with normalization $\int_{-\infty}^{\infty} n(z)dz = N$ (N is the total number of electrons in the bunch). As an example, for a uniform longitudinal bunch profile, one has $n(z) = N/\left(2\sqrt{3}\sigma_z\right)$ for $|z| \leq \sqrt{3}\sigma_z$ and $n(z) = 0$ for $|z| > \sqrt{3}\sigma_z$. If the constant wake function in eq.7 is used, then eq.9 yelds a linear wake-induced energy change along the bunch coordinate:

$$\delta_w(\bar{z}) = -\frac{2 N r_e L}{\gamma a^2} \left(1 + \frac{\bar{z}}{\sqrt{3}\sigma_z}\right) \tag{10}$$

where we have used the identity $Z_0 c \epsilon_0 = 1$; $r_e = 2.82 \cdot 10^{-15}$ m is the classical electron radius. It is straightforward to calculate the standard deviation of the wake-induced relative energy loss:

$$\sqrt{< \delta_w(\bar{z})^2 >} = -\frac{2}{\sqrt{3}} \frac{N r_e L}{\gamma a^2} \tag{11}$$

For a S-band, 1 GeV linac with inner iris radius of 10 mm and bunch charge of 200 pC, the total loss is of the order of a few MeV. Since the uncorrelated energy spread is a few order of magnitudes smaller than this, the energy loss translates into correlated energy spread. In the linear approximation, it could be removed by running off-crest some accelerating structures at the end of the linac in order to compensate this additional energy chirp.

3.3 High order energy chirp

Generally, a linear description of the longitudinal beam dynamics is not accurate enough. In fact, a nonlinear energy chirp usually affects the longitudinal phase space. It reduces the effective compression factor, enlarge the FEL spectral output bandwidth (via quadratic component of energy chirp) [47] and create current spikes at the bunch edges during compression (via cubic component of energy chirp) [34, 48], which lead to further detrimental effects on energy spread and emittance due to enhanced CSR field and wake fields. To investigate this nonlinear particle dynamics, we start with the expression for the bunch length transformation through magnetic compression at 2nd order:

$$z = z_0 + R_{56}\delta + T_{566}\delta^2 \tag{12}$$

R_{56} (T_{566}) is the integral of the first (second) order dispersion function along the chicane, taken with the signed curvature of each dipole. It governs the linear (quadratic) path-length dependence from the particle energy. In the following, we choose a longitudinal coordinate system such that the head of the bunch is at z<0. A chicane has $R_{56} < 0$ and $T_{566} > 0$ with this convention. A linac with energy gain $eV \sin\phi$ (V and ϕ are the RF accelerating peak voltage and phase, respectively) imparts to the beam the following linear and quadratic energy chirp, $\delta = \frac{\Delta E}{E_{BC}} \approx \delta_{0,u} + hz_0 + h' z_0^2$, with:

$$\begin{aligned} h &= \frac{1}{E_{BC}}\frac{dE}{dz} = \frac{2\pi}{\lambda_{RF}} \frac{eV\cos\phi}{E_0 + eV\sin\phi} \\ h' &= \frac{1}{2}\frac{dh}{dz} = -\left(\frac{2\pi}{\lambda_{RF}}\right)^2 \frac{eV\sin\phi}{E_0 + eV\sin\phi} \end{aligned} \tag{13}$$

E_0 and E_{BC} are the beam mean energy at the entrance and at the exit of the linac, respectively. The "linear compression factor" is defined as:

$$C = \frac{\sigma_{z0}}{\sigma_z} \approx \frac{1}{1 + hR_{56}} \tag{14}$$

In practice, compressions by a factor bigger than ~ 3 are dominated by nonlinear effects such as sinusoidal RF time-curvature (mostly giving a quadratic energy chirp) in the upstream linac and T_{566}. For simple magnetic chicanes with no strong focusing inside, the RF and the path-length effects $T_{566} \approx -3R_{56}/2$ always conspire with the same signed 2nd order terms to make the problem worse. By inserting eq.13 into eq.12, we obtain the bunch length transformation at 2nd order:

$$\begin{aligned} \sigma_z^2 = &\left(R_{56}\sigma_{\delta 0,u}\right)^2 + \left(1 + hR_{56}\right)^2 \sigma_{z0}^2 + \left(T_{566}\sigma_{\delta 0,u}^2\right)^2 + \\ &\left(h^2 T_{566} + h' R_{56}\right)^2 \sigma_{z0}^4 + \left(2hT_{566}\sigma_{z0}\sigma_{\delta 0,u}\right)^2 \end{aligned} \tag{15}$$

In order to linearize the 2nd order bunch length transformation, the use of a short section of RF *decelerating* field at a higher harmonic of the linac RF frequency [49, 50] is usually adopted, thereby maintaining the initial temporal bunch profile and avoiding unnecessary amplification of undesired collective effects. The necessary harmonic voltage is [50]:

$$eV_x = \frac{E_{BC1}\left[1 + \frac{\lambda_s^2}{2\pi^2}\frac{T_{566}}{|R_{56}|^3}\left(1 - \frac{\sigma_z}{\sigma_{z0}}\right)^2\right] - E_0}{\left(\frac{\lambda_s}{\lambda_x}\right)^2 - 1} \tag{16}$$

The square of the harmonic ratio $n^2 = (\lambda_s/\lambda_x)^2$ in the denominator suggests that higher harmonics are more efficient for 2nd order compensation, decelerating the beam less.

Unlike the quadratic chirp, the cubic energy chirp in a S-band linac is dominated by a contribution from the longitudinal wake potential (this may include both SC in the injector and geometric longitudinal wake field), rather than by higher order terms from the RF curvature. It has three main disrupting consequences: i) it reduces the efficiency of the magnetic compression for the bunch core, since during compression the edges "attract" particles from the core reducing the current in this region; ii) it induces current spikes at the edges that may be dangerous sources of CSR, with a direct impact on the transverse emittance and on the energy distribution; iii) wake field excited by a leading edge spike may cause additional energy spread in the low gap undulator vacuum chambers. The cubic chirp is always negative for a flat-top charge distribution [51]. After the interaction with longitudinal wake fields, its sign is reversed at the entrance of the second compressor, if present, so enhancing the energy-position correlation of the bunch edges with respect to the core. The edges are there over-compressed producing current spikes. On the contrary, a negative cubic chirp at the chicane provides under-compression of the edges. For these reasons the sign of the cubic term is related to the topology of the longitudinal phase space and to the final current profile.

For a given charge and bunch length, the interaction of the cubic chirp coming from the injector with the longitudinal wake field of the succeeding linac cannot be arbitrarily manipulated. However, the user has one more degree of freedom to manage the cubic chirp before reaching the magnetic compressor, that is by setting the harmonic cavity a few degrees away from the usual decelerating crest. Typical voltages of a fourth harmonic (X-band) RF structure adopted for compensating the quadratic energy chirp are in the range 20–40 MeV. The RF phase is usually shifted by a few X-band degrees from the decelerating crest to cancel the cubic energy chirp. Adjustments to the voltage and to the phase have to be studied with a simulator, depending on the cubic energy chirp coming from the injector and on the effective compression factor.

3.4 Current shaping

In some cases the knob of off-crest phasing the high harmonic structure to minimize the cubic energy chirp may be weak and a significant increase is needed in the amplitude of the structure voltage. One way to achieve this is to use a density distribution other than the standard parabolic one. This is one of the motivations leading to the technique of current shaping [34]. The basic premise for current shaping is that the output bunch configuration is largely pre-determined by the input bunch configuration and that therefore it is possible to find a unique electron density distribution at the beginning of the linac that produces a distribution at the end of the linac that is flat both in energy and in current. To find this distribution, one needs to reverse the problem, i.e. start at the end of the linac and move backwards towards the beginning of the linac. Eq.17 shows that for a given electron density λ_z and wake function w_z, the electron energy at the end of a section of the linac, defined as δ_f (with z_f being the electron coordinate taken with respect to the bunch center), can be determined using the electron energy δ_i and the coordinate z_i at the beginning of the section:

$$\delta_f(z_f) = \delta_i(z_i) + eU\cos(kz_i + \phi) - LQ\int_{z_i}^{+\infty} w_z(z_i - z')\lambda_z(z')dz' \quad (17)$$

where U, ϕ, L define the RF voltage, phase and length of the linac section, k is the wave number, e is the electron charge and Q is the bunch charge. For a relativistic beam, the electron distribution function λ_z does not change during acceleration, i.e $z_i = z_f$, and, therefore, eq.17 can be used to define $\delta_i(z_i)$ as a function of $\delta_f(z_f = z_i)$. Thus, beginning with a desirable electron distribution at the end of the linac section, one can find the distribution at the beginning of the linac section that will eventually make it. A similar situation arises in a bunch compressor if the CSR energy change is negligible with respect to that induced by the longitudinal wake field in the linac.Then, the electron coordinate at the beginning of the bunch compressor can be found using the electron coordinate at the end of the bunch compressor using eq.12.

The above considerations justify a concept of reverse tracking [34]. LiTrack [52], a 1-D tracking code, can be used to convolve the actual line-charge distribution with the externally calculated longitudinal wake function. A desirable distribution both flat in energy and current is set up at the end of the accelerator. Starting with this distribution and tracking it backward, the nearly linear ramped peak current shown in Figure 2 is obtained at the start of the accelerator. This

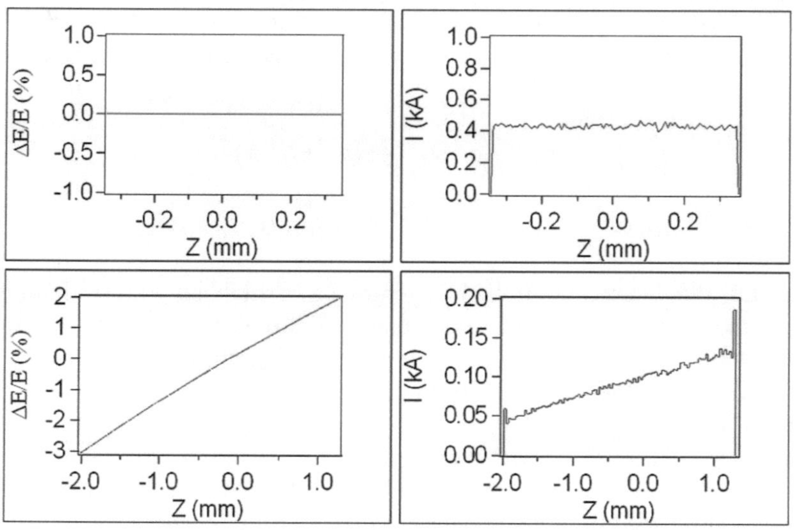

Fig. 2. Reverse tracking. It begins with "flat-flat" distribution at the end of the accelerator (top line) and moves towards beginning of the accelerator (bottom line). Published in [M. Cornacchia, S. Di Mitri, G. Penco and A. A. Zholents, Phys. Rev. Special Topics - Accel. and Beams, 9, 120701 (2006)].

result can be understood if one uses the wake function for an accelerating structure consisting of an array of cells, eq.8, and convolutes it with a linearly ramped current distribution. The wake potential is highly linear and this is why the final distribution is flat in energy.

Producing a linearly ramped electron bunch current at the exit of the injector is somewhat of a challenge because of the strong nonlinearity of the SC fields at low energy. The longitudinal blow-up of the electrons from the cathode to the first accelerating structure poses a limit to the ramping fraction of the bunch that meets the current linearity requirement. A fourth-degree polynomial distribution was found in [34] to offer the best cancellation of the high orders nonlinear contributions of the SC field, and thus increases the bunch fraction that follows a linear ramp. This cancellation helps preserving the linearity of the fields in the space-charge dominated part of acceleration.

4. Short-range geometric transverse wake field

4.1 Analytical Model

Similarly to the longitudinal case, the transverse wake function of a linac structure can be approximated with an analytical expression [53]:

$$w^1(z) = \frac{4s_1 Z_0 c}{\pi a^4}\left[1 - \left(1 + \sqrt{\frac{z}{s_1}}\right) e^{-\sqrt{z/s_1}}\right]\left[\frac{V}{C \cdot m^2}\right] \tag{18}$$

The transverse motion of a relativistic electron in the linac in the presence of the short-range geometric transverse wake field is described by an ordinary 2nd order differential equation in the complete form. The l.h.s. of this equation is the homogeneous equation for the betatron motion in the horizontal or vertical plane; the r.h.s. contains the convolution of the transverse wake function with the local current distribution and is also linearly proportional to the relative displacement of the particle from the axis of the accelerating structure [54, 55]:

$$\frac{1}{\gamma(\sigma)}\frac{\partial}{\partial\sigma}\left[\gamma(\sigma)\frac{\partial}{\partial\sigma}x\left(\sigma,\gamma\right)\right] + \kappa(\sigma)^2 x\left(\sigma,\gamma\right) =$$
$$\epsilon(\sigma)\int_{-infty}^{\zeta} w_n^1\left(\zeta-\zeta'\right)F(\zeta')\left[x\left(\sigma,\zeta'\right)-d_c(\sigma)\right]d\zeta' \tag{19}$$

where $\sigma = s/L$ is the distance from the linac entrance normalized with the total linac length L; $\zeta = z/l_b$ is the longitudinal bunch coordinate at location σ measured after the arrival of the bunch head, normalized with the full width bunch length; $F(\zeta) = I(\zeta)/I_{pk}$ is the local current normalized with the maximum peak current along the bunch; $\kappa = kL$ is the average normalized focusing strength k integrated along the linac length L; $w_n^1(\zeta)$ is the transverse wake function normalized with the wake amplitude; d_c is the transverse offset of the beam respect to the linac axis. Finally, $\epsilon(\sigma) = \epsilon_r\left(\gamma_0/\gamma(\sigma)\right)$ is the factor coupling the particle betatron motion (described by the homogeneous form of the previous equation) to the wake field driving term. It is given by [54]:

$$\epsilon_r = \frac{4\pi\epsilon_0}{I_A}\frac{w_n(1)I_{pk}l_b L^2}{\gamma(0)} \tag{20}$$

where I_A=17 kA is the Alfven current, $w_n(1)$ is the wake function normalized to its amplitude and computed for the particle at the bunch tail, I_{pk} is the peak current. Unlike the monopole nature of the longitudinal wake field pattern, the short-range geometric transverse wake field is excited by electrons traveling off-axis. When the electron bunch travels near the axis of the accelerating structures, the transverse wake field is dominated by the dipole field component. As a result, the bunch tail oscillates with respect to the head forming in the (z, x) and in the (z, y) plane a characteristic "banana shape" [56]. Persistence of the slice oscillations along the linac and their amplification may cause the conversion of the bunch length into the transverse dimension (beam break up). So, the displaced bunch tail adds a contribution to the projection of the beam size on the transverse plane, eventually increasing the projected emittance.

4.2 Emittance bumps

The FEL power relies on the energy exchange between the electrons and the light beam along the undulator chain; this interaction is made possible when the two beams overlap. Additionally for HGHG FELs, this is mandatory in the first undulator, where the external seeding laser has to superimpose on the electron bunch. The transverse kick induced by the dipole wake potential imposes an upper limit to the bunch length that is based on the single bunch emittance growth. In order to evaluate this limitation, we recall the approximate transverse emittance dilution through an accelerating structure of length L, due to a coherent betatron oscillation of amplitude Δ [57]:

$$\frac{\Delta\epsilon}{\epsilon} \approx \left(\frac{\pi r_e}{Z_0 c}\right)^2 \frac{N^2 \langle w \rangle^2 L^2 \beta}{2\gamma_i \gamma_f \epsilon} \Delta^2 \tag{21}$$

This is predominantly a linear time-correlated emittance growth and can be corrected. The wake field, $\langle w \rangle$, is expressed here as the approximate average transverse wake function over the bunch given by eqs.18, evaluated at the bunch centroid. Typical misalignment tolerances are in the range $\Delta = 10 - 100$ μm in order to ensure $\Delta\epsilon/\epsilon \leq 1\%$ per structure. If the electron bunch and the accelerating structure parameters do not completely fit into the approximated eq.6 the machine design and alignment tolerances are made more robust and reliable by particle tracking studies that include the geometric wake functions and all realistic alignment errors. Computer codes like `elegant`, PLACET, MTRACK and MBTRACK [58–61] adopt the Courant-Snyder variables to calculate the growth of the bunch slice coordinates caused by a random misalignment of various machine components in the presence of the geometric transverse wake fields. The effect of the wake field can therefore be integrated into the machine error budget.

We are going to show that control over the transverse wake field instability can be gained in a reliable way by applying local trajectory bumps, also called "emittance bumps" [62–66]. Special care is here devoted to the incoherent part of the trajectory distortion due to random misalignment of quadrupole magnets (150 μm rms), accelerating structures (300 μm rms), Beam Position Monitor (BPM) misalignment (150 μm rms) and finite resolution (20 μm) and beam launching error (150 μm, 10 μrad). The wake field effect in the presence of coherent betatron motion of the electron bunch is studied with the `elegant` code. The simulations show that a global trajectory correction provided through a response matrix algorithm is not sufficient to damp the transverse wake field instability; for this reason local trajectory bumps are applied to suppress it. The bumps technique looks for an empirical "golden" trajectory for which all the kicks generated by the transverse wake field compensate each other and the banana shape is finally canceled. In practice, the implementation of the local bumps foresees the characterization of the transverse beam profile as a function of the bunch longitudinal coordinate (banana shape), projected on screens separated by a proper phase advance (to reconstruct the head-tail oscillation). This could be done in a dedicated diagnostic section, downstream of the linac, by means of RF deflectors [67].

To enhance the wake field instability, we have designed the linac with 14 accelerating structures: the first 7 ones have 10 mm iris radius and bring the electron beam from the initial energy of 100 MeV to 600 MeV; the last 7 structures have the smaller iris radius of 5 mm and increase the energy to 1.2 GeV. The 800 pC, 10 ps long bunch is time-compressed twice, by a factor of 5 at 250 MeV and by a factor of 2 at 700 MeV. Thus, the high impedance (smaller iris) structures are traversed by the higher rigidity, shorter bunch. This scheme is expected to minimize the transverse wake field instability that, nevertheless, has still an impressive effect on the projected emittance, with respect to the early part of the linac. For illustration, only one set of errors – randomly chosen over a meaningful sample of error seeds – is shown in Figure 3. Simulations have been carried out with $2 \cdot 10^5$ particles divided into 30 longitudinal slices.

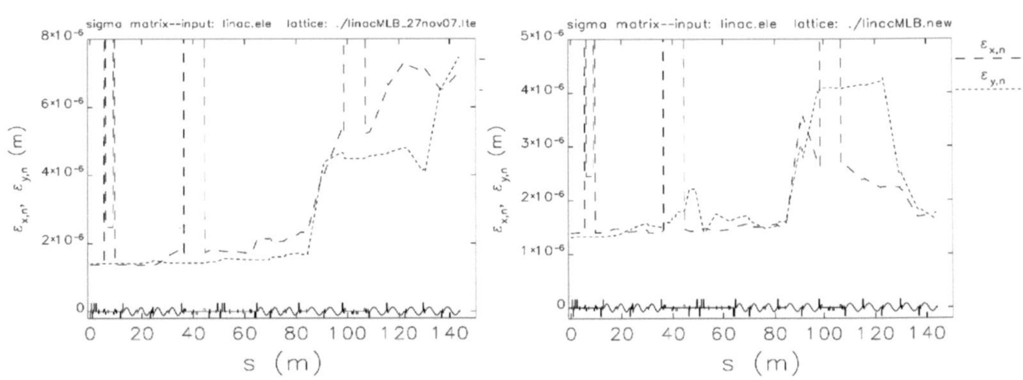

Fig. 3. Left: the projected emittances blow up as the beam enters into the small iris accelerating structures. The trajectory is corrected everywhere to 200 μm level. Right: suppression of the transverse wake field instability after some trajectory bumps have been done in the last linac section. Published in P. Craievich, S. Di Mitri and A. A. Zholents, Nucl. Instr. and Methods in Phys. Res. A 604 (2009)

4.3 Slice centroid Courant-Snyder amplitude

As a next step, the validity of the trajectory manipulation is checked in the presence of shot-to-shot trajectory jitter. This can be generated by beam launching error jitter, quadrupole magnet mechanical vibration and power supply current ripple, jitter of the residual dispersion induced by misaligned quadrupoles, energy jitter translating into trajectory jitter through residual dispersion. For a first rough estimation of the instability effect, let us reasonably assume that the transverse beam size, in each plane, is covered by (at least) four standard deviations (σ) of the particle position distribution. In order for the instability to be suppressed, we want the bunch tail do not laterally exceed the head by more than 1σ. In this case, the relative growth of the beam size is 25%. Equivalently, the relative emittance growth we could tolerate is 50%. Notice that if the instability is suppressed at the linac end, then the slice centroid transverse offset and divergence are small. Hence the bunch tends to maintain its shape in the (t, x) and (t, y) plane at any point of the line downstream. On the contrary, if the banana shape is pronounced, the slice optics in the bunch tail is mismatched to the magnetic lattice. Then, the bunch tail performs betatron oscillations around the head axis and the banana shape at any point downstream will depend on the Twiss parameters at the point of observation. For this reason, the Courant-Snyder amplitude of the slice centroid is now introduced [68] as a parameter to characterize the instability (same applies to the vertical plane):

$$\epsilon_{SC} = \gamma_x x_{cm}^2 + 2\alpha_x x_{cm} x'_{cm} + \beta_x x'^2_{cm} \tag{22}$$

ϵ_{SC} is a constant of motion in absence of frictional forces such as geometric wake fields and emission of radiation; this is just the case for the beam transport downstream of the linac, where also coherent and incoherent synchrotron radiation is neglected. ϵ_{SC} provides a measurement of the amplitude of motion that is independent of betatron phase. Its square root is proportional to the amplitudes of the slice centroid motion $x_{SC}(s)$ that describes the banana shape. In general, x_{SC} is the linear superposition of three main contributions: i) the betatron motion, $x_{S\beta}$, generated by focusing of misaligned quadrupoles; ii) the trajectory distortion, x_{ST}; iii) the transverse wake field effect, x_{SW}. Notice that $x_{offset} = x_{S\beta} + x_{ST}$ is approximately the same for all slices along the bunch. Regarding the instability, only the motion relative to the bunch head is of interest. Thus, we define a new slice centroid amplitude relative to the motion of the bunch head:

$$\epsilon_{SW,x} = \gamma_x (x_{SC} - x_{offset})^2 + 2\alpha_x (x_{SC} - x_{offset})(x'_{SC} - x'_{offset}) + \beta_x (x'_{SC} - x'_{offset})^2 \tag{23}$$

The effect of the trajectory jitter on the scheme for the suppression of the instability can be evaluated by looking to the shot-to-shot variation of the centroid amplitude $\epsilon_{SW,x}$ over the bunch duration. In fact, we require that the standard deviation (over all jitter runs) of the slice lateral deviation be less than the rms (over all particles) beam size $\sigma_x = \sqrt{\beta_x \epsilon_x}$: $\frac{\sigma_{x,SC}}{\sigma_x} \leq 1$. We manipulate this expression with the following prescriptions. First, x_{offset} is a constant. Second, the slice Twiss parameters are the same as the projected ones even in case of slice lateral displacement. Third, the slice Twiss parameters remain constant over all jittered runs. Then, we re-define the variable $\sqrt{\epsilon^i_{SW,x}} \equiv Q^i_x$ and the previous expression becomes an instability threshold given by the ratio between the standard deviation of Q^i_x and the square root of the rms projected (unperturbed) emittance:

$$\frac{\sigma_{Q,x}}{\sqrt{\epsilon_x}} \leq 1 \tag{24}$$

When eq.24 is applied to each slice of the bunch, it is possible to predict which portion of the electron bunch can be safely used for the seeded FEL operation even in the presence of trajectory jitter. When the condition 24 is widely satisfied for most of the bunch slices, that is if the machine error budget and jitter tolerances are respected, we do not expect any important effect of the jitter on the FEL performance.

5. Coherent synchrotron radiation

5.1 Analytical model

The effect of synchrotron radiation is here analyzed for a smooth electron density function, when the emission is at wavelengths of the order of the bunch length, l_b, and much longer than the typical wavelength of incoherent emission: $\lambda_{CSR} \geq l_b \gg \lambda_{incoh}$, where $\lambda_{incoh} =$

$(4\pi R/3\gamma^3)$, γ is the relativistic Lorentz factor and R is the bending radius. The coherent emission is characterized by an intensity spectrum that is proportional to the square of the number of particles N times the single particle intensity, unlike the incoherent emission that is simply linear with the number of particles:

$$\left(\frac{dI}{d\omega}\right)_{tot} = N(N+1)|F(\omega)|\left(\frac{dI}{d\omega}\right)_e \tag{25}$$

where $|F(\omega)|$ is the Fourier transform of the longitudinal particle distribution (form factor); it is of the order of 1 for very short bunches. When $\lambda_{CSR} \approx l_b$, a cooperative scale length of the process can be defined that describes the interaction of electrons and photons during the emission. This is the "slippage length" [25], $s_L = \frac{R\theta}{2\gamma^2} + \frac{R\theta^3}{24}$, where θ is the bending angle. In this case, the CSR emission depends on the details of the charge distribution, of the geometry of the electrons path and it causes a variation of the electron energy along the bunch (energy chirp). Owing to the fact that the energy variation happens in a dispersive region and that different slices of the bunch are subject to a different energy variation, they start betatron oscillating around new, different dispersive orbits during the emission, thus increasing the projection of beam size on the transverse plane. At the end of compression, the bunch will be suffering of an additional (nonlinear) energy chirp and of a projected emittance growth in the bending plane.

The energy variation along the electron bunch can be evaluated by means of the CSR wake potential. In the "steady-state" approximation, $R/\gamma^3 \ll l_b \leq s_L$, it can be expressed as follows [25]:

$$W_{CSR}^{SS}(z) = -\frac{1}{4\pi\epsilon_0}\frac{2e}{3^{1/3}R^{2/3}}\int_{-\infty}^{z}\frac{1}{(z-z')^{1/3}}\frac{d\lambda_z(z')}{dz'}dz' \tag{26}$$

The energy loss per unit length of the reference particle due to the radiation emission of the entire bunch is then $dE/dz = NeW_{CSR}^{SS}(z)$. In [25], the authors distinguish different regimes of CSR emission depending on relation between bunch length, bending magnet length and slippage length. So, using eq.26 in the *short bunch* ($l_b \leq s_L$), *long magnet* ($\gamma\theta \gg 1$) approximation for a Gaussian line-charge distribution, the induced rms relative energy spread [26] is (in S.I. units):

$$\sigma_{\delta,CSR} = 0.2459\frac{r_e N}{R^{2/3}\sigma_z^{4/3}}\frac{R\theta}{\gamma} \tag{27}$$

Eq.26 points out that the energy loss is proportional to the first derivative of the longitudinal charge distribution. So, a stronger CSR induced energy loss is expected, for example, from a Gaussian line-charge than from a uniform one with smooth edges. Also, a current spike in the bunch tail could drive a damaging CSR emission.

When the bunch length is much longer than the slippage length, the afore-mentioned steady-state regime provides incorrect results. Transient effects when the bunch enters and leaves the magnet have to be taken into account [25]. Moreover, the electron bunch moves inside the vacuum chamber that acts as a waveguide for the radiation. Not all spectral components of the CSR propagate in the waveguide and therefore the actual radiating energy is smaller than in a free space environment. For an estimation of the shielding effect of vacuum chamber, the recipe suggested in [27] is used:

$$\frac{\Delta E_{shielded}}{\Delta E_{free_s pace}} \simeq 4.2 \left(\frac{n_{th}}{n_c}\right)^{5/6} exp\left(-\frac{2n_{th}}{n_c}\right), n_{th} > n_c \tag{28}$$

Here $n_{th} = \sqrt{2/3}\,(\pi R/\Delta)^{3/2}$ is the threshold harmonic number for a propagating radiation, Δ is the vacuum chamber total gap, $n_c = R/\sigma_c$ is the characteristic harmonic number for a Gaussian longitudinal density distribution with the rms value of σ_c. The meaning of n_c is that the spectral component of the radiation with harmonic numbers beyond n_c is incoherent. Figure 4 shows the calculated effect of shielding for a vacuum chamber with Δ=8 mm. In case of very wide vacuum chambers (inner radius $\geq$ 30 mm), most of the CSR emission is not shielded when a bunch length of the order of 1 ps is considered.

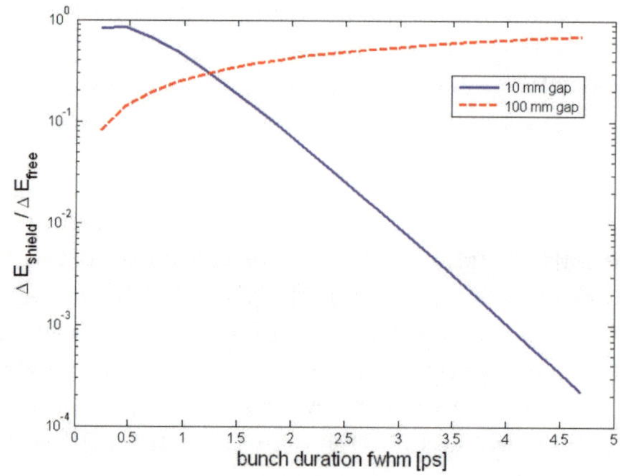

Fig. 4. Suppression of CSR by the vacuum chamber shielding.

5.2 Emittance growth

The energy loss induced by CSR is inversely proportional to the bunch length. Since in a magnetic chicane the bunch length reaches its minimum already in the third magnet, the global CSR effect is dominated by the energy spread induced in the second half of the chicane. Given the CSR induced energy spread $\sigma_{\delta,CSR}$, the beam matrix formalism [69] can be used to estimate the projected emittance growth induced by CSR in the transverse phase space:

$$\frac{\Delta\epsilon}{\epsilon_0} \simeq \frac{1}{2}\frac{\beta}{\epsilon}\theta^2\sigma^2_{\delta,CSR} \tag{29}$$

Due to the β-dependence of the emittance growth, an optics design with very small betatron function in the bending plane can help to reduce the CSR effect. This is especially true in the second half of the chicane, where the bunch length reaches its minimum. The physical meaning of this is given by recalling that, for any α, a small β-function corresponds to a high beam angular divergence. If this is large enough, the CSR kick is largely dispersed in the particle divergence distribution – the perturbed beam divergence is computed as the squared sum of the unperturbed beam divergence and the CSR kick, that can therefore be neglected – and no relevant CSR effect is observed in the bending plane. Typically, a horizontal betatron function at level of 1 m limits the relative projected emittance growth to below $\sim 10\%$.

This formalism, however, does not take into account the motion in phase space of the bunch slices that causes such emittance blow up. In fact, the CSR induced projected emittance growth is the result of the bunch slices misalignment in the transverse phase space. This misalignment is meant to be a spatial and an angular offset of each slice centroid respect to the others. This offset is correlated with the z-coordinate along the bunch. In principle, the emittance growth can be completely canceled out if this correlation is removed. The spatial (angular) offset evaluated at a certain point of the lattice is the product of η_x (η'_x) with the CSR induced energy change, integrated over the beam path. If a π betatron phase advance is built up between two points of the lattice at which the beam is emitting CSR in identical conditions, then we have the integral of an odd function over a half-period and its value is zero [70]. Such a scheme allows the design of even complex beam transport line (arc or dog-leg like) where a relatively large number of quadrupole magnets is dedicated to build a $-I$ transport matrix between successive dipole magnets. Large bending angles, usually translating into short transport lines, are therefore allowed, even in the presence of high charge, short bunches.

5.3 Numerical methods

We introduce here three particle tracking codes that can be used to support the analytical study of CSR instability. They are `elegant` [43], IMPACT [71] and CSRTrack3D [72]. The flexibility of these codes allows the investigation of the compression scheme and CSR effects independently from the analytical approximation for the magnet length ($\gamma\phi \gg 1$ or $\ll 1$) or bunch length ($\sigma_z \gg$ or $\ll R\phi^3/24$) [25]. Moreover, the codes allow the simulation of an arbitrary longitudinal current profile since they convolve the CSR wake function with the actual current profile at the entrance of the magnetic chicane. `elegant` implements a 1-D CSR steady-state and transient force approximation for an arbitrary line-charge distribution as a function of the position in the bunch and in the magnet; the charge distribution is assumed unchanged at retarded times [26]. The 1-D model ($\sigma_r \ll \sigma_z^{2/3}R^{1/3}$, where R is the orbit radius of curvature) does include neither the effects of the transverse distribution on the CSR fields nor the field variation across the beam. IMPACT computes quasi-static 3-D SC forces in the linac with the exception of CSR which is treated with the same 1-D algorithm as in `elegant`. CSRTrack3D treats sub-bunches of variant shape traveling on nonlinear trajectories in the compressor. Figure 5 shows the slice emittance distribution (in the bending plane) after that a 800 pC, 10 ps long bunch has been compressed by a factor of 10 in a symmetric magnetic chicane ($R_{56} = -49$ mm) at the energy of 250 MeV [73]. The good agreement between IMPACT (courtesy of J. Qiang, LBNL) and `elegant` demonstrates that SC forces in the range 100–250 MeV, simulated in IMPACT but *not* in `elegant`, do not affect the compression substantially. At the same time, CSRTrack3D (courtesy of K. Sonnad, LBNL) predicts some slice emittance bumps due to CSR, but not critical.

Assuming that the injector is able to produce a beam whose parameters satisfy the FEL requirements, the beam transport and manipulation in the main linac should not degrade the area in the phase space by more than $\sim 20\%$. Simulations indicate that this threshold can be satisfied for the longitudinal core of the bunch, while it is harder to apply it when

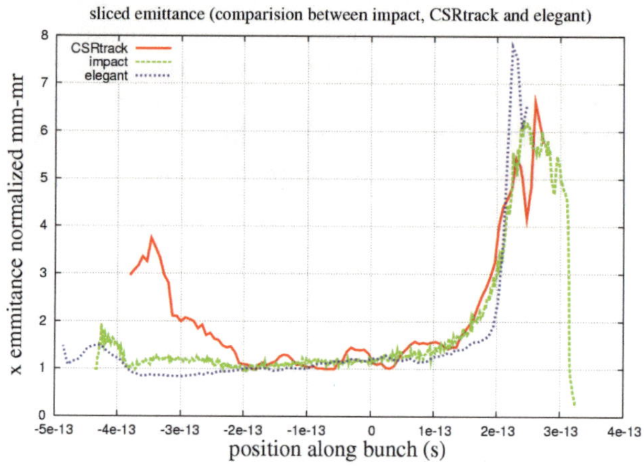

Fig. 5. Codes benchmarking slice emittance of a 800 pC, 10 ps electron bunch time-compressed $C = 10$ in a magnetic chicane, at 250 MeV. Published in S. Di Mitri et al., Nucl. Instr. and Methods in Phys. Res. A 608 (2009).

also the bunch edges are included. These regions are characterized by a lower charge density, therefore they are subjected to a different dynamics at very low energy, where the beam is generated in the presence of important SC forces that strongly depend on the charge density. The different dynamics of the bunch edges with respect to the core leads to a mismatch of the local distribution function (defined in the transverse and in the longitudinal phase space) with respect to the rest of the bunch. Moreover, the finite length of the bunch enhances a nonlinear behaviour of the space charge electric field at the bunch edges that introduces in turn a local nonlinear energy chirp, which leads to local over-compression and optics mismatch. Thus, we expect a stronger effect of the CSR instability in those regions. At the same time, the very ends of the bunch usually contain a smaller number of particles than the bunch core. This implies a bigger uncertainty in the computation of the beam slice parameters due to numerical sampling errors. For all these reasons, particle dynamics in the bunch head and tail is usually studied only with particle tracking codes and the final beam quality is referred to $\sim 80\%$ of the beam population contained in the bunch core. We finally notice that a large slice emittance at the bunch ends is not a limiting factor for a seeded FEL because those portions of the electron bunch are not foreseen to interact with the external seed laser. Also for a SASE FEL, we expect the amplification process would be greatly suppressed in this area.

6. Microbunching instability

6.1 Analytical model

CSR emission is only one aspect of a more complex dynamics called microbunching instability. This is driven by the interaction and reciprocal amplification of the CSR and Longitudinal Space Charge (LSC) field. The latter determines the variation of particles' longitudinal momentum. When the beam exits the photoinjector, the SC oscillation period is typically of the order of meters and any beam density modulation is practically frozen. Thus, without loss of generality, the microbunching instability is assumed to start at the photoinjector exit from a pure density modulation caused by shot noise or unwanted modulation in the photoinjector laser temporal profile. Such density modulation amplitudes are of the order of 0.01% in the sub-micron range and reach $\sim$ 1% at longer wavelengths [74]. As the beam travels along the linac, the density modulation leads to an energy modulation via the LSC wake. This is equal to the free space-charge wake for the wavelength of interest:

$$\lambda_m \ll \frac{2\pi d}{\gamma} \tag{30}$$

d being the transverse size of the vacuum chamber and γ the Lorentz factor. The expression for the LSC impedance is [24]:

$$Z(k) = \frac{iZ_0}{\pi k r_b^2}\left[1 - \frac{k r_b}{\gamma} K_1\left(\frac{k r_b}{\gamma}\right)\right] \tag{31}$$

where $Z_0 = 377\Omega$ is the free space impedance, r_b is the radius of the transverse cross section for a uniform distribution and K_1 is the modified Bessel function of the second kind.

According to the theory developed in [40], the current spectrum is characterized by a bunching factor:

$$b(k) = \frac{1}{Nec}\int I(z)e^{-ikz}dz \tag{32}$$

where N is the total number of electrons. $b(k)$ couples with the LSC impedance along a path L to produce energy modulation of amplitude [40]:

$$\Delta\gamma(k) \approx -\frac{I_0 b(k)}{I_A}\int_0^L \frac{4\pi Z(k,s)}{Z_0}ds \tag{33}$$

where $I_A = 17$ kA is the Alfvén current. We now consider that the bunch length is compressed in an achromatic magnetic chicane characterized by a momentum compaction $R_{56,1}$. For a generic initial energy distribution $V_0(\delta\gamma/\gamma)$ at the entrance of BC1, the resultant density modulation can be expressed through the bunching factor at the compressed wavelength [40]:

$$b_1(k_1) = \left[b_0(k_0) - ik_1 R_{56,1}\frac{\Delta\gamma(k_0)}{\gamma}\right]\int d\left(\frac{\delta\gamma}{\gamma}\right) V_0\left(\frac{\delta\gamma}{\gamma}\right) e^{\left(-ik_1 R_{56,1}\frac{\delta\gamma}{\gamma}\right)} \tag{34}$$

where $k_1 = 2\pi/\lambda_1 = k_0/(1+hR_{56,1})$ is the wave number of the modulation after compression; it is equal to the initial wave number k_0 times the linear compression factor $C = 1/(1+hR_{56,1})$, h being the linear energy chirp. The bunching evolution in a two-stage compression is obtained by iterating the previous expression:

$$b_2(k_2) = \left\{\left[b_0(k_0) - ik_1 R_{56,1}\frac{\Delta\gamma(k_0)}{\gamma}\right]\int d\left(\frac{\delta\gamma}{\gamma}\right) V_0\left(\frac{\delta\gamma}{\gamma}\right) e^{\left(-ik_1 R_{56,1}\frac{\delta\gamma}{\gamma}\right)} - ik_2 R_{56,2}\frac{\Delta\gamma(k_1)}{\gamma}\right\} \times$$
$$\times \int d\frac{\delta\gamma}{\gamma} V_1\left(\frac{\delta\gamma}{\gamma}\right) exp\left(-ik_2 R_{56,2}\frac{\delta\gamma}{\gamma}\right) \tag{35}$$

where the suffix $_2$ refers to the BC2 element. So, according to eq.33 the energy modulation amplitude in front of BC1 is:

$$\Delta\gamma(k_0) = \frac{I_0 b_0(k_0)}{I_A} \int_0^{BC1} \frac{4\pi Z(k_0, s)}{Z_0} ds \tag{36}$$

while that in front of BC2 is:

$$\Delta\gamma(k_1) = \frac{I_0 b_1(k_1)}{I_A} \int_{BC1}^{BC2} \frac{4\pi Z(k_1, s)}{Z_0} ds \tag{37}$$

The bunching described by eq.34 assumes a very simple form for an initial Gaussian energy distribution:

$$b_1(k_1) = \left[b_0(k_0) - ik_1 R_{56,1} \frac{\Delta\gamma(k_0)}{\gamma} \right] exp\left[-\frac{1}{2} \left(k_1 R_{56,1} \frac{\sigma_\gamma}{\gamma} \right)^2 \right] \tag{38}$$

The present analysis is in the linear approximation because it assumes that the microbunching instability starts from a small energy or density modulation, $|CkR_{56}\Delta\gamma/\gamma| \ll 1$. The spectral dependence of the microbunching instability gain in the density modulation can be expressed as the ratio of the final over the initial bunching. In the case of magnetic compression, if the initial bunching term can be neglected with respect to the chicane contribution, the instability is said to be in the "high gain regime", $G(k) \gg 1$. So, the gain in the density modulation after linear compression, due to an upstream energy modulation and for a Gaussian energy distribution, is given by:

$$G(\lambda) = \left| \frac{b_f(\lambda_f)}{b_i(\lambda_i)} \right| = k_f R_{56} \frac{\Delta\gamma}{\gamma} exp\left(-\frac{1}{2} k_f R_{56} \frac{\sigma_{\gamma,i}}{\gamma} \right) \tag{39}$$

As a numerical example, we assume an initial shot noise with a constant spectral power and calculate the initial bunching according to the formula:

$$|b|^2 = \frac{\sigma_I^2}{I_b^2} = \frac{2e}{I_b} \Delta\nu \tag{40}$$

where $\Delta\nu$ is the bandwidth. Then, we convolute it with spectral gain function $G(\lambda)$ to obtain:

$$\left(\frac{\sigma_E}{E_0} \right)^2 = \frac{2ec}{I_b} \int G(\lambda)^2 \frac{d\lambda}{\lambda^2} \tag{41}$$

Here we used a substitution $\Delta\nu = c\Delta\lambda/\lambda^2$. The slice energy spread in the electron bunch after magnetic compression can be calculated by assuming that the energy spread induced by the microbunching instability will eventually become uncorrelated energy spread. This gives us a large value, $\sigma_E \approx 4$ MeV, which for a 1.5 GeV FEL is one order of magnitude larger than the specification we have mentioned in Section 1.

6.2 Landau damping

The exponential term of eq. 38 shows that the particle longitudinal phase mixing contributes to the suppression of the instability if the initial uncorrelated energy spread σ_γ/γ is sufficiently larger than the energy modulation amplitude $\Delta\gamma/\gamma$. In case of non-reversible particle mixing in the longitudinal phase space, this damping mechanism is called energy Landau damping. The "laser heater" was proposed in [16] in order to have an efficient control over the uncorrelated energy spread with the ability to increase it beyond the original small level. The laser heater consists of an undulator located in a magnetic chicane where a laser interacts with the electron beam, causing an energy modulation within the bunch on the scale of the optical wavelength. The corresponding density modulation is negligible and the coherent energy/position correlation is smeared by the particle motion in the chicane.

In order to demonstrate the effect of the laser heater, we compute the spectral gain function for a few different setting of the laser heater and plot them in Figure 6. The parameters in Table 1 have been used for the computation. It is seen here that the larger the energy spread added by the laser heater the more efficient is the suppression of the gain at the high frequency end of the spectra. We also compute the uncorrelated energy spread at the end of the linac as a function of the energy spread added by the laser heater only with the beam and accelerator parameters listed in Table 1. The analytical result is shown in Figure 6. The calculation is simplified by the fact that the interaction between the laser and the electron beam is weak because the required energy spread is small. In this case the changes in laser and beam dimensions along the interaction region can be neglected. Even the slippage effect is negligible because the slippage length is small with respect to the electron and laser pulse length. The heating process is therefore well described by the small gain theory with a single mode [75].

Parameter	Value	Units
Uncorrel. Energy Spread (rms)	2	keV
Initial Beam Energy	100	MeV
Beam Energy at BC1	320	MeV
R_{56} of BC1	-26	mm
Lin. Compression Factor in BC1	4.5	
Peak Current after BC1	350	A
Linac Length up to BC1	30	m
Lin. Compression Factor in BC2	2.5	
Beam Energy at BC2	600	MeV
R_{56} of BC2	-16	mm
Peak Current after BC2	800	A
Linac Length up to BC2	50	m
Linac Length after BC2	70	m

Table 1. Parameters used to compute the microbunching instability gain.

As an alternative to the beam heating, energy modulation and transverse emittance excitation induced by CSR can be moderated, in principle, with an appropriate design of the compressor lattice. Although transverse microbunching radiative effects excite emittance directly [76, 77], an indirect emittance excitation via longitudinal-to-transverse coupling typically dominates

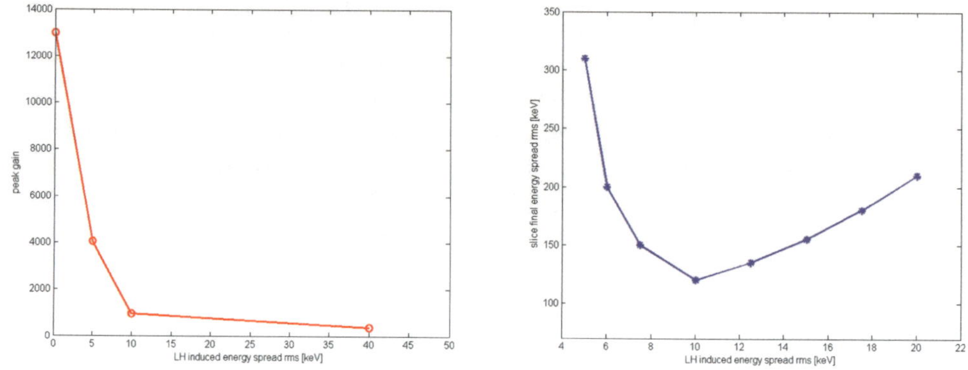

Fig. 6. Left: spectral gain function for several beam heating levels. Right: Final uncorrelated energy spread vs. energy spread added by the laser heater. For beam heating weaker than that minimum, the instability is not suppressed and the final uncorrelated energy spread grows because of the energy modulation cumulated at the linac end at very short wavelengths. For stronger beam heating, instead, the final uncorrelated energy spread is dominated by that induced by the laser heater. Owing to the (approximate) preservation of the longitudinal emittance during bunch length compression, the final energy spread is linearly proportional to the initial one.

them. This coupling is characterized by the function:

$$H = \gamma_x \eta_x^2 + 2\alpha_x \eta_x \eta_x' + \beta_x \eta_x'^2 \tag{42}$$

where γ_x, α_x and β_x are the Twiss functions and η_x, η_x' are the dispersion function and its derivative, all in the horizontal bending plane. Using H, we write for the emittance contribution due to CSR:

$$\Delta\epsilon_x \approx H\delta^2 \tag{43}$$

where δ is the spread of the energy losses caused by CSR. It is obvious from eq.43 that the lattice with small H gives less emittance excitation. Since the strongest CSR is expected in the third and fourth bending magnet of the chicane where the electron bunch is the shortest, we pursue the compressor design with reduced H in this magnet. Now we would like to give the argument why we may not want to get the smallest possible H . While moving through the chicane bending magnets, the electrons with different amplitudes of the betatron oscillations follow different paths with path lengths described by the following equation:

$$\delta l = \int_0^s \frac{x(s')}{R} ds' = x_0 \int_0^s \frac{C(s')}{R} ds' + x_0' \int_0^s \frac{S(s')}{R} ds' \tag{44}$$

Here x_0, x_0' are the electron spatial and angular coordinate at the beginning of the chicane and $C(s), S(s)$ are the cos-like and sin-like trajectory functions. It can be shown that the rms value of Δl taken over the electrons in any given slice of the electron bunch is related to the electron beam emittance through the function H, i.e.:

$$\Delta l_{rms} \approx \sqrt{H\epsilon_x} \tag{45}$$

Thus, the lattice with large H spreads slice electrons more apart than the lattice with small H and washes out the microbunching more effectively. In fact, without accounting for this effect, the gain of the microbunching instability would be significantly overestimated. This effect is very similar to the effect of the Landau damping due to the energy spread. Because of the last argument, it is desirable to design the magnetic compressor such as the magnitude of H in the last bend of the chicane can vary at least within a factor of four. It will give some flexibility to maneuver between such tasks as containing the emittance excitation due to CSR that benefits from smaller H and containing energy spread growth due to the microbunching instability that benefits from larger H.

6.3 Numerical methods

Simulation of the microbunching instability with particle tracking codes requires a large number of macroparticles. The microbunching amplitude, b, due to shot noise in an electron beam with peak current I_b within the bandwidth $\Delta\lambda$ can be estimated:

$$b = \sqrt{\frac{ec}{I_b \Delta\lambda}} \tag{46}$$

For $I_b = 75$ A and $\Delta\lambda = 10$ μm this formula gives $b = 2.52 \cdot 10^{-4}$. Typically, the microbunching due to granularity of the distribution of macro-particles is much larger. For example, we calculate for a 6 ps long electron bunch (fwhm) with 10^6 macroparticles, $b = 1.3 \cdot 10^{-2}$, which is approximately 50 times larger than the real shot noise.

There are several solutions to overcome the sampling noise problem. Following [73], we mention three of them: i) a smoothed initial particle distribution is taken as start for `elegant` particle tracking code; the particle binning is then filtered during the simulation. Several tens of million particles representing a 0.8 nC, 10 ps long bunch were tracked on parallel computing platforms to resolve the final modulation at wavelengths of 1-10 μm [78]; ii) IMPACT Particle-In-Cell (PIC) code tracked up to 1 billion particles, thus reducing the numerical sampling noise by brute force. The convergence of the final result for the increasing number of macroparticles was demonstrated in [79]; iii) a 2-D direct Vlasov solver code can be used that is much less sensitive to numerical noise than PIC codes. The 4-D emittance smearing effect is simulated by adding a filter, as shown in [80, 81]. The latter technique follows the evolution of the distribution function using Vlasov's kinetic equation. Ideally, this is absolutely free from computational noise, although some noise can be introduced on which, due to the final size of the grid, the initial distribution function is defined. However, in practice, this noise can be easily kept below the sensitivity level. It has been demonstrated that the tracking codes results and the analytical evaluation converge with small discrepancy when applied to the beam dynamics in a 1.5 GeV linac, in the presence of a moderate two-stage magnetic compression. In the case of comparison of the simulation results with the linear theory, it becomes apparent that a true result will likely be different because of the anticipation that the linear model should fail at the high frequency end of the noise spectra. Nevertheless, even in the analytical case the result gives a correct assessment of the magnitude of the effect. These techniques have been developed and compared for the first time during the design of the FERMI@Elettra FEL [37]. In that case, `elegant` demonstrated that such a linac-based, soft X-ray facility is very sensitive to small initial density modulations and that the instability enters into the nonlinear regime as the beam is fully compressed in BC2 [78]. The longitudinal phase space becomes folded and sub-harmonics of the density and energy modulation appear. Consequently, the uncorrelated energy spread produced in the injector region has to be increased with a laser heater. For the same case study, IMPACT and the Vlasov solver predicted [80] that a minimum beam heating of 10 and 15 keV rms, respectively, is necessary to suppress the microbunching instability in the one- and two-stage compression scheme. This led to a final slice energy spread of 110 and 180 keV rms, respectively, with a nominal uncertainty of about 15% from code to code.

In spite of the results obtained so far, the microbunching instability study still presents some challenges. In spite of the the Vlasov solver agreement with the linear analytical solution of the integral equation for the bunching factor for a compression factor of 3.5, as shown in [82], entrance into the nonlinear regime is predicted by that code when the compression factor reaches 10 [80]. Unfortunately, the analytical treatment of the nonlinear regime remains a work in progress [83] and no nonlinear analytic treatment of the microbunching instability exists at present for codes benchmarking. Second, the initial seed perturbations for the instability are currently not well determined, both in configuration and in velocity space. Moreover, complications from the bunch compression process, which can lead to "cross-talk" amongst different modulation frequencies, make it difficult to extract the frequency-resolved gain curve. Finally, a fully resolved 3-D simulation of microbunching instability can only be accomplished with massive parallel computing resources that are impractical for the machine fine tuning. As mentioned before, only IMPACT implements 3-D SC forces, while `elegant` and Vlasov solver adopt a 1-D LSC impedance. However, the substantial agreement between the codes suggests that the 3-D SC effect (which is expected to mitigate the microbunching instability) is probably masked by the differences in the computational methods and in the treatment of the numerical noise.

7. Machine configurations and start-to-end simulations

In spite of the specific features that each new FEL source is showing in its conceptual design, flexibility is still a key word for all existing projects, because it allows facility upgrades, new beam physics and back solutions in case of unexpected behaviours. So, if multiple FEL scheme are usually studied for the same source, the driving linac allows different optical and compression schemes for electron beam manipulation. As an example, a moderate compression factor up to 30 in a $\simeq$ 1 GeV linac could be achieved either with a one-stage or a two-stage magnetic compression scheme. However, the two schemes lead to some differences in the final current shaping, transverse emittance and energy distribution, mainly due to a different balance of the strength of collective effects such as geometric wake fields, CSR emission and microbunching instability, as discussed in [84]. The one-stage compression scheme optimizes the suppression of the instability with respect to the two-stage compression for two reasons: firstly, the phase mixing is more effective in BC1 due to the larger R_{56} and to the larger relative energy spread. Secondly, the absence of the high energy compressor does not provide the opportunity to transform the energy modulation accumulated by LSC downstream of BC1 into current modulation. Another positive aspect of the one-stage compression, performed early enough in the linac, is that of minimizing the effect of the transverse wake field, since the induced wake potential is reduced by a shorter bunch length.

The drawbacks are that a short bunch is affected by longitudinal wake field along a longer path than in the two-stage option, where the path to a short final bunch proceeds in two stages. The wake field corrupts the longitudinal phase space by increasing the energy spread, by reducing the average beam energy and by inducing nonlinearities in the energy distribution. We have seen a manipulated current profile has been studied in [34] to overcome this problem. From the point of view of the stability, the two-stage compression has the intrinsic advantage of self-stabilizing the shot-to-shot variation of the total compression factor, C. Let us assume an RF and/or a time jitter makes the beam more (less) compressed in BC1; a shorter bunch then generates stronger (weaker) longitudinal wake field in the succeeding linac so that the energy chirp at BC2 is smaller (bigger). This in turn leads to a weaker (stronger) compression in BC2 that approximately restores the nominal total C.

A specific application of the magnetic compression *in order to suppress* the microbunching instability was presented in [84]. After removing the linear energy chirp required for the compression at low energy (BC1), an additional and properly tuned R_{56} transport matrix element (BC2) is able to dilute the initial energy modulation and to suppress the current spikes created by the microbunching instability without affecting the bunch length. In this case the energy and density modulation washing out is more efficiently provided by two magnetic chicanes having R_{56} of the *same sign*. In fact, the energy modulation smearing is induced by a complete rotation of the longitudinal phase space; the two chicanes must therefore stretch the particles in the same direction (see Figure 7).

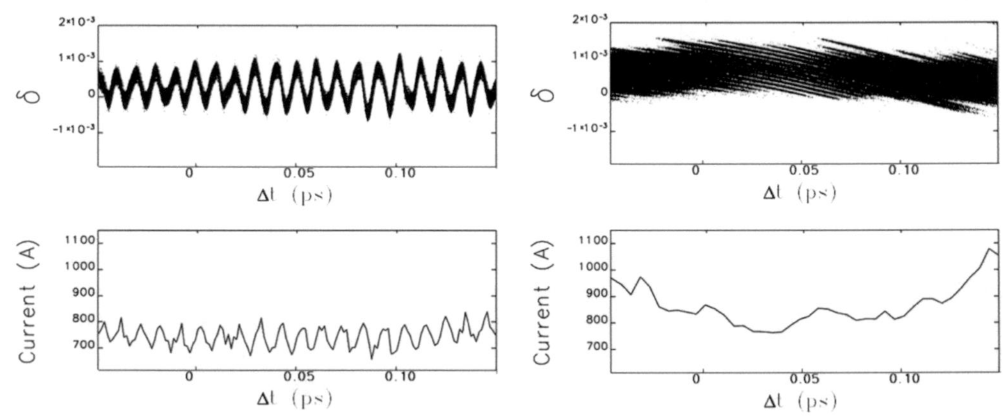

Fig. 7. Particle distributions of the bunch core *upstream* (left) and *downstream* (right) of BC2. An initial modulation amplitude of 1% is introduced at 30 μm wavelength, corresponding to an initial bunching factor of $7 \cdot 10^{-2}$. After BC2, the bunching factor calculated for 3 μm wavelength (C = 10) shrinks to $3 \cdot 10^{-5}$. The final projected normalized horizontal emittance for 60% of the particles in the transverse phase space is 2 mm mrad. Published in S. Di Mitri, M. Cornacchia, S. Spampinati and S. V. Milton, Phys. Rev. Special Topics - Accel. and Beams, 13, 010702 (2010).

Thus we see that a multi-stage compression scheme opens different possibilities to the final beam quality. The eventual machine configuration can be chosen depending on the actual FEL requirements in terms of electron beam quality. Once the configuration is fixed, start-to-end, time-dependent simulations are performed to evaluate the global facility performance in the presence of static imperfections and shot-to-shot jitter sources. This is done by chaining SC codes such as GPT and Astra to a linac code such as `elegant` or LiTrack, then these to FEL codes like Genesis [85] or Ginger [86]. GPT, LiTrack and Genesis have been used in [73] to calculate the sensitivity of the injector, main linac and FEL output, respectively, to the jitter of the photo-cathode emission time, charge, RF voltage and phase, bunch length, emittance and mean energy. First, it was calculated how large the jitter could be in each parameter independently to cause an rms variation of 10% peak current, 0.1% mean energy and 150 fs arrival time. Then, these uncorrelated sensitivities were summed to generate a linac tolerance budget. Jitter analysis of some slice electron beam parameters was also implemented, with an important parameter being the quadratic energy chirp that affects the FEL output bandwidth. Finally, the tolerance budget was used to simulate shot-to-shot variations of the machine parameters and to perform a global jitter study.

8. Notes on the particle-field interaction

In Section 1 we have mentioned the short range SC forces as one of the Coulomb inter-particle interactions that limits the applicability of the Liouville's theorem to the particle motion. We want to make here a more precise statement. In general, Liouville's theorem still applies in the 6-D phase space in the limit of very small correlations established by the space charge forces between particles, so that each particle moves in the same way than all the others, in the collective (also named "mean") field generated by all the others. Quantitatively, this situation is satisfied if the number of particles in the Debye sphere surrounding any particle is large, that is $\lambda_D \gg n^{-1/3}$, where n is the density of charged particles in the configuration space and λ_D is the Debye length that is the ratio of the thermal velocity, $(KT/m)^{1/2}$, to the plasma frequency $\omega_p = (q^2 n/m\epsilon_0)^{1/2}$, q and m being the particle charge and mass. Then, a smoothed out potential due to all particles may be calculated from the density distribution in the configuration space and its contribution included in the Hamiltonian system of forces. This procedure leads to the derivation of the Maxwell-Vlasov equation, which self-consistently describes the behaviour of an assembly of charged particles.

By definition, the SC forces describe a Coulomb interaction within a bunch. Their extension to a train of bunches is straightforward. Typically, being the distance between different bunches of the train much larger than the Debye length, each bunch is treated as independent from the others. This is not the case for the geometric wake fields. If the relaxation time of the wake field is shorter than the repetition time of the accelerator, then the electro-magnetic field associated with two succeeding bunches do not interfere and the single bunch wake field is said to be in the *short range* regime. This is the case already treated in this Chapter. As opposite, the *long range* wake field is usually present in rings, recirculating linacs and single-pass linacs dealing with a bunch train. In this regime, different bunches "communicate" through the narrow-band (high Q, quality factor) impedances. That is, wake fields deposited in various high-Q resonant structures can influence the motion of following bunches and can cause the motion to become unstable if the beam currents are too high. To effectively couple the bunch motion, high order modes must have a damping time $\tau \approx 2Q/\omega$, where ω is the mode resonant frequency, longer than the bunch spacing. For modes with $Q \leq 100$, this restricts the frequencies to less than 10 GHz. The frequency limit is lower for smaller Q.

9. References

[1] C. Pellegrini, Nucl. Instr. and Meth. A 445 (2000)
[2] H. Haus, Quantum Electronics, IEEE Journal of 17 (1981)
[3] G. Dattoli, A. Marino, A. Renieri and F. Romanelli, IEEE J. Quantum Electron. QE-17, (1981)
[4] E. L. Saldin, E. A. Schneidmiller and M. V. Yurkov, Nucl. Instr. And Methods in Phys. Research., Sect. A, 490 (2002)
[5] R. Coisson and F. De Martini, Quantum Electronics 9 (1982)
[6] R. Bonifacio et al., Nucl. Instr. Methods in Phys. Research A 296 (1990)
[7] L.H. Yu, Phys. Rev. A 44 (1991)
[8] I. Ben-Zvi, K. M. Yang, L. H. Yu, Nucl. Instr. Methods in Phys. Research A 318 (1992)
[9] L.H. Yu et al., Science 289 (2000)
[10] L.H. Yu et al., Phys. Rev. Lett. 91, 074801 (2003)
[11] G. De Ninno et al., Phys. Rev. Lett. 101, 053902 (2008)
[12] A. J. Lichtenberg, Phase Space Dynamics of Particles, Wiley, New York (1969)
[13] B. E. Carlsten, Nucl. Instr. Methods in Phys. Research A 285 (1989)
[14] M. Ferrario et al., Phys. Rev. Letters, 99 234801 (2007)
[15] M. Ferrario, V. Fusco, C. Ronsivalle, L. Serafini and C. Vaccarezza, New J. Phys. 8 (2006)
[16] E. L. Saldin, E. A. Schneidmiller and M. Yurkov, TESLA-FEL2003-02 (2003)
[17] L. Palumbo, V. G. Vaccaro and M. Zobov, LNF-94/041 (P)
[18] K. L. F. Bane and M. Sands, SLAC-PUB-95-7074 (1994)
[19] L. I. Schiff, Rev. Sci. Instr. 17 (1946)
[20] J. S. Nodvick and D. S. Saxon, Phys. Rev. 96 (1954)
[21] R. L. Warnock and P. Morton, Part. Accel. 25 (1990)
[22] J. B. Murphy, S. Krinsky, and R. L. Gluckstern, Proc. of Part. Accel. Conf. 1995, (1995)
[23] S. Heifets, G. Stupakov and S. Krinsky, Phys. Rev. Special Topics - Accel. and Beams 5, 064401 (2002)
[24] Ya. S. Derbenev, J. Rossbach, E. L. Saldin, and V. D. Shiltsev, DESY-TESLA-FEL-95-05 (1995)
[25] E. L. Saldin, E. A. Schneidmiller, and M. V. Yurkov, Nucl. Instr. Methods in Phys. Research A 398 (1997)
[26] M. Borland, Phys. Rev. Special Topics - Accel. and Beams, 4, 070701 (2001)
[27] R. Li, C. L. Bohn and J. J. Bisognano, Particle Accelerator Conference (1997)
[28] K. J. Kim, Nucl. Instrum. Meth. A 250, (1986)
[29] R. Bonifacio, C. Pellegrini and L. M. Narduci, Opt. Comm. 50 (1984)
[30] R. Scheffield, Physics of Particle Accelerators, AIP Vol. 184 (1989)
[31] D. Palmer, *The Next Generation photo-injector*, Stanford University dissertation (1998)
[32] S.B. van der Geer et al., in Comp. Accel. Physics, http://www.pulsar.nl/gpt (2002)
[33] K. Floettmann, http://www.desy.de/ mpyflo/ (1990)
[34] M. Cornacchia, S. Di Mitri, G. Penco and A. A. Zholents, Phys. Rev. Special Topics - Accel. and Beams, 9, 120701 (2006)
[35] P. Emma, Proc. of Part. Accel. Conf., Dallas, Texas, USA, WAG01 (1995)
[36] TESLA TDR, DESY 2000-011 (2001)
[37] C. Bocchetta et al., *FERMI@Elettra Conceptual Design Report* (2007)
[38] G. Stupakov, Z. Huang, Phys. Rev. Special Topics - Accel. and Beams, 11, 014401 (2008)

[39] T. Shaftan et al., Nucl. Instr. Methods in Phys. Research A, 528 (2004)
[40] Z. Huang and K.-J. Kim, Phys. Rev. ST Accel. Beams, 5, 074401 (2002)
[41] Z. Huang et al., Phys. Rev. Special Topics - Accel. and Beams, 7, 074401 (2004)
[42] M. Trovo' et al., in Proc. of FEL Conf. 2005, THPP059, Stanford, CA, USA (2005)
[43] M. Borland, APS LS-287 (2000)
[44] K. L. F. Bane, in Workshop on the Phys. and Appl. of High Bright. Electr. Beams, Erice, Italy (2005)
[45] K. L. F. Bane et al., SLAC-PUB-7862 (1994)
[46] P. Craievich, T. Weiland and I. Zagorodnov, Nucl. Instr. and Methods in Phys. Research A 558 (2006)
[47] W. Fawley, G. Penn, ST/F-TN-06/07 (2006)
[48] P. Emma, A. C. Kabel, EPAC'04, MOPKF081, Lucerne, Switzerland (2004)
[49] D. Dowell et al., in Proc. of the 1995 Part. Accel. Conf., WUPB20, Dallas, Texas, USA (1995)
[50] P. Emma, LCLS-TN-01-1 (2001)
[51] M. Cornacchia et al., ST/F-TN-06/15 (2006)
[52] K. Bane, P. Emma, PAC'05, FPAT091, Knoxville, Tennessee (2005)
[53] K. L. F. Bane et al., SLAC-PUB-9663 (2003)
[54] J. Delayen, Phys. Rev. Special Topics - Accel. and Beams, 6, 084402 (2003)
[55] J. Delayen, Phys. Rev. Special Topics - Accel. and Beams, 7, 074402 (2004)
[56] A. W. Chao, B. Tichter and C.-Y. Yao, SLAC-PUB-2498 (1980)
[57] P. Emma, LCC–0021 (1999)
[58] G. Guignard, CERN-SL/91-19 (AP)
[59] G. Guignard and J. Hagel, Proc. of EPAC'96, WEP015G, Sitges, Barcelona, Espana (1996)
[60] E. T. D'Amico et al., Proc. of PAC'01, RPAH082, Chicago, Illinois (2001.)
[61] J. Resta-Lopez et al., Proc. of EPAC'08, MOPP027, Genoa, Italy (2008)
[62] J. Seeman et al., SLAC-PUB-5705 (1992)
[63] R. W. Assmann et al., Proc. of LINAC'96, TUP54, Geneva, Switzerland (1996)
[64] G. Guignard et al., Proc. of PAC'97, 8W015, Vancouver, Canada (1997)
[65] P. Tenenbaum, SLAC-TN-04-038 (2004)
[66] P. Eliasson and D. Schulte, Phys. Rev. Special Topics - Accel. and Beams, 11, 011002 (2008)
[67] P. Craievich, S. Di Mitri, M. Ferianis, M. Veronese, M. Petronio and D. Alesini, Proc. of DIPAC 2007, TUPC10, Venice, Italy (2007)
[68] P. Craievich, S. Di Mitri and A. A. Zholents, Nucl. Instr. and Methods in Phys. Res. A 604 (2009)
[69] A. W. Chao and M. Tigner, Handobook of Accelerator Physics and Engineering, World Scientific, 3rd edition (2006)
[70] D. Douglas, JLAB-TN-98-012 (1998)
[71] J. Qiang, S.Lidia, R. D. Ryne and C.Limborg-Deprey, Phys. Rev. Special Topics - Accel. and Beams, 9, 044204 (2006)
[72] M. Dohlus and T. Limberg in Proc. of the FEL Conf. 2004 (M0C0S05)
[73] S. Di Mitri et al., Nucl. Instr. and Methods in Phys. Res. A 608 (2009)
[74] T. Shaftan and Z. Huang, Phys. Rev. Special Topics - Accel. and Beams, 7, 080702 (2004)
[75] G. Dattoli and A.Renieri, Laser Handbook, Vol. 4 (1985)
[76] Ya.S. Derbenev, V. D. Shiltsev, FERMILAB–TM–1974, SLAC-PUB 7181 (1996)
[77] R. Li, Particle Accelerator Conference, New York, (1999)

[78] M. Borland, Phys. Rev. Special Topics - Accel. and Beams, 11 030701 (2008)
[79] I. Pogorelov et al., in Proc. of ICAP'06, WEPPP02, Chamonix, France (2006)
[80] M.Venturini et al., Phys. Rev. Special Topics - Accel. and Beams, 10 104401 (2007)
[81] M.Venturini, R.Warnock and A. A. Zholents, Phys. Rev. Special Topics - Accel. and Beams, 10 054403 (2007)
[82] M. Venturini et al., LBNL Report, Berkeley, LBNL-60513 (2006)
[83] M.Venturini, in Proc. of the Second Workshop on Microbunching Instability, LBNL, /https://www.elettra.trieste.it/FERMI/ (2008)
[84] S. Di Mitri, M. Cornacchia, S. Spampinati and S. V. Milton, Phys. Rev. Special Topics - Accel. and Beams, 13, 010702 (2010)
[85] S. Reiche et al., in Proc. of PAC 2007, TUPMS038, Albuquerque, NM, USA (2007)
[86] W.M.Fawley, LBNL-49625 (2002)

3

Hole Coupled Infrared Free-Electron Laser

Prazeres Rui
CLIO / Laboratoire de Chimie Physique
France

1. Introduction

Infrared FELs are used worldwide as user facilities. The FELs are especially efficient in infrared since they produce high power laser pulses, and are tunable across a large wavelength range, typically 1 to 2 decades, such as the CLIO mid-infrared laser facility which is covering from 3μm to 150 μm (Glotin et al., 1992; Prazeres et al., 2002; Ortega J.-M. et al., 2006). We are principally interested here in far infrared lasers. Several facilities are now operating in the world within this range:

- "CLIO", at Laboratoire de Chimie Physique, Orsay, France
- "FELIX", at FOM Institute, Nederland
- "FELBE", at Helmholtz-Zentrum Dresden-Rossendorf, Germany
- "FIR FEL", at IR-FEL Center, Tokyo University of Science, Japan
- "UCSB FEL", at University of California, Santa Barbara, U.S.A.

The amount of FEL optical gain which is required to achieve laser operation, i.e. to compensate the optical cavity losses, is easier to obtain in the infrared range than in visible or UV. Indeed, the FEL gain is strongly dependent on the electron beam energy γmc^2: in first approximation, the gain is proportional to $1/\gamma^3$. On the other hand, the FEL wavelength varies as $1/\gamma^2$. Therefore, large wavelength operation of the FEL requires low electron beam energy, which produces a larger amount of gain. Also, and this is a consequence of above comments, the requirements of quality (energy spread, emittance) of the electron beam are much less severe for infrared operation than for visible or shorter wavelength. This is why the X-ray FEL only have been developed recently, whereas the first FEL was operating in infrared (λ=3.4μm) in 1977 (Deacon et al., 1977). However, some other difficulties arise in infrared range: optical diffraction, absorption in materials or in air. Also, light detectors are less sensitive, cumbersome and more difficult to use - requiring liquid nitrogen, and liquid helium for far infrared. These difficulties have consequences on the design of an optical cavity for a far infrared FEL.

2. FEL optical cavity

The length of the optical cavity of a FEL must be rather large (a few meters) because it must include the undulator, magnetic dipoles and quadrupoles, electron beam deviation coils and various diagnostics. As a consequence, the Rayleigh length of the optical beam inside the cavity must be adapted, in order to avoid important optical beam divergence. The transverse profile of laser must be less than the diameter of the cavity mirrors, and less than the aperture of all insertion devices, including the undulator vacuum chamber. It is important to keep the optical cavity losses at minimum value in order to get a maximum of laser power. The following calculation gives an approximate expression of the cut-off wavelength (upper limit) as a function of the geometrical configuration of the optical cavity. The Rayleigh length Z_R of a spherical cavity is (for a pair of identical mirrors):

$$Z_R = \frac{\pi . w_o^2}{\lambda} = \frac{1}{2}\sqrt{(d(2R_c - d)} \tag{1}$$

where d is the cavity length, R_c is the mirrors radius of curvature and w_o is the minimum of the waist $w(z) = w_o\sqrt{1+(z/z_R)^2}$. The waist w_o is in the center (z=0) of the cavity. The cut-off wavelength λ_x can be defined by the condition where the laser mode hits the undulator vacuum chamber, i.e. when:

$$FWHM = 2.35.\left(\frac{W(L_w/2)}{2}\right) = b \tag{2}$$

where FWHM is the transverse size of the laser mode at the vacuum chamber extremity ($z=L_w/2$), L_w and b are respectively the length and the transverse aperture of the undulator vacuum chamber. This gives the cut-off wavelength λ_x:

$$\lambda_x = \frac{\pi . b^2}{Z_R} . \frac{0.72}{1+\left(L_w/2Z_R\right)^2} \tag{3}$$

For example, for CLIO with Z_R=1.2m, L_w=2m and b=14mm, we obtain $\lambda_x \cong 220\mu m$. In this case, the extremity of the vacuum chamber acts as a waveguide for the intracavity laser mode.

2.1 Hole coupling

Due to the lack of wide band multilayer transmitting mirrors or beam splitters, the only practical solution, to extract the laser power from the optical cavity, is to use a hole of a few millimeters in one mirror. In a preliminary configuration of the CLIO FEL in 1992 (Glotin et al., 1992), we used an intracavity Brewster plate to extract the laser power. Various materials was used: ZnSe from λ=3µm to 18µm, and KRS5 up to 40µm. However, this solution has two main consequences: the band-pass is limited due to absorption in the Brewster plate, and the variation of index of the plate creates a desynchronization of the FEL cavity length tuning according to wavelength.

Using a hole in the center of the output mirror avoids such problems, and this is presently the most commonly used system on FEL facilities (CLIO, FELIX, FELBE,...). Nevertheless, this is not a perfect solution, because it may alter the laser spot - as shown below. Moreover, the output coupling depends on wavelength. Indeed, the laser beam spot, on the output mirror, increases with wavelength, thus the extraction rate T_x decreases with wavelength. The extraction rate T_x is the ratio between (1) the laser output energy (through the hole) and (2) the intracavity laser energy. Therefore, in practice it is useful to have, inside the vacuum of the cavity, a mechanical system that allows changing the output mirror, with several hole dimensions.

In first approximation, T_x should be proportional to the hole area, but in fact it is not the case because the intracavity laser profile is modified by the hole - it is a dynamic process in the cavity which leads to a steady state laser mode taking into account the perturbation due to the hole. The figure 1 shows a numerical simulation that has been performed with our code "MODES" (Prazeres & Billardon, 1992; Prazeres, 2001; Prazeres et al., 2005). It displays the output laser power at λ=20µm, as a function of the hole diameter, using the parameters of CLIO. For small holes, less than 2mm, the perturbation is negligible in the cavity, and the laser spot is not modified. But for a diameter of 3mm (point B), there is a jump of intracavity laser mode which creates a drop of the extraction rate T_x and consequently of the output power. This is due to a splitting of the laser profile on the output mirror (inside the cavity), as it is shown in the images on side of the curve. Such mode profile exhibits a minimum on the hole surface (at the center of the image), and reduces the extraction rate T_x.

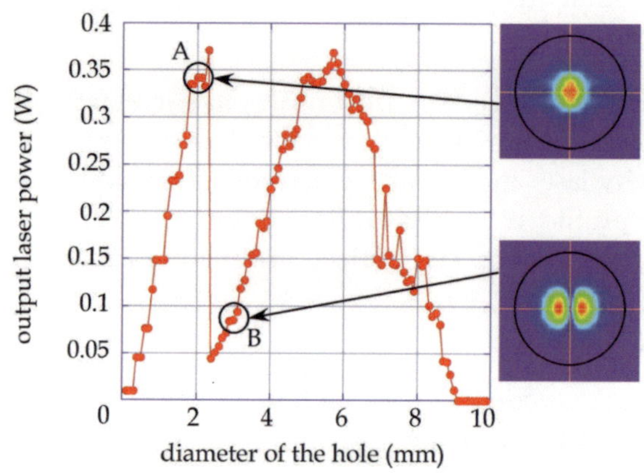

Fig. 1. Numerical simulation: output laser power vs. hole diameter (for CLIO data at λ=20µm). Images on side are the laser transverse profile inside the cavity, on output mirror. The black circles represent the mirror diameter of 38mm.

However, in practice, the tuning of the FEL operator is made so as to optimize the laser output power. It is a compromise between intracavity power and output coupling, and this procedure tends to select intracavity modes that are not necessarily centered on the hole. In this aim, the output mirror may be tilted by a few milliradians. This is shown in figure 2 which displays the output laser power as a function of mirror tilt on both axes (θ_x, θ_y). The point θ_x=θ_y=0 corresponds to the point B in figure 1, which the relevant laser profile exhibits a double spot in horizontal axis. Therefore, a horizontal tilt of the mirror allows the laser spot to fit upon the extraction hole. Indeed, the figure 2 shows that a horizontal tilt of 0.5mrd increases the laser power by a factor 5. In this case, as shown on the image on side, the laser spot fits better to the hole. This allows a better output coupling, and then a larger output power.

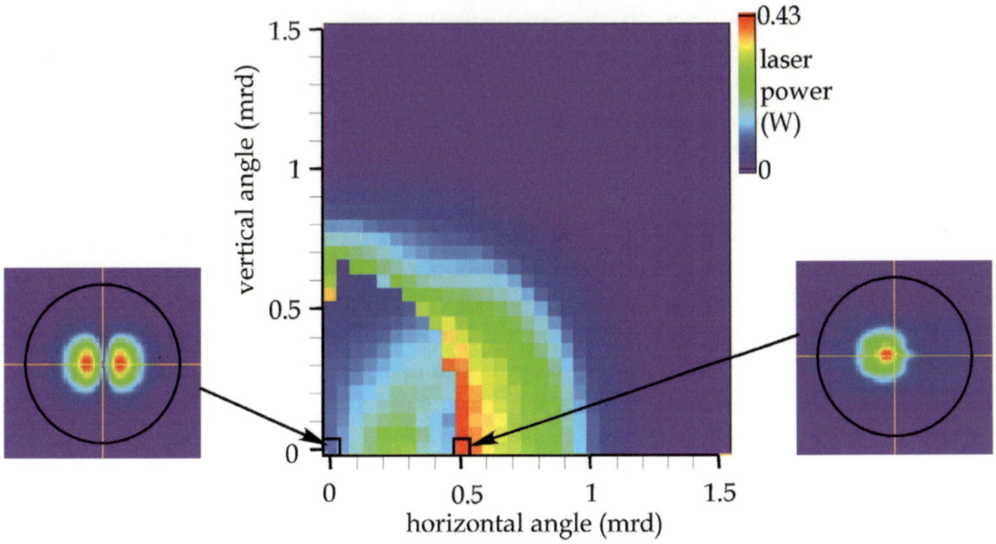

Fig. 2. Numerical simulation: laser power vs. angular tilt of the output mirror, at λ=20μm with a hole diameter of 3mm. Images on side are the laser transverse profile inside the cavity, on output mirror. The black circles represent the mirror diameter of 38mm.

As a consequence, it is not straightforward to calculate analytically the extraction rate T_x from the output coupling of a pure Gaussian mode. It is necessary to take into account the features of the intracavity laser mode. Also, when the intracavity laser profile is off-axis, or distorted, the output profile is strongly affected by the hole and does not keep the cylindrical symmetry of the system. These effects are analyzed here in section 5.

2.2 Undulator vacuum chamber

The transverse size of laser mode increases with wavelength, and beyond a certain limit it tends to undergo losses at the undulator vacuum chamber, which is generally the closer section of the FEL cavity. However, the transverse size of this chamber cannot be increased if one wants to maintain a sufficiently strong undulator magnetic field on-axis. Indeed, the upper limit of laser wavelength range corresponds to a maximum of magnetic field, which corresponds to the closer undulator gap. This determines the vacuum chamber transverse size along the magnetic field axis. On the other axis, there are no geometrical constraints because it is a free space - see fig.3. Therefore, the cross section of the vacuum chamber is, in principle, of rectangle shape with the shorter axis along magnetic field axis. However, due to pressure constraints upon the vacuum chamber, an elliptical shape offers a better rigidity. Such configuration is used on the CLIO FEL, as shown in figure 3. The minimum undulator gap is 18mm, and corresponds to an inner vacuum chamber size of 14mm on vertical axis. The inner size on horizontal axis is 35mm, it is large enough to avoid any diffraction problem. Note that in this chapter, we always consider a "usual configuration" of planar undulator, i.e. with magnetic field on vertical axis. Therefore, when speaking in this chapter about "horizontal walls" of the undulator vacuum chamber, these are always perpendicular to magnetic field, as shown in the example of figure 3.

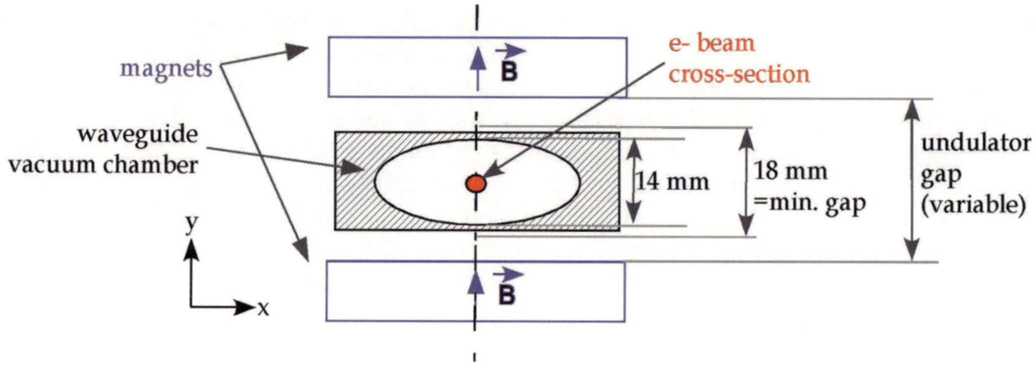

Fig. 3. Sectional drawing of the undulator vacuum chamber of the CLIO FEL

Some undulators are designed to be inside vacuum (Tanaka et al., 2005): the undulator gap can be as close as possible to the electron beam. This solution is used when a strong magnetic field is required, but the electron beam trajectory has to be very precise across the undulator in order to not hitting the magnets, which would create heavy damages.

The inner surface roughness of the undulator vacuum chamber is also an important parameter, because at large wavelength the laser mode may be guided by the horizontal chamber walls. Indeed, the horizontal walls are closer to the laser mode, whereas vertical walls are far enough to avoid any guiding effect. This is equivalent, in first approximation, to an infinite plate horizontal waveguide. This approximation is also valid in the case of an elliptical cross section waveguide, as shown in figure 3. Therefore, the inner surface of the vacuum chamber must be of mirror quality for infrared spectral range. As an example, the vacuum chamber of CLIO is done in extruded aluminium, which possesses a good conductivity and reflectivity. Now, if we consider a pure Gaussian laser mode propagating through such waveguide, we can have two borderline cases: (1) at short wavelength, the laser divergence in the vacuum chamber is small enough to avoid any clipping by the metallic walls, and no guiding occurs – it is equivalent to free space propagation ; and (2) at large wavelength, close to the cut-off λ_x in expression (3), the laser mode fills the vacuum chamber cross section, creating a guided propagation across the chamber. Between these two extreme cases, i.e. for intermediate wavelengths, the guiding effect may only partially occur, i.e. only on the extremities of the vacuum chamber. In this case, the simple model of infinite plate horizontal waveguide is fully applicable. This model is involved in our numerical code "MODES" (Prazeres & Billardon, 1992; Prazeres, 2001; Prazeres et al., 2005), which is used in this chapter to calculate the FEL features such as: cavity optical losses, output coupling, laser mode profile, laser power,...

2.3 Configuration of the optical cavity

Two main configurations are possible for the optical cavity of an infrared FEL: (1) the waveguide fills the whole length of the optical cavity, from upstream mirror to downstream mirror, or (2) the waveguide only fills the undulator section, the other parts of the cavity being in free-space propagation. The 1st type of cavity has been used in the UCSB FEL (University of California Santa Barbara [UCSB], Center for Terahertz Science and Technology): the waveguide covers from upstream cavity mirror up to downstream mirror.

In this case, the waveguide cross section is rectangular, and both cavity mirrors are of cylindrical type: the mirrors are flat in vertical plane because of the guiding effect, and concave in horizontal plane because of the free-space propagation - we are here in the approximation of infinite plate waveguide. The 2nd type of cavity, that we can nominate "partially guided", is used on other FELs such as CLIO. The figure 4 represents a layout of the optical cavity of CLIO. In this case, both cavity mirrors are in free-space propagation, and are of concave type. There is also an example of "hybrid configuration", on the ELBE FEL (Helmholtz-Zentrum Dresden-Rossendorf [HZDR]), where the upstream cylindrical mirror is in contact with the waveguide, whereas the concave downstream mirror is in free-space.

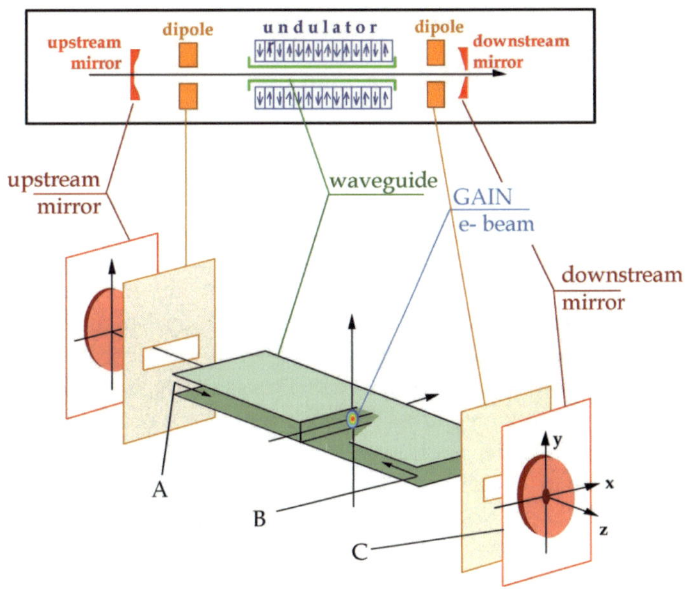

Fig. 4. Schematic layout of the CLIO Free-Electron Laser optical cavity

2.4 Radius of curvature of the cavity mirrors

When the laser wavelength is far from the cut-off λ_x in expression (3), the influence of the waveguide is negligible and a pair of spherical mirrors can be used for the cavity ($R_X=R_Y$). But for the "partially guided cavity" configuration, and at large wavelength, i.e. of the order of the cut-off wavelength λ_x, the waveguide section modifies strongly the laser mode. In this case, toroidal mirrors are more adapted for the cavity. The horizontal radius of curvature R_X is the same as for a free-space cavity, for both mirrors, because it corresponds to a free-space propagation. But for vertical axis, the calculations have shown (Prazeres et al., 2002) that it is necessary to focus the laser beam on the waveguide entrance. In this aim, the vertical radius of curvature R_Y must be approximately equal to the distance between the mirror and the waveguide entrance, for both mirrors. This is well described in the numerical simulation, in figure 5, using the codes "MODES" with parameters of CLIO. It shows, at various wavelengths, the laser power as a function of the vertical radius of curvature of both mirrors R_{1Y} and R_{2Y}, where R_1 and R_2 are respectively upstream and downstream mirrors. We have R_{1X}=3m and R_{2X}=2.5m for horizontal axis. The "S" labeled square dots corresponds to a pair of spherical mirrors: $R_{1X}=R_{1Y}$ and $R_{2X}=R_{2Y}$.

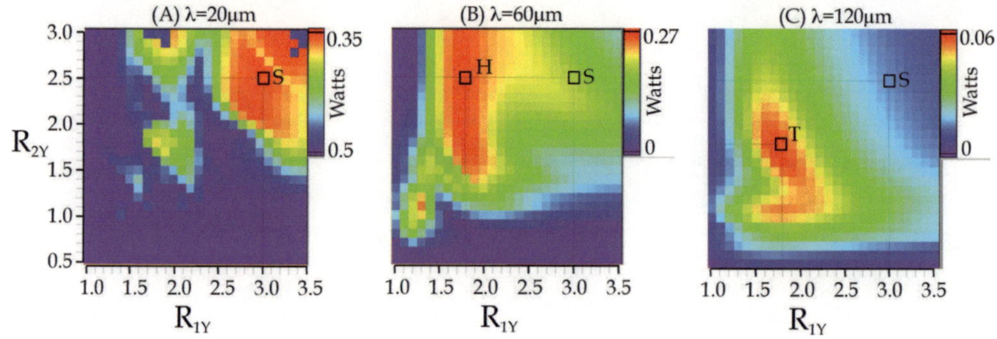

Fig. 5. Numerical simulation at (A) λ=20μm, (B) λ=60μm and (C) λ=120μm: laser power vs. radius of curvature in vertical plane, of both cavity mirrors. Abscissa corresponds to upstream mirror, and ordinate to downstream mirror.

The best configuration at λ=120μm is a pair of toroidal mirrors, with $R_{1Y}=R_{2Y}$=1.8m in vertical, and R_{1X}=3m and R_{2X}=2.5m in horizontal. It is represented by a "T" labeled dot. However, in practice, when using a pair of toroidal mirrors, the main difficulty appears during the procedure of alignment of the cavity: when using an HeNe laser, the red spot is deformed and spread out in vertical plane. This makes the alignment much more difficult as when using spherical mirrors. A good compromise is to use a "hybrid cavity", at least for the intermediate wavelength range: a toroidal upstream mirror, and a spherical downstream mirror that has a hole in center for output coupling. The "hybrid cavity" is represented by a "H" labeled dot on the figure, and it corresponds to the optimum power at λ=60μm. Consequently, and as spoken above, a mechanical system in vacuum for changing the output mirror may be very useful to swap rapidly the cavity mirrors. Such system is used in several FELs: CLIO, FELIX,...

3. Shift of undulator resonance in the waveguide

The waveguide, which is used in the undulator section, has an influence on the FEL process. Indeed, the phase velocity, inside the waveguide, is dependant on the eigenmode number, and then the resonant condition of the undulator is modified (Nam et al., 2000). This resonant condition requires that: during the travel of an electron through a period λ_o of the undulator, the radiation, which is produced by the electron, crosses the same distance plus one wave period λ. In free-space, this condition can be written:

$$v_z t = l_o \text{ and } ct = l_o + l \quad (4)$$

where v_z is the longitudinal velocity of the electron in the undulator. This is equivalent to:

$$\omega = c\beta_z (k_o + k) \quad (5)$$

where $\beta_z = v_z/c$ is the normalised longitudinal velocity of the electron, $k_o = 2\pi/\lambda_o$ is the undulator wavenumber and and $k=\omega/c$ is the wavenumber. This gives the resonant frequency ω_R in free-space:

$$\omega_R = k_o \frac{c\beta_z}{1-\beta_z} \tag{6}$$

corresponding to the well known expression of resonant wavelength in free-space:

$$\lambda_R = \lambda_o \frac{1-\beta_z}{\beta_z} = \frac{\lambda_o}{2\gamma^2}\left(1+\frac{K^2}{2}\right) \tag{7}$$

where K is the undulator parameter and γ is the normalized electron energy.

In the waveguide (Bonifacio & De Salvo Souza, 1986), the resonant condition is written:

$$\omega = c\beta_z (k_o + k_{//}) \tag{8}$$

and the dispersion relation is:

$$\omega = c\sqrt{k_\perp^2 + k_{//}^2} \tag{9}$$

The wavenumber $k = \sqrt{k_{//}^2 + k_\perp^2}$ is a combination of (1) the longitudinal wavenumber $k_{//}=\omega/v_\varphi$, where v_φ is the phase velocity in the waveguide, and of (2) the transverse wavenumber $k_\perp = q\pi/b$, where q is the eigenmode number and b is the waveguide aperture along vertical axis (in the infinite plane parallel configuration). From these two expressions, resonant condition and dispersion relation, we obtain (Nam et al., 2000) two solutions ω_1 and ω_2 corresponding to the resonant frequencies in the waveguide:

$$\omega_{1,2} = \frac{\omega_R}{1+\beta_z}\left(1 \pm \beta_z\sqrt{1-X}\right) \tag{10}$$

where ω_R is the resonant frequency in free-space, $X = \left(k_\perp / \beta_z \gamma_{//} k_o\right)^2$ is the waveguide parameter and $\gamma_{//} = 1/\sqrt{1-\beta_z^2}$ is the longitudinal normalized energy of the electrons.

The situation where X=0 corresponds to the free-space configuration. The upper frequency ω_1 is equal to the "standard" resonant frequency ω_R of the undulator, and the lower frequency $\omega_2 \cong ck_o/2$ corresponds to a backward wave with the resonant wavelength $\lambda_2 \cong 2\lambda_o$ in the centimeter spectral range.

When X increases, X>0, the two frequencies ω_1 and ω_2 are shifting, and are converging one towards each other. The extreme situation where X=1 corresponds to the "zero slippage condition", where the wave group velocity $v_g = c^2/v_\varphi$ equals the electron bunch longitudinal velocity v_z. In this case, both frequencies are merging into a single resonant frequency $\omega_1 = \omega_2 \cong \omega_R/2$.

When considering the FEL spectral range $\lambda<100\mu m$, and using an overmoded waveguide with an aperture b>1cm, then the waveguide parameter X is close to zero: X<<1. Therefore, it is possible to make a first order linear approximation of the upper resonant frequency ω_1:

$$\omega_1(q) \cong \omega_R - \frac{q^2 \pi c \lambda_o}{4b^2} \qquad (11)$$

corresponding to a resonant wavelength λ_1:

$$\lambda_1(q) \cong \lambda_R + \frac{\lambda_R^2 q^2 \lambda_o}{8b^2} \qquad (12)$$

As an example, for CLIO, the cut-off wavelength of the waveguide is $\lambda_c=2b/q$ = 28mm for the first mode q=1. The wavelength is in the range: $5\mu m<\lambda<150\mu m$. The relative wavelength shift $(\lambda_1(q)-\lambda_R)/\lambda_R$ is shown in the table 1, as a function of the mode q, for different wavelengths.

$(\lambda_1(q)-\lambda_R)/\lambda_R$ (in %)	q=1	q=3	q=5
$\lambda_R = 50\mu m$	0.15 %	1.4 %	4 %
$\lambda_R = 100\mu m$	0.32%	3%	8%

Table 1. Percentage of shift of the undulator wavelength resonance

This shift has to be compared to the linewidth of the undulator radiation: $\Delta\lambda/\lambda_R \cong 1/N$, where N is the number of undulator periods. We have for CLIO $\Delta\lambda/\lambda_R$ = 2.5 %, with N=40. Therefore, the wavelength shift of the resonance is negligible for the mode q=1, but is important for $q \geq 5$.

In principle, this may have an important consequence on the FEL gain. Indeed, the spectral range of the FEL gain is centred on the resonance wavelength λ_R and it is limited, in first approximation, to the resonance wavelength linewidth, which is of the order of $\Delta\lambda/\lambda_R \cong 1/N$. When a waveguide is present in the undulator section, the laser mode is a sum of eigenmodes, with their own resonant frequency $\lambda_1(q)$. Therefore, at large values of resonance shifts $\lambda_1(q)-\lambda_R > \Delta\lambda$, for the eigenmode q, no gain occurs. In other words, only the low order eigenmodes will be amplified by the FEL process because their wavelength remains inside the gain band-pass.

However, the coupling of eigenmodes compensates this effect of gain selection. Indeed, outside of the waveguide, i.e. in free space areas, the transverse distribution of laser amplitude A(x,y) is modified by the divergence and by the diffraction, and it is focused by the mirror on the waveguide entrance. This makes a recombination, a coupling of the eigenmodes, inside the waveguide, at each cavity round trip of the laser pulse. On the other hand, the FEL gain is limited, in transverse plane, to the electron beam cross-section. Therefore, only the central part of the eigenmode is amplified. This makes a distortion of the eigenmode profile, which creates a mode coupling and feeds the high order modes. As a consequence, and if we consider the worst case, where only the first mode q=1 is amplified, the mode coupling allows to share the benefice of the gain among the eigenmodes. In addition, the high order modes, which are not interacting with the gain process, are not participating to the FEL gain saturation process. But these high order modes are still stored in the optical cavity, and are contributing to the laser pulse energy, although not contributing to the gain saturation process. The numerical simulations, which have been performed (Prazeres et al., 2009) using the CLIO FEL configuration, have shown that such effect of eigenmode selection by gain gives an error of less than 10% to 20% on the output power, and it does not change significantly the laser mode distribution.

4. "Spectral Gaps" phenomenon with a waveguided FEL

In a "partially guided" configuration, using in addition a hole coupling system, we have shown that a phenomenon called "spectral gaps" (Prazeres et al., 2009) may produce a strong drop of power at some wavelengths of the laser. This effect is always present, independently of FEL tunings, of hole coupling size or of cavity mirror radius of curvature. It is very inconvenient for users since holes are observed during the wavelength scan of the FEL. This effect is due to the influence of the waveguide in the optical cavity.

4.1 Power measurements (on CLIO)

As an example, we show here the measurements on the CLIO FEL. The figure 6 shows the output laser power of CLIO as a function of wavelength, for a 15 MeV electron beam. The points correspond to measurements, and the continuous line to a numerical simulation using the code "MODES". Spherical mirrors are used here for the cavity, with a hole of 2 mm for output coupling. These measurements have been recorded in a purged chamber in order to avoid laser absorption in air.

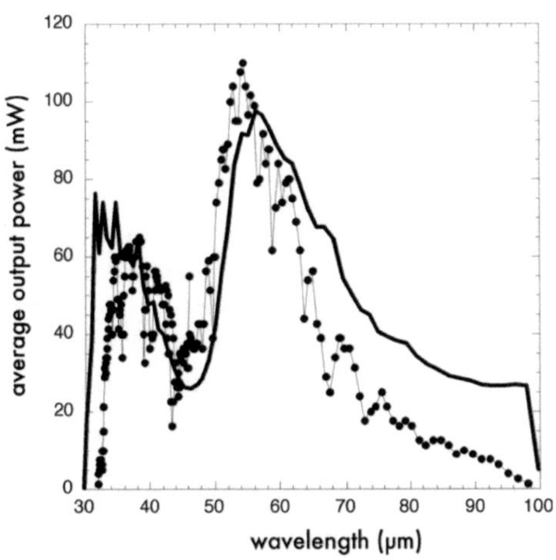

Fig. 6. Laser power vs. centroid wavelength, for spherical mirrors cavity, using hole coupling of 2mm (dotted line: measurement, continuous line: simulation)

A Spectral Gap appears at λ=45μm and corresponds to a power decrease by a factor of about 2. The simulation fits very well measurements, including the absolute value of the power, with a set of electron beam parameters: bunch charge Q=0.5nC, energy spread σ_γ/γ=0.5%, pulse length σ_z=4ps, emittance 150 π mm.mrd.

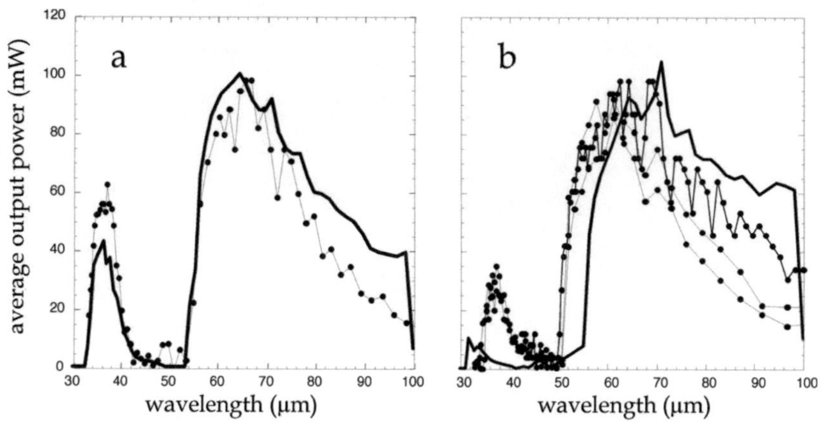

Fig. 7. Laser power vs. centroid wavelength, for hybrid cavity (a) and toroidal cavity (b), and using a hole coupling of 3mm (continuous line is a numerical simulation)

The figure 7a corresponds to another set of measurements, using a hybrid cavity (toroidal upstream mirror and spherical downstream mirror) with a 3mm hole coupling. The Spectral Gap is even more noticeable in this case. The figure 7b as been obtained with a full toroidal cavity, i.e. toroidal on both cavity mirrors. These measurements show that Spectral Gaps is an intrinsic effect, independent from the optical cavity configuration.

4.2 Simulation for a spherical cavity

In order to explain the Spectral Gap phenomenon, we used the numerical code "MODES" to compute the cavity losses L and the hole coupling extraction rate T_X as a function of wavelength, and the laser mode profile at various places in the optical cavity. The figures 8 and 9 show the results in the case of a spherical mirrors cavity, i.e. related to power variations of figure 6.

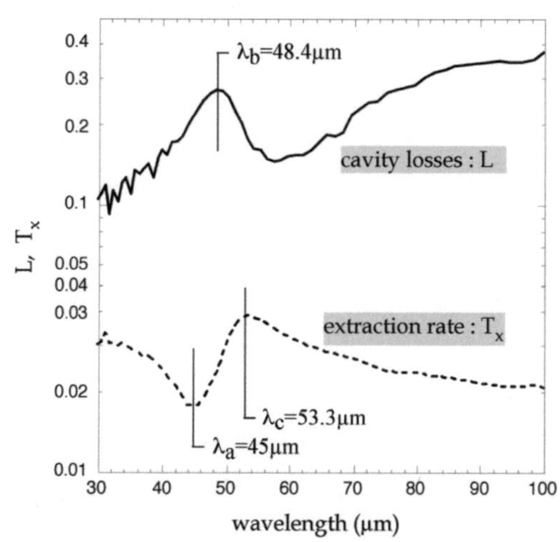

Fig. 8. Simulation of cavity losses L (continuous line) and hole coupling extraction rate T_X (dotted line) as a function of wavelength, for spherical cavity

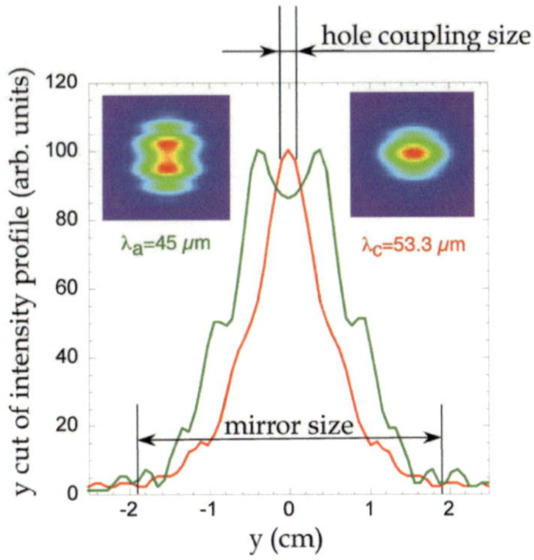

Fig. 9. Numerical simulation in the case of spherical cavity: laser mode transverse profiles on the output mirror, for 2 different wavelengths

The reason of the Spectral Gap is clearly visible here: the extraction rate T_X exhibits a minimum at wavelength λ_a=45µm - note that vertical axis is in log scale. The figure 9 displays the transverse profile of the laser mode, on the output mirror (point C in fig.4), for wavelengths λ_a=45µm and λ_c=53.3µm. At wavelength λ_a, the profile is wider and exhibits a small minimum of intensity in center, which accounts for the decrease of extraction rate T_X. This is a first explanation of the output power decrease at 45µm.

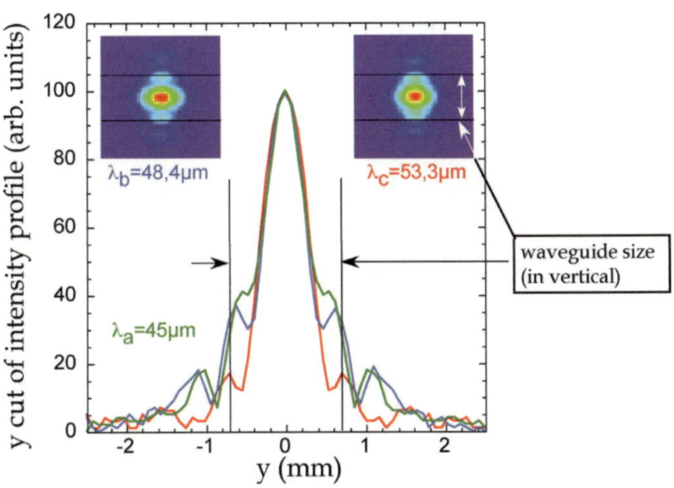

Fig. 10. Numerical simulation in the case of spherical cavity: laser mode transverse profiles on downstream waveguide entrance (point B in fig.4), for 3 different wavelengths

Indeed, this is not the only cause of Spectral Gaps: despite of the lower extraction rate, the cavity losses L are also maximum at wavelength λ_b=48.4µm - see figure 8. Note that mirror reflectivity is independent of wavelength, and Ohmic losses in the waveguide are negligible. A detailed analysis of the laser mode profile, at the waveguide entrance, is displayed in figure 10. It shows that high losses, at wavelength λ_b, are due to a vertical broadening of the transverse size of laser mode, which does not fit the waveguide aperture. It occurs only in the vertical plane, where the laser mode is guided.

At λ_b=48.4µm, about 20% of laser energy is lost on each entrances of the waveguide. It represents only 8% at λ_b=53.3µm. These losses occur twice in a complete cavity round trip of the laser pulse. This explains the peak of losses in the curve of figure 8, which is also responsible for the Spectral Gap observed in figure 6.

4.3 Simulation for a toroidal cavity

As shown in figure 7b, the Spectral Gap phenomenon with a toroidal cavity is even stronger than in the case of spherical cavity (figure 6). The figure 11 shows the cavity losses L, and the hole coupling extraction rate T_X, as a function of wavelength, for a toroidal cavity.

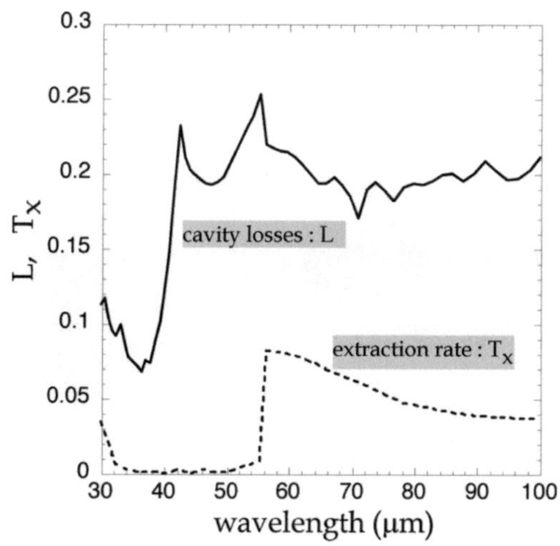

Fig. 11. Simulation, total cavity losses L and hole coupling extraction rate T_X as a function of wavelength, for toroidal cavity

The decreasing of the extraction rate T_X is much stronger here: there is a large range showing nearly zero extraction, between 32 and 55 µm. Also, within this range, the losses are varying strongly, and exhibit an important minimum at 37 µm. Such small losses are creating a large intracavity power P_{in}, as shown in figure 12 which displays both intracavity power P_{in} and output power P_{out}. The large value of P_{in} compensates partially the low extraction rate T_X, and leads to a significant output power P_{out} in the range 30-40µm. This effect is shown in simulations (fig.12), but it is even clearly observed in the measurements in figure 7b.

In real experimental conditions, a small misalignment of the cavity mirrors, or of the electron beam, may allow the intracavity power of the laser mode to be better out-coupled by the hole. This "default" is not taken into account in the simulations, and it could explain why the measured power overcomes the simulation in figure 7b in the range 30-40µm. This range corresponds to a critical situation where the intracavity power is large, whereas the extraction rate is reduced: the extracted power comes from this compromise.

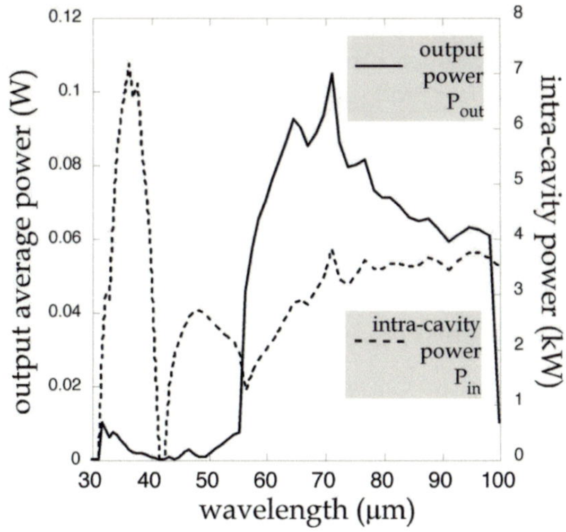

Fig. 12. Simulation, output power P_{out}, and intracavity power P_{in}, as a function of wavelength, for toroidal cavity.

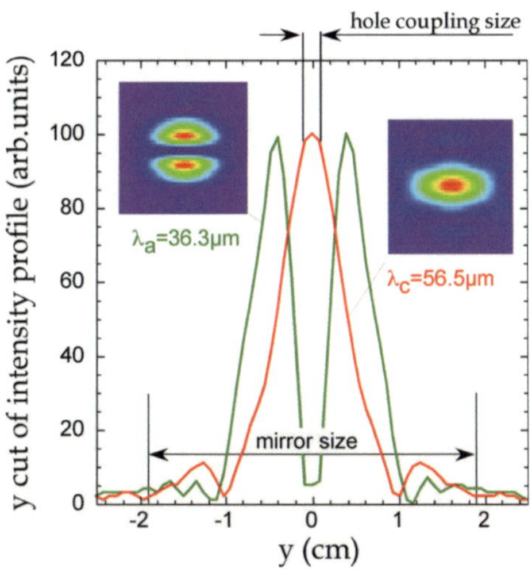

Fig. 13. Numerical simulation in the case of toroidal cavity: laser mode transverse profiles on the output mirror, for two different wavelengths.

The low extraction rate T_X for $\lambda < 55$ µm is explained by the image of laser mode on output mirror, displayed in figure 13: a double spot profile avoids the extraction hole close to λ=36µm. And the mode jumping, from 2 spots to 1 spot, at 56 µm, explains the sharp variation of T_X.

Also, the strong variations of cavity losses in figure 11 are due to variations of the mode profile at the waveguide entrance. Figure 14 displays the mode profiles at the downstream waveguide entrance (point B in fig.4), at 36µm and 55µm, corresponding respectively to minimum and maximum of cavity losses L. At λ=36.3 µm, the sum of optical losses, at both entrances, represents only 2.6%, which is rather low as compared to other wavelengths.

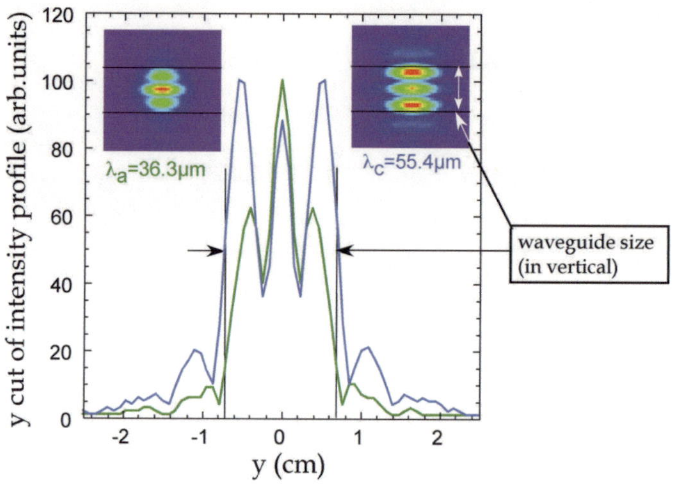

Fig. 14. Numerical simulation in the case of toroidal cavity: laser mode transverse profiles on downstream waveguide entrance, for two different wavelengths

As a conclusion, the explanation of Spectral Gap phenomenon is quite complex, because it is due to two effects: (1) a substantial increasing of cavity losses by clipping of the laser mode profile at the waveguide entrances, and (2) a minimum of the extraction rate T_X due to exotic transverse profiles on the hole coupling mirror. And both effects are strongly dependant on slight variations of the transverse mode profile. They are due to the presence of the waveguide. Indeed, we will see below that it is depending on the dimensions of the waveguide aperture.

4.4 Measurements on other FELs

The optical cavity of the FELIX infrared FEL (FOM Institute for Plasma Physics Rijnhuizen) is of same type as for CLIO, but with minor differences: the downstream cavity mirror is in contact with the waveguide, i.e. there is no free space between that mirror and the waveguide entrance. But there is a free space of 1.66 m in front of the upstream mirror. The distance between cavity mirrors is 6 meters, and the waveguide length is 4.34 m. The figure 15 shows a comparison between simulation, with "MODES" code, and measurement of the output power. Two measurements are displayed here: (1) for a perfect alignment of the cavity, and (2) for an alignment optimized at λ=35 μm. It shows that the Spectral Gap phenomenon exists on FELIX. The depth of Spectral Gaps may change with FEL alignment, but their positions remain identical.

The numerical simulation exhibits the two main Spectral Gaps, at 38 μm and 53 μm. A detailed simulation (Prazeres et al., 2009) makes a confirmation that the Spectral Gap is still due to two effects which are described above: (1) an increasing of optical losses by clipping into waveguide entrances, and (2) a reduction of the extraction rate T_X due to a split of the laser profile on the output mirror. The relevant laser profiles are shown in figure 15. The Spectral Gap occurs at wavelengths corresponding to a jump of laser mode into a complex shape.

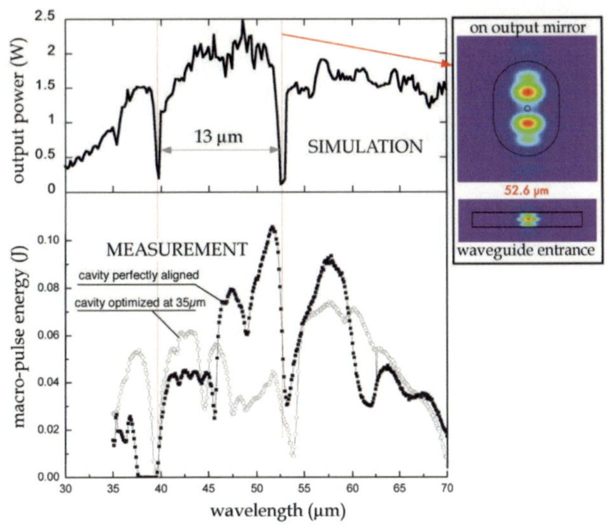

Fig. 15. Results for the FELIX infrared FEL

The configuration of the U100 FEL (Helmholtz-Zentrum Dresden-Rossendorf [HZDR], ELBE laboratory), is very similar to the FELIX FEL: the waveguide is in contact with downstream mirror. But the optical cavity is much larger, 11.5 meters long, and the waveguide also, 7.9 m. Figure 16 shows the result of a numerical simulation with "MODES" code. It displays the output power as a function of wavelength. Four Spectral Gaps are observed and they seem to be periodically spaced, with about a 7μm period. Unfortunately, the measurements of output power on ELBE are very lacunar, and they exhibit air absorption lines in the spectrum. A comparison with measurements is not available.

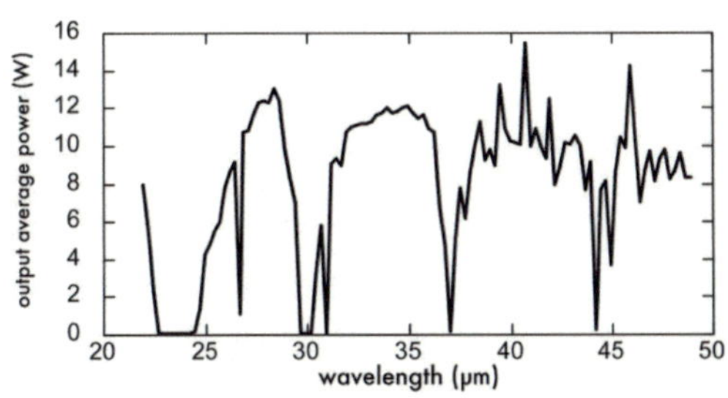

Fig. 16. Simulation of output power on the U100 FEL in ELBE, at 32MeV, with 2mm hole coupling

4.5 Analytical description of the Spectral Gaps

The Spectral Gaps (SG) phenomenon is due to a particular combination of eigenmodes TEq and TMq, produced in the waveguide, which creates a mode profile producing large losses or low extraction rate. These eigenmodes are only dependant on modes q because the waveguide is equivalent to infinite parallel plates. The figure 17 shows an example of the energy distribution of eigenmodes in the case of CLIO at λ=50 μm, corresponding to a mode profile which gives a two peaks structure on output mirror, as displayed in figure 13.

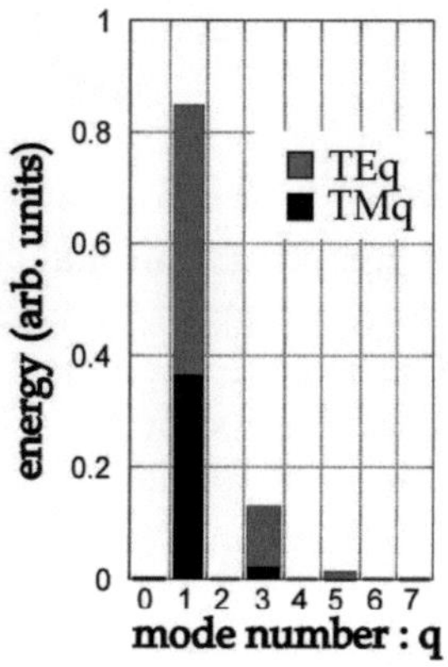

Fig. 17. Energy distribution of eigenmodes on TEq and TMq in the case of CLIO, with toroidal mirrors, at λ=50μm

Even modes are not present here, because they have a zero amplitude on axis and are not amplified by the electron beam that is centred on axis. This figure shows that the most important modes are q=1 and q=3.

It is likely that the phenomenon of SG is linked to the relative phase of the dominant cavity modes at the ends of the waveguide. Indeed, this phase will determine the diffraction losses at the entries of the waveguide and the mode structure at the extraction hole. Therefore, when sweeping the FEL wavelength, one expects that a phase difference of 2π in a cavity round trip will account for the wavelength difference between two successive SG.

The phase velocity in the waveguide for the mode q is:

$$v_\varphi = \frac{c}{\sqrt{1-\left(\lambda/\lambda_c\right)^2}} \cong c\left(1+\frac{\lambda^2 q^2}{8b^2}\right) \qquad (13)$$

where $\lambda_c=2b/q$ is the cut-off wavelength, and b is the waveguide aperture in vertical axis. The approximation corresponds to an overmoded waveguide with $\lambda<<\lambda_c$. The propagation time of mode q, in a waveguide of length L_w, is $t_q=L/v_q$. The relevant dephasing of mode q is:

$$\Phi_q = \frac{2\pi}{T} t_q \cong \frac{2\pi L_w}{\lambda}\left(1-\frac{\lambda^2 q^2}{8b^2}\right) \qquad (14)$$

the phase difference $\Delta\phi = \phi_1 - \phi_3$ between modes q=1 and q=3 is:

$$\Delta\Phi = \Phi_1 - \Phi_3 = \frac{2\pi L_w \lambda}{b^2} \qquad (15)$$

and for a complete cavity round trip:

$$\Delta\Phi_{RT} = \Phi_1 - \Phi_3 = \frac{4\pi L_w \lambda}{b^2} \qquad (16)$$

This calculation does not make any hypothesis on the absolute phase distribution of the modes. It only gives the phase shift between mode 1 and mode 3, for one cavity round trip. Now, when choosing λ corresponding to a SG, then the "next" SG in the FEL spectrum, at wavelength $\lambda' > \lambda$, must keep the same phase structure. This is the unique hypothesis that we do here: the phase structure is the same for all SGs (for a given FEL configuration, of course). Therefore, for:

$$\Delta\Phi_{RT}(\lambda') = \Delta\Phi_{RT}(\lambda) + 2\pi \qquad (17)$$

one obtains:

$$\delta\lambda = \lambda' - \lambda = \frac{b^2}{2L_w} \qquad (18)$$

where $\delta\lambda$ is expected to be the wavelength difference (λ - λ') between two successive SG in the FEL power spectrum. Note that the 2π periodicity in expression (17) corresponds to a 4π periodicity when using a pair of higher modes q=3 and q=5. Therefore, as stated in the above hypothesis, the whole structure of modes TEq and TMq is kept constant between two consecutive SGs at λ and λ'.

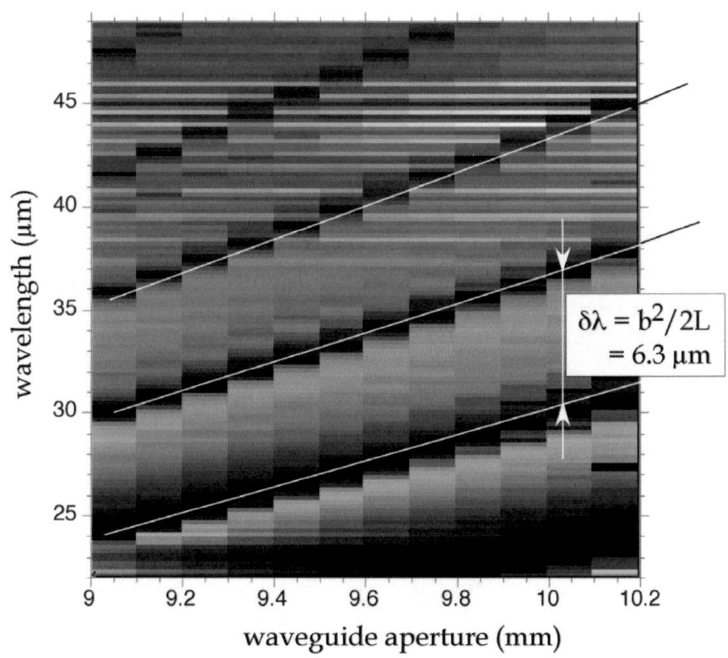

Fig. 18. Simulation – laser power vs. wavelength and waveguide aperture "b"

In order to check this simple analytical model, we show in figure 18 a simulation of FEL power as a function of wavelength, for various dimensions of waveguide aperture "b". The parameters used in this calculation correspond to the ELBE free-electron laser, which the geometrical configuration gives a short wavelength period $\delta\lambda$ for SG phenomenon.

The lines correspond to SGs, and the distance between two successive SG is 6.3 μm as predicted by the simple analytical expression (18) with b=1cm and L_w=7.92 m. Therefore, the simple model and the simulation agree very well, even if the available data do not allows to verify it experimentally.

Now for FELIX, the expression (18) gives $\delta\lambda$=11.5 μm, with b=1 cm and L_w=4.34 m. The simulation curve in figure 15 shows that the distance between two successive SG is $\delta\lambda$=13 μm, which is not far from analytical and experimental values.

For the case of CLIO, the expression (18) gives $\delta\lambda$=50 μm. The first SG in spectrum is measured at λ=45 to 50 μm. Therefore, the next SG should be observed close to λ=100 μm, which is not the case in the measurements. Indeed, we make the assumption here that such simple analytical model must be realistic when the wavelength period $\delta\lambda$ of SG is short as compared to the FEL wavelength full range. But a "large" period of $\delta\lambda$=50μm corresponds quite to the full spectral range, and the features the FEL are not constant between these two extreme points.

Nevertheless, the expression (18) provides an order of magnitude of the number N_{SG} of SGs in a given FEL spectral range (λ_{max}- λ_{min}):

$$N_{SG} \cong (\lambda_{max} - \lambda_{min}).2L_w / b^2 \quad (19)$$

Therefore, in the design of a far-infrared waveguide FEL, it is important to chose the waveguide geometry, with L_w and b, in order to reduce N_{SG} at minimum or at less than one.

5. Analysis of laser beam transverse profile

The hole coupling is a convenient system for laser extraction from the cavity. However, the intracavity optical mode is influenced by the hole, and this modifies the profile of the extracted light. The optimization of the FEL output power corresponds to a compromise between hole coupling extraction T_x and intracavity laser power P_{in}. Indeed, the "natural" tendency of the intracavity laser mode is to reduce the optical losses L, i.e. to create a mode with a minimum of hole coupling T_x, leading to a maximum intracavity power P_{in}. Whereas the desire of FEL user is to get a maximum of output power $P_{out}=T_x*P_{in}$ which is proportional to the hole coupling T_x. Therefore, the machine parameters are adjusted in this aim. This compromise is mainly dependent on the relative values of 2 parameters: the laser wavelength and the extraction hole size. For "small" hole size, or large wavelength, the laser profile is not strongly affected, but the output coupling T_x is small. On the other hand, for "large" hole size, the intracavity laser profile tends to exhibit, on the output mirror, a minimum on Z axis which tends to reduce the output coupling T_x - see section 2.1. The better compromise is in between these two configurations. It gives a maximum of output FEL power P_{out}, but it still produces a distorted transverse profile of the output laser beam.

5.1 Laser profile measurements

In order to have a diagnostic of this effect, it would be interesting to have an image of the intracavity laser mode. But this is not always possible because of the mechanical configuration of the FEL cavity. Nevertheless, we have done measurements, on CLIO, of the laser profile outside of the cavity. These measurements have been done using a 2D infrared camera, which is placed outside of the optical cavity, and on the focal plane of a focusing mirror. The detector is a 2D matrix of pyroelectric detectors. The image, which is projected on the surface of detector, corresponds to the magnified image of the hole on output downstream mirror. The measured profile is shown on figure 19, for different wavelengths. At short wavelength (λ < 10 µm), as displayed in fig. 19a for λ=6.5µm, the profile always exhibits a typical "crescent moon" shape. Even when trying to modify all parameters of the FEL, we always obtain this typical profile. This shows that the intracavity laser mode is not centered on Z axis, and only one off-axis part of the profile is extracted by the hole. As the wavelength is increased, as shown in fig.19b for λ=17µm, the profile tends to fill the hole, as expected. The horizontal lines are due to diffraction in the laser beam line - outside of the cavity.

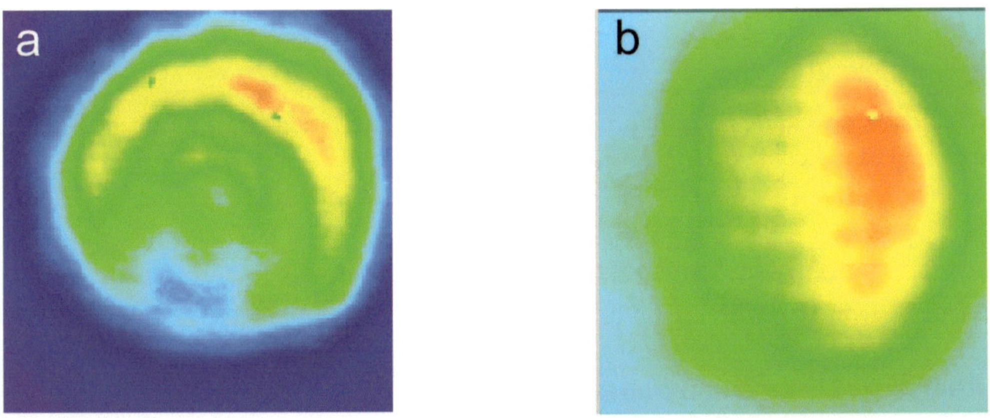

Fig. 19. Measurement of laser mode transverse profiles, in the experimental room, at λ =6.7 µm (a), and λ =17 µm (b)

Note that there is a competition, in the hole-coupled FEL configuration, between the hole extraction process and the optical gain process. The optical gain of the FEL is limited in space: it is limited along Z axis in the middle part of the undulator, and it is limited in (X,Y) plane by the electron beam cross-section, which RMS dimensions are close to 1 or 2mm. This forces the laser beam to stay on Z axis in the middle of the undulator. Whereas the hole on output mirror avoids that laser spot stays on Z axis at the extremity of the cavity. These two constraints force the laser beam to be misaligned in the cavity. As a consequence, the tuning of angular tilt of cavity mirrors, or the electron beam alignment in undulator section, have a strong influence on the intracavity laser mode, and they are crucial parameters for the output laser power P_{out}. As described here in section 2.1, a slight misalignment of the optical cavity breaks the axial symmetry of the laser mode and allows optimizing the output laser power P_{out}.

5.2 Numerical simulations

Indeed, numerical simulations have been performed with the "MODES" code, and they have shown that the output power may be increased by using a tilt θ on the output downstream mirror (while keeping upstream mirror at normal incidence). The relevant laser profile is displayed in figure 20, which shows a numerical simulation of the amplitude distribution A(x,y). It takes into account the same experimental conditions as for the measurements: it calculates the wave propagation along the optical beam line and through the focusing mirror up to the detector plane. The wavelengths are respectively λ=6.5µm (a) and λ=17µm (b). These profiles have been calculated using a tilt on output downstream mirror. This tilt is oriented of 45°, with $\theta_x=\theta_y$ =0.4mrd, in order to get θ=0.56mrd at 45° in the XY plane. These images may be compared to the image on figure 19, which have been measured in same conditions with the pyroelectric camera. The profile on fig.20a exhibits a "crescent moon" shape, which reproduces reasonably well the measurement. The profile on fig.20b is larger and fills the hole size, as shown in measurements on fig.19b.

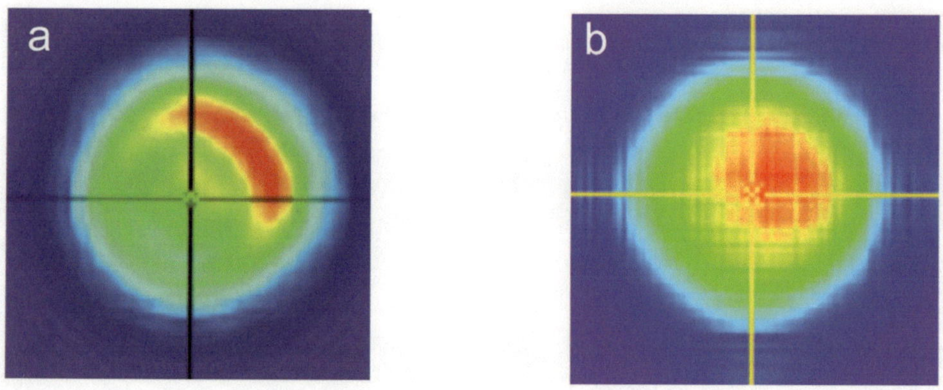

Fig. 20. Numerical simulation of laser mode transverse profile, calculated outside of cavity and within the experimental conditions, on the detector plane, at λ =6.7 µm (a), and λ =17 µm (b)

The figure 21 shows a simulation within same conditions, but it displays the profile inside of the cavity, on the output downstream mirror, at λ=6.7µm (a) and λ =17µm (b). The diameter of mirror (38mm) is represented by a circle, and the extraction hole (2mm) by a black dot in center. The laser profile exhibits, on fig.21a, a single spot, which is not centered on the hole. This explains the "crescent moon" shape of the extracted laser profile. At 17µm, the laser distibution is closer to the center of the hole, and this is sufficient to create a rather homogeneous output laser profile on fig.20b.

In order to show a comparison with on-axis alignment of the cavity, the figure 22 displays the result of the numerical simulation calculated for θ = 0, i.e. for normal incidence at the downstream mirror. In this case, the intracavity mode exhibits a double spot, in figure 22a, with a minimum on extraction hole and almost zero output power. In figure 22b, corresponding to the profile calculated outside of the cavity, on the detector plane, the mode also exhibts a double symmetrical spot. But in practice, it is not possible to observed such a laser profile, because of weakness of the power. This is another argument to say that the laser mode is never aligned on Z axis (Z axis being defined by the two centers of cavity mirrors).

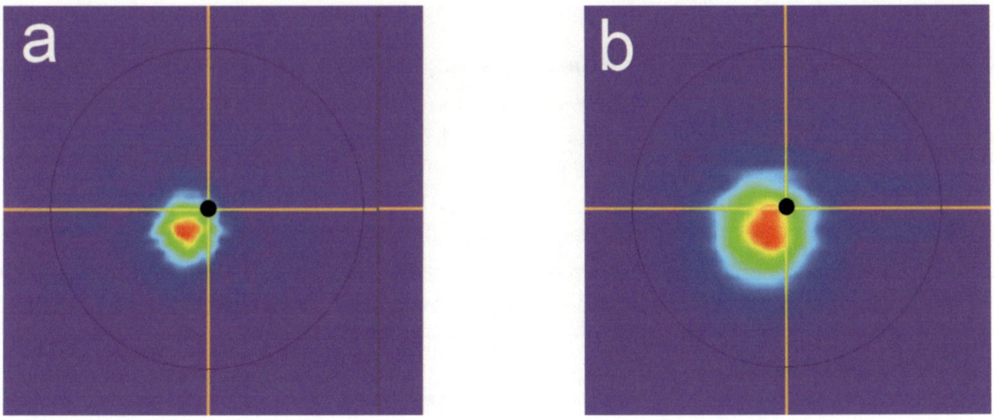

Fig. 21. Numerical simulation of laser mode profile inside of cavity and on downstream mirror, (a) at λ =6.7μm, (b) at λ =17μm

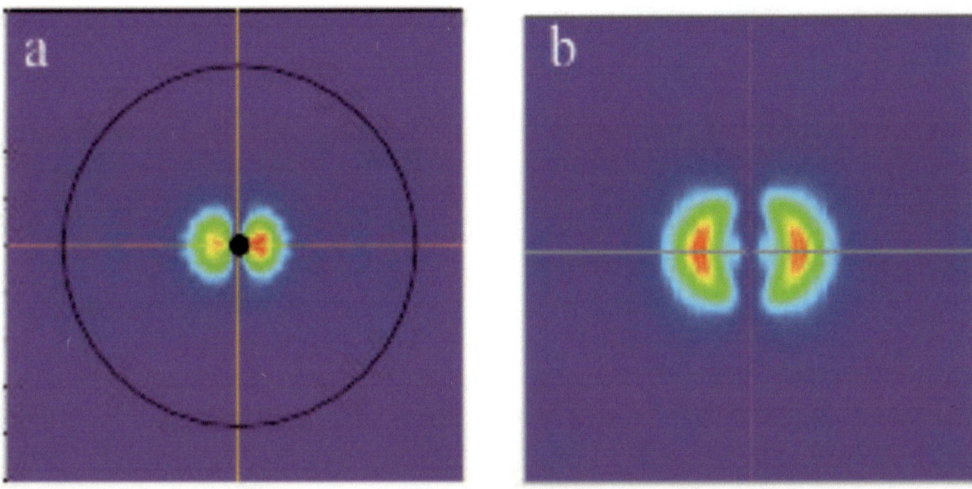

Fig. 22. Numerical simulation of laser mode profile, for λ=6.7μm, at normal incidence at the downstream mirror, (a) on cavity output mirror, (b) in experimental room

The competition phenomenon between the hole extraction process and the optical gain process, which is explained above, has been observed in the numerical simulations. The electron beam is centered on Z axis, and forces the laser mode to be aligned on the extraction hole (as an "automatic hole alignement process"). This effect is very important to get important output laser power, as it force the system to keep a large extraction rate T_x. In order to point out this effect, a numerical simulation has been done within same conditions as above, but using a flat electron beam profile, i.e. a large cross-section. This produces a uniform amplification of the laser cross-section, and avoids the effect of "automatic hole alignement". Within these (non realistic) conditions, the numerical simulation gave almost zero extraction rate T_x, and zero output power P_{out}, for any angular tilt θ of the downstream mirror. Therefore, if the laser beam is not pinched by the electron beam in the cavity, it always finds a position which avoids the hole. In other words, this shows that the small electron beam cross-section forces the laser beam to pass through the hole. Whereas the "natural tendancy" of the laser mode would be to avoid the hole.

6. Conclusion

The far infrared is the most effective operating range for the free-electron lasers: large gain available, large output power. It represents also a most reasonable cost since using a "low" energy electron beam, and therefore it can be a rather compact device. However, some inconveniences require peculiar attention during the designing of such FEL. The optical diffraction imposes to use a waveguide in the optical cavity. Then, if the waveguide does not fill the whole cavity, a phenomenon of "Spectral Gaps" may occur, which creates a drop of power at some wavelengths within the FEL tuning range. In order to avoid these Spectral Gaps, the length of the waveguide has to be as short as possible, and its aperture must be as large as possible. The first constraint imposes a limit in the number of undulator periods, whereas the second one limits the maximum magnetic field on axis. A compromise has to be find with these parameters, as a function of the requirements for the FEL performances. On the other hand, the lack of wide band optics, in infrared, requires to use a hole coupling extraction in the optical cavity. However, this system produces generally an exotic laser transverse distribution ("crescent moon" shape) on the output laser spot. This effect may be very inconvenient for the users of the laser beam. It may be erased by reducing the hole dimension on the extraction mirror, but in this case the output laser power decreases.

The laser mode inside the optical cavity is generally not Gaussian because of the influence of the waveguide and of the hole coupling. It is a complex combination of divergence in free-space, guiding of eigenmode and diffraction by the hole. Therefore, a simple analytical model cannot be satisfying to describe the features of the laser mode. Contrariwise, a numerical code like "MODES" gives results that are in rather good agreement with the various measurements on CLIO, and even taking into account the absolute value of laser power.

7. Acknowledgment

We are thankful to Lex Van der Meer, and more generally to the FELIX free-electron laser group, for their contribution by sending us experimental data for the FEL

8. References

Bonifacio, R. & De Salvo Souza, L. (1986). Tuning and slippage optimization in a high-gain FEL with a waveguide. *Nucl. Instr. & Meth. in Physics Reaserch*, Vol.A276, pp. 394-398

Deacon, D.A.G.; Elias, L.R.; Madey, J.M.J; Ramian, G.J.; Schwettmann, H.A. & Smith, T.I. (1977). First Operation of a Free-Electron Laser. *Physics Review Letters*, Vol.38, pp. 892-894

Glotin, F.; Berset, J.M.; Chaput, R.; Kergosien, B.; Humbert, G.; Jaroszynski, D.; Ortega, J.M.; Prazeres, R.; Velghe, M. ; Bourdon, J.C.; Bernard, M.; Dehamme, M.; Garvey, T.; Mencik, M.; Mouton, B.; Omeich, M.; Rodier, J.; & Roudier P. (1992). First lasing of the CLI0 FEL, *Proceedings of EPAC 1991 3rd European Particle Accelerator Conference,* pp. 620, Berlin, Germany, December 1991

NAM, S.K.; Yi, J. & KIM, K.B. (2000). *The small-signal gain in a waveguide free electron laser.* Journal of physical society, *Vol.37, No3, (September 2000), pp. 236-240*

Oepts, D. & van der Meer, A.F.G. (2010). Start-up and Radiation Characteristics of the FELIX Long Wavelength FEL in the Vicinity of a Tuning Gap. Proceedings of 32th International Free-Electron Laser Conference, pp. 323-327, Malmö, Sweden, August 23-27, 2010

Ortega, J.M.; Glotin, F.; Prazeres R. (2006). Extension in far-infrared of the CLIO free-electron laser. Proceedings of international Workshop on Infrared Microscopy and Spectroscopy with Accelerator-Based Sources (WIRMS 2005), Rathen, Germany, June 2005. Infrared Physics & Technology, Vol.49, pp. 133-138.

Prazeres, R; Glotin, F & Ortega, J.M. (2009). Analysis of periodic spectral gaps observed in the tuning range of free-electron lasers with a partial waveguide. *Physical Review Special Topics -Acceletors and Beams.* Vol.12, 0110701

Prazeres, R.; Glotin, F. & Ortega, J.M. (2005). Measurement and calculation of the "electron efficiency" on the "CLIO" free-electron laser. *European Physical Journal - Applied Physics,* Vol.29, pp. 223-230

Prazeres, R.; Glotin, F.; Rippon, C. & Ortega J.M. (2002). *Operation of the "CLIO" FEL at long wavelengths and study of partial guiding in the optical cavity.* Nucl. Instr. & Meth. in Physics Reaserch, *Vol.A483, pp. 245–249*

Prazeres R. (2001). A method of calculating the propagation of electromagnetic fields both in waveguides and in free space using the Fast Fourier Transform. *European Physical Journal - Applied Physics,* Vol.AP16, pp. 209-215

Prazeres, R. & Billardon, M. (1992). Numerical calculation of transverse optical modes for characterization of an optical cavity. *Nucl. Instr. & Meth. in Physics Reaserch,* Vol.A318, pp. 889-894

Tanaka, T.;Hara, T.; Tsuru, R.; Iwaki, D.; Marechal, X.; Bizen, T.; Seike, T. & Kitamura, H. (2005). In-Vacuum undulators, Proceedings of 27th International Free-Electron Laser Conference, *pp. 370-377, Stanford, USA, August 21-26, 2005*

4

Compact XFEL Schemes

Emmanuel d'Humières and Philippe Balcou
Université de Bordeaux – CEA - CNRS – CELIA, Bordeaux
France

1. Introduction

The quest for X-ray lasers has long been a major objective of laser science, starting from the early proposals of Duguay and Rentzepis, and of Jaeglé (2006), up to the modern large-scale projects of extreme UV and X-ray free electron lasers in Europe, Japan, Korea and the US.

X-ray radiation is one of the most efficient tools to explore the properties of matter in multidisciplinary domains. Over the past fifty years three generations of synchrotrons have been developed delivering x-ray beams with always shorter wavelength and higher brightness to a rapidly growing users community. These large scale instruments have lead to important discoveries and outstanding applications. However, while femtosecond x-ray pulses are now essential to open new possibilities of research and applications (Ultrafast Phenomena proceedings 1992-2002; Zewail 2000; Bloembergen et al., 1999; Rousse et al., 2001), the shortest pulse duration at synchrotron is a few tens picosecond. To face these technological limits, several methods have been proposed and demonstrated, based, for example on electron bunch slicing or Thomson scattering off a part of the bunch (Schoenlein et al., 1996 ; Schoenlein et al., 2000). However, these mechanisms are limited by their very low efficiency. Major progresses have been made with the fourth generation of synchrotron: the free electrons lasers (FEL). Even larger than a synchrotron, X-ray FELs can produce femtosecond x-ray pulses billion times more intense than any conventional source (Emma et al., 2010).

This continued effort, involving several scientific communities from laser physics, plasma physics, and accelerator physics, has already resulted in many experimental demonstrations of lasing, or of laser-like radiation, mostly in the extreme ultra-violet, or very soft Xray ranges. In the case of Xray lasers based on the interaction between intense laser pulses and plasmas, numerous lasing lines have been brought to saturation. High-harmonic generation, a process by which a femtosecond intense laser is converted directly into an XUV coherent beam, has bloomed into an extremely effective method for applications to time-resolved studies, down to the attosecond scale. In the field of laser-plasma interactions, incoherent X-ray sources obtained by shining an intense laser on plasmas or on free electrons have already allowed scientific premieres, such as sub-picosecond time-resolved diffraction studies of non thermal melting. Last but not least, the short-wavelength free electron lasers have recently achieved lasing, especially with the first X-ray light from the American LCLS (Emma et al., 2009). The European XFEL, and the Japanese SCSS projects will soon follow.

The impressive push to all those new technologies is not related to a problem of availability of X-ray sources, as the latter are readily available to the scientific and industrial communities thanks to last generation synchrotron sources. However, a X-ray laser beam would possess inherently many novel properties, especially in terms of very short (femtosecond) time duration, and focusability to very small spots, thereby opening several new fields of research: single protein crystallography, strong field science in the X-rays, Warm Dense Matter, and even many medical applications, especially in oncology (Carroll 2003). The huge potential of all these applications was shown by the various scientific surveys performed to present the scientific cases of the major XFEL projects. Increasing the energy range of X-ray lasers beyond the current limit of roughly 10 keV for XFELs, would open even more numerous applications, and reaching the γ-ray range could even start brand new fields, such as nuclear laser spectroscopy.

However, the current XFEL projects are so large, both in size and budget, that they are bound to remain Large Scale Infrastructures, with a real issue of beam-time availability, and will not have the possibility to disseminate in university-scale research centers, industrial laboratories, or hospitals. The other technologies mentioned, such as laser-plasma X-ray lasers, high harmonic generation, Thomson (Inverse Compton) scattering on relativistic electrons, etc, are either too limited in flux, or simply restricted to the extreme ultra-violet wavelengths, and, while very useful for many studies, cannot be considered for some of the highest profile applications of XFELs.

Being able to combine the large flux and brilliance, and the operation in the real X-ray range of XFELs on one hand, and the compacity and reduced cost of XUV / X-ray sources derived from short pulse intense lasers on the other hand, would be an outstanding breakthrough in X-ray science. Facilities to generate X-ray pulses during the interaction of intense lasers and matter are compact, less costly, but yield mostly incoherent light, with low brightnesses. The development of coherent compact X-ray sources would allow to finally meet the current demand in X-ray sources for the applications discussed above.

Different setups are now considered, LINAC accelerated electrons coupled with a magnetic or an optical wiggler and laser wakefield accelerated electrons with a magnetic or an optical wiggler. Laser wakefield acceleration of electrons is increasingly considered as a potential compact substitute of conventional accelerator technology. It has made tremendous progresses in the last ten years thanks to the advent of short and high power Titane:Saphire systems. Two prospective schemes have recently being proposed with this goal, both based on the impressive progress of Laser WakeField Acceleration (LWFA) of electrons. On one hand, a few groups have proposed to use electron packets, accelerated in the LWFA or bubble regime up to energies around 1 GeV, and inject them into an undulator to obtain the micro-bunching effect of free electron lasers (Gruner et al., 2007). A second proposal revisits an old idea, which has remained so far inapplicable: the laser-undulator free electron laser (Petrillo et al., 2008). Both schemes are extremely interesting in the soft X-ray range around 1 keV. Both require however in a stringent way extremely challenging parameters of mono-energeticity and emittance of the laser-accelerated electron bunches. To the best of our knowledge, no compact scheme for an XFEL has yet been proposed, robust enough to be adapted to realistic conditions of relativistic electron bunches.

More recently, a new scheme using an all-optical setup has been proposed [P. Balcou et al. EPJD 2010]. It uses two counter-propagating lasers to create the wiggler and the X-ray radiation is created through a Raman scattering process between the electron bunching, the wiggler lasers and the generated X-rays.

We explore in this Chapter different opportunities to create a compact X-ray FEL, by coupling the physics of free electron lasers, of laser-plasma XUV lasers, and of extreme non-linear optics. In Section 2 the key physical mechanisms involved in these schemes are presented. In Section 3 several compact XFEL schemes proposed in the last twenty years are described. Section 4 is devoted to the recent proposal of a Raman compact XFEL and to a discussion on the possible parameters of an all-optical compact XFEL using this scheme. Section 5 closes this Chapter by discussing the perspectives of compact XFEL schemes.

2. Key physical mechanisms involved in these schemes

In this Section, the key physical mechanisms involved in the various compact XFEL schemes presented in this Chapter are discussed. First, the various techniques used to produce collimated energetic electron beams are reviewed. Then, magnetic and optical wigglers are presented and finally, the physics of electron beams trapped in an optical wiggler is discussed.

2.1 Energetic electron beam production

For the different compact XFEL schemes described below, a bunch of free electrons, with a kinetic energy in the range from 10 to 50 MeV, and hence a Lorentz factor from 20 to 100, needs to be produced using either a linear accelerator or a small high power short laser system. This element is typical of Thomson (inverse Compton) scattering experiments, or of free electron lasers, except for the use of smaller electron kinetic energies.

2.1.1 Linear particle accelerators

A linear particle accelerator (or LINAC) is a method of particle acceleration using oscillating electric potentials accelerating charged particles on a linear path. It was invented in the late 1920s. The design of a LINAC depends on the type of particles which it is supposed to accelerate and greatly varies in size depending on the energy the accelerated particles can reach (from a few meters to a few kilometers at SLAC in California).

In 1924, a theoretical paper by G. Ising, Stockholm, describes a method for accelerating positive ions (canal rays) by applying the electrical wavefront from a spark discharge to an array of drift tubes via transmission lines of successively greater lengths.

In 1928, an experimental paper (including the theory of betatron) by R. Wideroe, Switzerland, describes the successful acceleration of Potassium ions to 50 keV. The Potassium ions travel from one gap to the next one in one-half a radio frequency (RF) period. Since higher frequency oscillators did not exist at the time, lighter particles traveling faster could not be accelerated.

From 1931 to 1935, K. Kingdon (G.E.), L. Snoddy (Univ. of Virginia) et al., accelerate electrons from 28 keV to 2.5 MeV by applying progressive wavefronts to a drift tube array.

In 1947 and 1948, at Stanford, W. Hansen, E. Ginzton, W. Kennedy et al. build the Mark I disk-loaded LINAC yielding 4.5 MeV electrons in a nine-foot structure powered to 1 MW at 2856 MHz. It is the first of a series: Mark II (40 MeV), Mark III (1.2 GeV), and SLAC (30 GeV). Parallel efforts take place in Great Britain, France and the USSR, and at MIT and Yale in the USA.

In 1973, P. Wilson, D. Farkas and H. Hogg at SLAC invent the RF energy compression scheme called SLED which in the next five subsequent years gets installed on the 3-km LINAC, boosting its energy up to 30 GeV electrons.

A LINAC is composed of a particle source, a high voltage source for the initial injection of particles, a vacuum chamber containing electrically isolated cylindrical electrodes energized by sources of RF energy (Fig. 1). The frequency of the driving signal and the spacing of the gaps between electrodes are designed so that the maximum voltage difference appears as the particles crosses the gap. The particle velocity is therefore increased when it passes the gap and it is accelerated.

The disadvantages of LINACs are their length for devices designed to reach high energies, the associated power requirements and the fact that to reach high energies the device needs to be operated in bursts as the accelerating cavity walls can not sustain continuous heating.

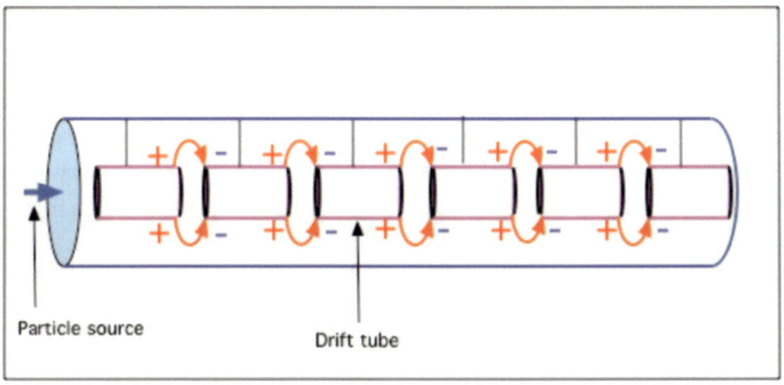

Fig. 1. Linear Particle Accelerator schematic.

Linear accelerators have made major contributions to physics research including neutron sources, colliding electron-positron beams, X-ray FELs, and heavy-ion rare-isotope beams. In addition electron linacs are used in hospitals around the world generating X-rays for radiation therapy, an application that represents one of the most significant spins-offs of high-energy and nuclear physics research.

Table 1 summarizes some industrial applications of RF LINACs with the associated characteristics. The LINACs are considered small devices.

		Favorable RF frequency
Electron beam processing - Sterilization - Polymer Reforming (< 10 MeV, > 10 kW)	High beam power	L-band S- band
Cargo inspection (3-9 MeV, ~1 kW)	↑ ↓	C-band
Radiotherapy (6-9 MeV, < 1 kW)	More compact	X-band

Table 1. Summary of industrial applications of RF LINACs.

The accelerating structure depends on the type of LINAC. The main types of LINACs are (Loew and Talman 1983):

- DC linacs, like Van de Graafs, in which the structure consists of a column of electrodes. These electrodes sustain a DC electric field which accelerates a continuous stream of particles. DC LINACs are limited to a few tens of MeV.
- Induction LINACs in which the accelerating electric fields are obtained, according to Faraday's law, from changing magnetic fluxes. These changing magnetic fluxes are generated by large pulsed currents driven through linear arrays of magnetic toroids. Induction LINACs are generally used in medium energy high-current pulsed applications.
- RF LINACs, either low frequency (UHF), microwave frequency (L, S, C, or X-band), laser frequency; CW or pulsed; traveling-wave or standing wave; room temperature or superconducting. In all these cases, the structure is a conducting array of gaps, cavities or gratings along which RF waves with an electric field parallel to the beam can be supported and built up through some resonant process. RF LINACs are used for a wide range of applications from injectors, to entire high-energy accelerators, medical accelerators.

For RF LINACs, the frequency of the driving oscillator is a crucial parameter. If a particle spends one RF cycle traveling between gaps, and the gap is small compared to the drift tubes. The distance between gaps, L, is then $L = \beta\lambda$, where $\beta = v/c$, c is the velocity of light, and λ is the RF wavelength. Note that the RF frequency is proportional to β. For a velocity ~c and a distance between gaps of a few cm, the frequency of the RF source ~GHz. High energy electron LINACs were, therefore, not possible until the development of high power microwave RF sources (Wilson, 2008). The relativistic relation between particle kinetic energy in electron volts, E_K, the rest energy E_0 and the normalized velocity is $E_K = E_0[(1 - \beta^2)^{-1/2} - 1]$. Here $E_0 = m_0c^2/e$, where e is the elementary charge and m_0 the electron rest mass. The difference between the relativistic and nonrelativistic regimes is approximately marked by a particle having an energy equal to its rest energy. For an electron this is an energy of 511 keV, while for a proton it is about 1 GeV. If electrons are injected at ~10 MeV, they can therefore be assumed to travel at c for design purposes.

2.1.2 Laser wakefield acceleration

In this section, a brief presentation of the principle and of the evolution of Laser Wakefield Acceleration of electrons is given. Different regimes of laser wakefield acceleration exists depending on the laser pulse duration, pulse width and intensity. The bubble regime of laser electron acceleration presented at the end of this section allows to produce quasi-monoenergetic electron beams with characteristics that make them very promising for XFEL applications.

Laser particle acceleration in vacuum is limited to the energy an electron can gain during a half laser period. With nowadays and even envisioned laser parameters, it will be difficult to reach energies higher than hundreds of MeV. Only a few experimental demonstrations of laser electron acceleration in vacuum exist, and the electron energies are lower than one MeV (Malka et al., 1997). It is possible to reach higher electron energies through resonance between the particle velocity and the wave phase velocity. The electromagnetic wave has a super-luminous phase velocity and is not suited for particle acceleration. On the contrary, its group velocity in a plasma is lower than the velocity of light:

$$v_g = c\left(1 - \frac{n_e}{n_c}\right).$$

It depends on the ratio of the plasma electron density and the critical density: $n_c = \frac{m_e \omega^2 \varepsilon_0}{e^2}$.

Choosing the plasma density, it is easy to control the laser pulse group velocity and therefore the phase velocity of the plasma wave excited in its wake.

Plasma waves are adapted to accelerate electrons. The original idea was proposed by Tajima and Dawson in 1979 (Tajima and Dawson 1979). This mechanism consists of three steps – the excitation of a strong plasma wave using the laser pulse, the injection of electrons in the accelerating phase of the wave and the acceleration of electrons on a sufficient distance. The demonstration of plasma waves generated by laser pulses was obtained experimentally at the end of the eighties and electron beams accelerated in plasma waves nowadays reach energies higher than a GeV (Leemans et al., 2006).

Linear regime - The plasma wave is excited during the propagation of a laser pulse in an underdense plasma, $n_e << n_c$, by the ponderomotive force (see Fig. 2). The plasma density defines the plasma wave pulsation, ω_{pe}, whereas the pulse velocity defines its phase velocity, $v_{ph}=\omega_{pe}/k_p=v_g$. The wave vector can therefore be written:

$$k_p = \frac{\omega_{pe}}{c}\left(1 - \frac{\omega_{pe}^2}{\omega^2}\right)^{-1/2} = \frac{\omega}{\gamma_p \beta_p c}$$

Where we have introduced the relativistic factor of the plasma wave, $\gamma_p = \left(1-\beta_p^2\right)^{-1/2} = \omega / \omega_{pe}$, and $\beta_p = \sqrt{1 - n_e / n_c}$. Moreover, in a low density plasma, the plasma wave is strongly relativistic, $\gamma_p >> 1$, and its wavelength is γ_p times higher than the laser wavelength.

The amplitude of the plasma wave depends on the laser pulse amplitude and on its duration τ_0. The most advantageous case is for $\tau_0\omega_{pe} \sim 1$ when the electrostatic potential of the plasma wave is $\psi_p \sim a^2 m_e c^2$, where $a=eE/(m_e c\omega)$. The amplitude of the electron density perturbation, $\delta n_e / n_e \sim a^2$, is therefore directly proportional to the laser intensity. For relativistic laser pulses, the plasma wave becomes strongly non-linear, the electrons accumulate in front of the laser pulse whereas an excess of ions if formed behind the laser pulse.

To have a more complete vision, it is necessary to take into account the laser pulse radius in the direction perpendicular to the propagation axis. If the laser radius w_0 is small compared to the electron inertia length, $w_0 < c / \omega_{pe} = 1 / k_p$, the laser pulse diffracts rapidly on the Rayleigh length $z_R = \frac{1}{2} k w_0^2$ which is not sufficient to create an efficient wakefield. On the contrary, for large pulse radii, $k_p w_0 > 1$, filamentation and self-guiding of the laser pulse can occur.

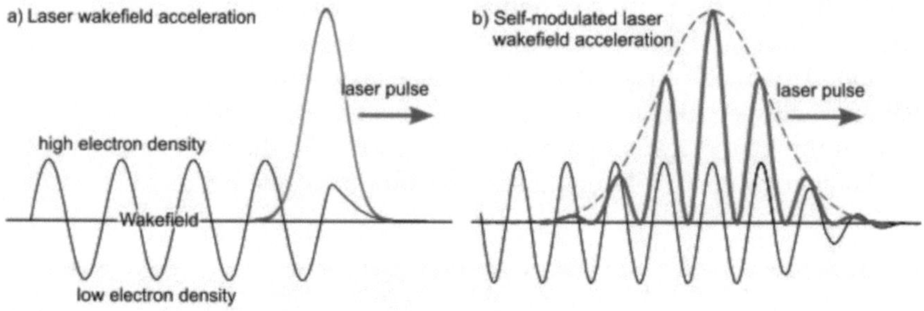

Fig. 2. Laser eneveloppe and electron density schematic in the laser wakefield regime (a) and in the self-modulated wakefield regime.

The relativistic self-guiding threshold power is given by the formula $P_c = 8\pi n_c m_e c^5 / \omega_{pe}^2$ (Gibbon 2005). The condition on laser power P=P_c in dimensionless units corresponds to the condition $a^2 k_p w_0^2 = 8$ and therefore the wakefield excitation occurs at the same time as the laser pulse self-guiding. Self-guiding is indeed desirable to guide the pulse on a distance of several Rayleigh lengths. The guiding criteria qualitatively comes from the condition that the radial pronderomotive force, $f_p \sim a m_e c^2 / w_0$, is balanced by the attraction force acting on electrons due to the ions electrostatic field, $eE_r \sim m_e \omega_{pe}^2 w_0$. The equilibrium criterion is then $k_p w_0 \sim 2\sqrt{a}$. The factor 2 in this formula was obtained in (Lu et al. 2007) using numerical simulations.

Beat wave - Initially, laser pulses had pulse durations longer than the plasma period. To couple more efficiently the laser pulse with the plasma wave, a technique had to be developed to generate an electromagnetic wave at the plasma frequency. This mechanism requires two counterpropagating pulses with pulsations ω_1 and ω_2 chosen so their difference corresponds to the plasma pulsation ω_2-ω_1=ω_{pe}.

The superposition of these two pulses therefore produces a beat wave at ω_{pe} that excites resonantly the plasma wave. The amplitude of the plasma wave reaches about 30% of the initial density in this regime, which limits the accelerating field to a few GV/m.

In 1993, Clayton et al. (Clayton et al., 1994) obtained a final energy of 9.1 MeV for injected electrons energies of 2.1 MeV. Experiments in this regime were also conducted at UCLA (Everett et al., 1994) (gain of 30 MeV), at Ecole Polytechnique (Amiranoff et al., 1995) and at Osaka (Kitagawa et al., 1992) for instance.

The physical mechanisms limiting this technique are the ions movement, which needs to be taken into account for such high pulse durations, the relativistic dephasing of the plasma wave for higher laser intensities and the growth of instabilities.

Non-linear regimes - Self-modulated wakefield - With the advent of laser systems with high laser intensity, shorter pulse durations (500fs) and high energy (100J), plasma non-linear effects could be studied. The cumulated effects of self-guiding and self-modulation of the laser envelope with the perturbation of the initial electron density generate a laser pulse train which resonates with the plasma wave. These effects are presented in Fig. 2. The self-modulated wakefield mechanisms has first been studied theoretically (Sprangle et al., 1992; Antonsen and Mora, 1992; Andreev et al., 1992). These studies show that when the pulse duration his higher than the plasma period and when the laser power is higher than the self-guiding critical power, a unique laser pulse is modulated at the plasma wavelength during its propagation. This mechanism, designated as Raman scattering and that describes the decomposition of an electromagnetic wave in a plasma wave and another electromagnetic wave shifted in frequency, leads to a modulation similar to the ones obtained by the wave beating mechanism using two laser pulses and produces energetic electrons (Joshi et al., 1981).

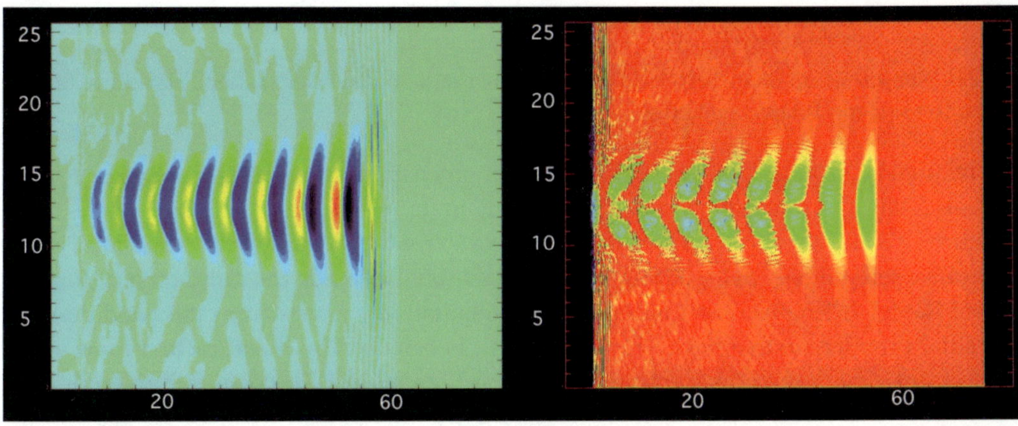

Fig. 3. 2D Particle-In-Cell simulation results of the propagation of a high intensity laser in a low density plasma. (a) Electric field in the propagation direction. (b) Electron density map. Axis are in wavelengths. Laser intensity is 4.3×10^{18} W/cm^2 and pulse duration is 10 fs. Laser FWHM is 3 wavelengths. Plasma density is 0.02533 n_c and plasma length is 80 wavelengths.

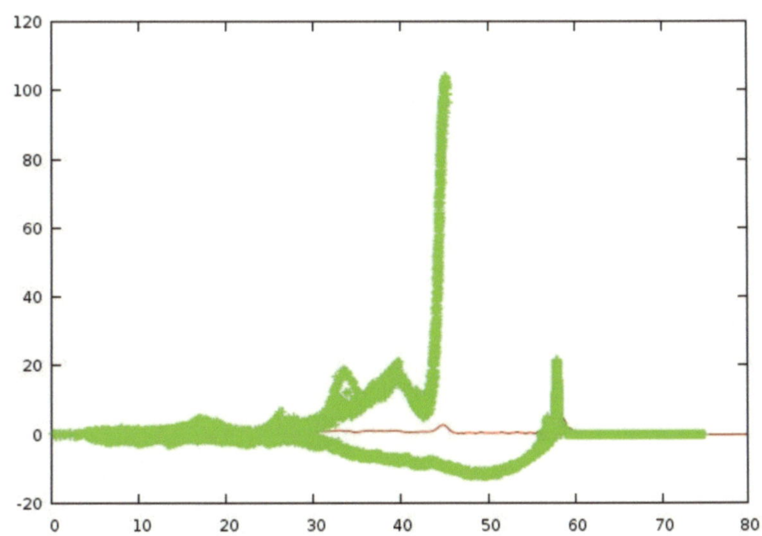

Fig. 4. 1D Particle-In-Cell simulation results of the propagation of a high intensity laser in a low density plasma. Electron phase space at t=198 fs. Laser intensity is 1.7×10^{19} W/cm^2 and pulse duration is 10 fs. Plasma density is 0.02533 n_c and plasma length is 80 wavelengths. Y-axis corresponds to $\beta\gamma$ and x-axis is in wavelengths.

Forced wakefield - Thanks to the advent of the Chirped Pulse Amplification technique [Strickland and Mourou 1985], the new properties of high intensity laser interaction with matter, typically experimented on big laser facilities, became accessible to smaller laser systems adapted to the academic community. These systems, most using Titane Sapphire crystals, fit in a room of a few square meters and deliver an energy of about 2-3 J in 20 fs on target. This corresponds to lasers of the 100 TW class with focalized intensities of the order of a few 10^{19} W/cm^2. Numerous publications have shown that these moderate energies installations, working at a high repetition rate of 10 Hz and of a reasonable price can produce energetic electron beams of very high quality. For example, using the "Salle Jaune" laser at LOA in France, electrons were accelerated up to 200 MeV in 3 mm of plasma (Malka et al., 2002). The accelerating mechanism is called forced wakefield to distinguish it from the self-modulated wakefield mechanism (see Fig. 2). Indeed, using short duration laser pulses, plasma heating in the forced wakefield regime is less important than in the self-modulated regime. This allows reaching higher plasma wave amplitudes and therefore high electron energies. Thanks to shorter interaction duration between the laser and the accelerated electrons, the electron beam quality is also enhanced. The measure of the normalized transverse emittance gives comparable values as conventional accelerators with similar energies (rms normalized emittance of 3π mm·mrad for electrons with an energy of 55 ± 2 MeV) (Fritzler et al., 2004). Electron beams with maxwellian spectral distributions, generated using ultra short laser pulses, were obtained in several laboratories around the world: at LBNL (Leemans et al., 2004), at NERL (Hosokai et al., 2003), and in Europe at LOA in France (Malka et al., 2001) or at MPQ in Germany (Gahn et al., 1999) for example. Figure 3 shows the results of a 2D Particle-In-Cell simulation of the propagation of a high intensity pulse in a low density plasma leading to the creation of a plasma wave and strong electrostatic fields. Figure 4 shows the electron phase space in a 1D simulation with similar interaction parameters leading to the production of a broad electron energy spectrum with maximum energies higher than 50 MeV.

The previous regimes are limited by non-linear effects of which wavebreaking is one of the most important. The maximum electric field that a plasma wave can sustain is limited by wave breaking. This wave breaking takes place when the electrons participating to the plasma wave are trapped in the wave itself and then accelerated. This leads to the loss of structure of the electrons generating the electric field of the wave, and therefore to the damping of its amplitude. For a relativistic plasma wave, the electric field when wave breaking starts is (Arkhiezer and Polovin, 1956): $E_{break} = \sqrt{2(\gamma_p - 1)}E_0$, where $E_0 = m_e c\omega_{pe} / e$. This formula is obtained in the cold plasma limit. Thermal effects will launch wave breaking before the cold wave breaking limit (Rosenzweig, 1988; Katsouleas and Mori, 1988).

Bubble regime - More recently, theoretical works based on 3D Particle-In-Cell simulations have shown the existence of a robust acceleration scheme called the bubble regime (Pukhov and Meyer-ter Vehn, 2002). In this regime, the laser dimensions are shorter than the plasma wavelength in the longitudinal direction but also in the transverse directions. The laser pulse therefore resembles a light sphere with a radius smaller than 10 μm. If the laser energy contained in this volume is sufficiently high, the laser ponderomotive force efficiently expels radially plasma electrons, creating an electronless cavity behind the laser pulse, surrounded by a dense electron zone. The total expulsion of electrons was predicted in numerical simulations (Pukhov and Meyer-ter-Vehn 2002) and two years later, it was observed in experiments (Faure et al. 2004). The electrons expelled by the laser glide around the cavity (see Fig. 5) and some of them enter the cavity from the back. Some electrons are injected in the cavity and accelerated along the laser axis, producing an electron beam with radial and longitudinal dimensions smaller than the laser dimensions (see Fig. 5). The details of the electron injection process are not yet completely understood, but it is observed in several simulations that are in good agreement with experiments. A new electron injection technique using a secondary laser has recently been demonstrated experimentally (Faure et al., 2006).

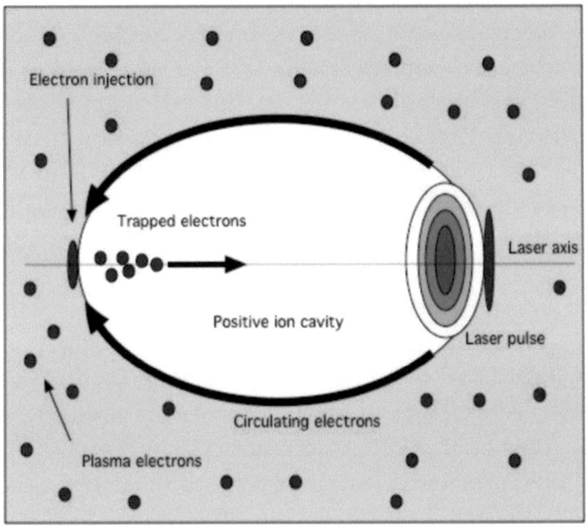

Fig. 5. Schematic of the bubble regime of laser electron acceleration in low density plasmas.

The signature of this regime is a quasi-monoenergetic electron distribution. This property differs significantly from previous results on laser electron acceleration. It comes from the combination of several factors:

- Electron injection in the cavity is different from the injection linked to wavebreaking in the self-modulated and forced wakefield regimes. The injection does not originate from the "breaking" of the accelerating structure. It is located at the back of the cavity, which gives similar properties to injected electrons in the phase space.
- Acceleration occurs in the stable accelerating structure during the propagation as long as laser intensity is sufficiently high.
- Electrons are trapped behind the laser, which limits the interaction with the laser transverse electric field. Trapping stops when the charge contained inside the cavity compensates the ion charge.
- Rotation of electrons in the phase space also contributes to reduce the spectral width of electron bunches (Tsung et al., 2004).

As soon as an electron enters the cavity, it is attracted to the center by the ion electrostatic field. At the same time, the electric field focalizes the electron towards the laser axis. The value of this accelerating force has already been estimated above:

$$eE_r \sim m_e \omega_{pe} c \sqrt{a}$$

A magnetic field also exists inside the cavity. It originates from the fact that the cavity glides on fixed ions creating a current density $-en_e c\beta_p$ in the direction opposite to the propagation direction. The amplitude of this field is proportional to the distance to the center of the cavity and its maximum value is:

$$eB_\varphi \sim m_e \omega_{pe} \sqrt{a}$$

so that the Lorentz force is equal to the electrostatic force.

We can now estimate the maximum energy that an electron can gain inside the cavity. In the reference frame of the cavity, it is stretched in the propagation direction by the factor γ_p. Therefore, the electron energy gain in the cavity $\varepsilon_e' \sim m_e c^2 \gamma_p a$ is the product of the electric force $\sim m_e \omega_{pe} c \sqrt{a}$ and the acceleration length $\gamma_p \sqrt{a} / k_p$. It is then necessary to multiply this result by the factor γ_p to come back to the ion reference frame. The final energy of an accelerated electron is therefore:

$$\varepsilon_e \sim m_e c^2 \gamma_p^2 a = m_e c^2 a \frac{\omega^2}{\omega_{pe}^2}$$

It is proportional to the laser field amplitude and moreover it is inversely proportional to the plasma density.

It is therefore better to work with moderate laser intensities, a~5 and to use low density plasmas with densities $\leq 10^{18}$ cm^{-3} to reach GeV electron energies. Nevertheless, the increase of the electron energy also requires a longer acceleration length. The length travelled by the electron in the laboratory reference frame during its acceleration is:

$$l_{acc} \sim \frac{R}{1-\beta_p} \sim \frac{c}{\omega} \gamma_p^3 \sqrt{a}$$

It strongly increases with the energy gain and self-guiding of the laser pulse beyond a few mm seems difficult.

Several laboratories have obtained electron bunches with quasi-monoenergetic spectra: in France (Faure et al. 2004) with laser pulses duration shorter than the plasma period, but also using laser pulses longer than the plasma period in the United Kingdom (Mangles et al. 2004), and in the US (Geddes et al., 2004), then in Japan (Miura et al. 2005) and in Germany (Hidding et al., 2006). There is great interest in such beams for applications: it is now possible to transport and refocalize this beam using magnetic elements. With a maxwellian spectrum, it would have been necessary to select an energy range for transport, which would have strongly decreased the electron flux.

Future of laser electron acceleration - Some solutions are now proposed to reach energies of several GeVs. It is possible to increase the acceleration length by guiding the laser pulse either using a capillary filled with gas (Leemans et al. 2006) or by tuning properly laser and plasma parameters (Hafz et al., 2008). Nevertheless, acceleration using multiple stages seems necessary to go further. The first demonstrations of electron acceleration in two stages were reported (Liu et al., 2011 ; Pollock et al., 2011). Counter-propagating lasers to separate the wake creation and the injection mechanism lead to a better reproducibility with a good quasi-monoenergeticity (J. Faure et al. 2006, Davoine et al 2009). This scheme is designated as controlled optical injection. Bunch duration was also measured at LOA, France (O. Lundh 2011). Their analysis shows that the electron beam, produced using controlled optical injection, contains a temporal feature that can be identified as a 15 pC, 1.4–1.8 fs electron bunch (root mean square) leading to a peak current of 3–4 kA depending on the bunch shape.

For the XFEL compact schemes application, the maximum electron energy is not the most important parameter. Reaching better emittance with smaller energy spread is the goal. Table 2 summarizes some important parameters of electron beams accelerated either by LINACs or by laser wakefield acceleration.

	Typical energy	Energy spread	Beam intensity	Emittance
LINAC (CERN LINAC96 compendium)	30-100 MeV	0.5-3 %	0.1-70 A	0.2-100 mm·mrad
Laser electron acceleration (X. Davoine et al. 2009)	60 MeV	1-2 %	25 kA	1 mm·mrad

Table 2. Summary of some important parameters of electron beams accelerated either by LINACs or by laser wakefield acceleration

As shown in Table 2, the main advantage of laser wakefield acceleration is in the very high achievable beam intensities. It is important to note that this table does not include parameters of very high energy LINACs which are out of the scope of this Chapter.

2.2 Wigglers (magnetic and optical)

2.2.1 Magnetic wigglers

Coherent emission by an electron beam in a FEL relies upon the self-consistent interaction between the electrons and radiation as the electron beam undulates in the magnetic undulator field. The undulator and resonant electromagnetic fields form a ponderomotive potential which co-propagates at the mean electron velocity along the axis of the undulator (Winick et al., 1981). The relativistic electron beam is bunched by this potential and forms a periodic density modulation at the resonant radiation wavelength (see Fig. 6). Emission from the electron beam is therefore coherent and may be many orders of magnitude greater than the incoherent emission from a similar, but unbunched, beam.

1D models of the FEL interaction between a pulse of electrons and a linearly-polarized radiation field in a planar undulator FEL are widely used. In deriving the equations that describe the FEL interaction, it is possible to neglect explicit three-dimensional effects such as electron beam emittance and radiation diffraction. These effects may be re-introduced into the reduced one-dimensional model by using further approximations. When three-dimensional effects are included in FEL models they generally tend to degrade the quality of the FEL interaction by reducing saturation powers and decreasing the radiation gain per unit length. The one-dimensional FEL model is therefore the 'best-case' model for a given set of parameters.

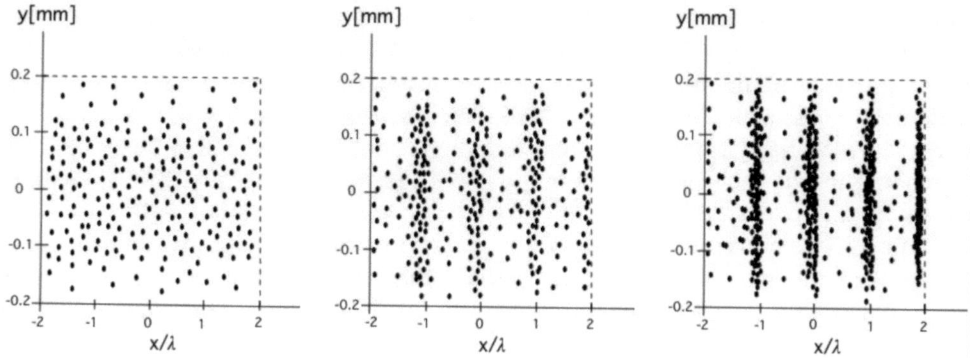

Fig. 6. Schematic of the bunching of electrons in a magnetic undulator.

A magnetic wiggler is characterized by its undulator parameter $K=eB_0\lambda_u/(2\pi m_e c)$. λ_u is the wiggler wavelength and B_0 is its magnetic field amplitude (Lau et al., 2003). Synchrotron radiation is emitted inside a cone with opening angle $1/\gamma$. If $K \leq 1$, the electron trajectory is inside the radiation cone. Therefore, the photons emitted by an electron at various positions interfere with each other. The radiation is therefore monochromatic. the wavelength of the undulator light is given by $\lambda = \lambda_u/2\gamma^2(1 + K^2/2)$. The essential feature of the high-gain FEL is that a large number of electrons radiate coherently. In that case, the intensity of the radiation field grows quadratically with the number of particles: $I \sim N^2$. More information on magnetic wigglers can be found in other chapters of this book.

2.2.2 Optical wigglers

Many authors have shown that the action of a laser field, counter-propagating with respect to the electrons, is basically equivalent to that of the magnetic field in a wiggler, and thus can be used to obtain an FEL effect (Sprangle 2009b ; Dobiasch et al., 1983 ; Gea-Banacloche et al, 1987; Gallardo et al., 1988). In fact, the laser parameter a is found to be almost interchangeable with the FEL wiggler parameter K. Many realistic effects that have been studied in the conventional synchrotron/FEL community may be immediately applied to the laser synchrotron. However, the strength parameter K depends strongly on the undulator period. Replacing the magnet period of few cm, by a laser wavelength around 1 µm, results in practice in a very low value of K, and hence in low gain, so that a very high number N of oscillations is required.

Another way to produce an optical wiggler is to use multiple lasers to generate an optical lattice. An optical lattice is formed by the interference of counter-propagating laser beams. The beam interaction produces a spatially periodic potential. One of its uses is to trap neutral cooled atoms at the locations of potential minima in atomic physics (three laser beams for the optical molasses technique) and to produce Bose-Einstein condensates. The advantage on magneto-optical traps is the periodicity induced in the cold atomic gases, making it alike a solid crystal.

Using two counter-propagating laser beams, it is possible to create a ponderomotive potential array. Kapitza and Dirac (1933) have shown that electrons interacting with a light standing wave can diffract from this light lattice – thus undergoing the reverse process of light diffraction on a matter density grating. In the low intensity limit, the interaction with the light is a small perturbation to the electron free motion, that induces a momentum transfer of $\pm(h/\pi)k$, where h is the Planck constant and k is the wavevector of either beam forming the standing wave (D.L. Freimund et al., 2001). Conversely, at high intensities of the order of 10^{13} W/cm2 or more for near infrared lasers, the electron dynamics is modified considerably by the action of the light lattice. Free electrons interacting with a spatially non uniform laser field are indeed submitted to a significant ponderomotive force, i.e., a drift force tending to expel the electrons from the regions of highest intensity (Kibble 1968). Free electrons are in effect embedded into a spatially varying potential induced by light, resembling a series of parallel half-pipes. This situation is extremely similar to that of optical molasses, well-known in the field of cold atoms, that play for instance a role in Sisyphus cooling. Giant momentum exchanges of several 10^3 $(h/2\pi)k$ or more can be induced; at such levels, the quantum description of electrons becomes useless, and a classical description is valid. The existence of the strong field Kapitza-Dirac effect has been demonstrated experimentally by Bucksbaum et al. (1988), and briefly rewieved by Hartemann (2000). Interaction of relativistic electrons and a standing wave in the strong field KD regime has only been briefly considered by Fedorov et al. (1988).

2.2.3 Raman FEL lasers: From Raman to Compton regimes of FEL emission

The low-gain, tenuous-beam limit is relevant to free-electron laser configurations in which the electron beam current is low and the gain of the signal in a single pass is less than unity. In the linear theory of this regime, the beam plasma frequency $\omega_b \ll \omega$ and collective effects due to beam space-charge waves are negligible. The high-gain regime is applicable to intense-beam FELs. The fields in this regime exhibit exponential growth of the fluctuation fields. The wiggler field provides for the coupling between the beam space-charge wave and either polarization state of the electromagnetic field, as well as a growth mechanism for the electromagnetic waves in the absence of the beam-plasma waves (Freund and Antonsen, 1996). In the former case, coherent amplification occurs by three-wave coherent Raman scattering in which the wiggler represents the pump wave, the beam-plasma wave mode represents the idler, and the output signal is the daughter wave. The latter case is coherent Compton scattering in which the wiggler scatters off the electron beam. There are two principal regimes of interest in the solution of the dispersion equation corresponding to the low- and high-density regimes. The high-gain Compton regime is achieved when the ponderomotive potential is larger than the space-charge potential of the beam-plasma waves. The opposite case in which the space-charge potential is larger than the ponderomotive potential is the Raman regime. In the Compton regime, the electron beam interacts with the ponderomotive potential formed by the beating of the wiggler and radiation fields. For high currents, the electrostatic potential due to the beam space-charge waves is dominant, and the interaction leads to stimulated Raman scattering of the space-charge wave off the wiggler. An intermediate regime exists in which both mechanisms are operative.

2.3 Behaviour of free electrons in relativistic laser fields

2.3.1 Electron motion in an electromagnetic field

This motion is described by Newton's equation with Lorentz force. Electric and magnetic field are given by Maxwell's equations. To use these equations in an invariant form, it is better to use the vector potential $\vec{A}$ in Coulomb's gauge. The equation of motion of the electron is then:

$$d_t\vec{p} = e\partial_t\vec{A} + e(\vec{v}\cdot\nabla)\vec{A} - e\nabla(\vec{v}\cdot\vec{A}) \tag{1}$$

Considering a plane electromagnetic wave propagating in the z direction, $\vec{A}_\perp(z,t)$, we can decompose this equation in two components. In the direction perpendicular to the propagation direction, $d_t\vec{p}_\perp = e\partial_t\vec{A}_\perp + ev_{//}\partial_z\vec{A}_\perp \equiv ed_t\vec{A}_\perp(z_e(t),t)$, where $z_e(t) = \int_t v_{//}dt$. This relation gives the conservation of generalized momentum $\vec{P}_\perp = \vec{p}_\perp - e\vec{A}_\perp$ and for an electron initially at rest, $\vec{p}_\perp = e\vec{A}_\perp(z_e(t),t)$. It is a direct consequence of the fact that the wave is homogeneous in the x and y directions.

In the direction parallel to the propagation direction, Eq. 1 gives: $d_tp_{//} = -\frac{e}{m_e\gamma}\vec{p}_\perp\cdot\partial_z\vec{A}_\perp$, and using the conservation of generalized momentum:

$$d_tp_{//} = -\frac{e^2}{2m_e\gamma}\partial_z\vec{A}_\perp{}^2 \tag{2}$$

γ is the relativistic factor of the electron. The kinetic energy of the electron is $\varepsilon = (\gamma-1)m_ec^2$ and the velocity of the electron is related to its energy by $\vec{v} = \partial_{\vec{p}}\varepsilon$, leading to:

$$d_t\varepsilon = e\vec{v}\cdot\partial_t\vec{A} \tag{3}$$

Using Eq. 2 and Eq. 3 leads to:

$$d_t\left[(\gamma-1)m_ec^2 - p_{//}c\right] = \frac{e^2}{2m_e\gamma}(\partial_t + c\partial_z)\vec{A}_\perp{}^2 \tag{4}$$

For a progressive wave propagating at the velocity c, the vector potential is a function of the variable t-z/c and the right side of Eq. 4 is equal to zero. The conservation of the parallel momentum is therefore obtained and for a particle initially at rest:

$$p_{//} = m_ec(\gamma-1) \tag{5}$$

A relation between parallel and perpendicular components of the momentum is therefore also obtained:

$$p_{//} = \frac{\vec{p}_\perp^{\,2}}{2m_ec} \tag{6}$$

After the wave has passed, the electron is again at rest. The sign of the parallel momentum is always positive meaning that the electron is pushed by the wave in its propagation direction.

From Eq. 5 and Eq. 6, it is also possible to deduce the divergence angle of the electron and its link with the particle energy:

$$\tan\theta = \frac{p_\perp}{p_{//}} = \sqrt{\frac{2}{\gamma-1}} \tag{7}$$

Relativistic electrons are therefore ejected in the propagation direction, whereas non-relativistic electrons are ejected in the polarization direction.

To study in more detail the trajectories of the electrons in the wave, we can use the proper time $\tau = t - z_e(t)/c$ to get a linear equation for the orbit of the particle:

$$\frac{d\vec{r}_e}{d\tau} = \gamma\vec{v} = \frac{\vec{p}}{m_e} \tag{8}$$

For a linearly polarized wave, $\vec{A}_\perp(z,t) = (m_e c / e) a \cos(\omega\tau)\vec{e}_x$, one can obtain:

$$kx_e(t) = a\sin(\omega\tau),\ kz_e(t) = \frac{a^2}{8}(2\omega\tau + \sin(2\omega\tau)).$$

The simple motion in the perpendicular plane comes with a more complicated motion in the propagation direction of the wave. It is periodical but it is not a sinusoid. It also contains several harmonics of the wave frequency. In the non-relativistic regime, the main harmonic is dominant. In the relativistic regime, the number of harmonics becomes important.

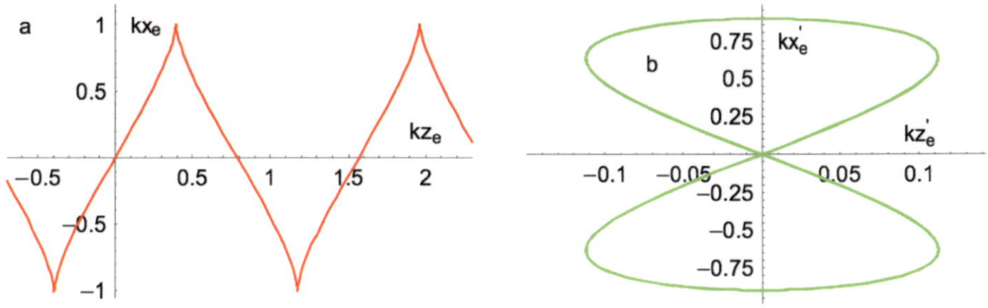

Fig. 7. Orbit of an electron in the (x,z) plane due to a linearly polarized progressive electromagnetic wave: (a) in the laboratory reference frame and (b) in the reference frame moving at the particle drift velocity. The dimensionless amplitude of the wave is a=1.

An example of the orbit of an electron in a linearly polarized wave is shown in Fig. 7. In the reference frame of the laboratory, maximum velocity is achieved when x_e is close to zero, and the orbit is peaked for large values of x_e, when acceleration is at its maximum. The shape of this orbit is more simple in the reference frame moving at the drift velocity of the article v_d in the z direction as it becomes closed. It is the well-known "figure eight". v_d is given by $a^2c/(a^2+4)$.

2.3.2 Relativistic ponderomotive force

The force on the right side of Eq. 2 is proportional to the square of the laser pulse amplitude. It therefore contains an average component and an oscillating component with a period of π/ω, the laser pulse half-period. The average part of this force is the ponderomotive force. It acts on all charged particles (but mostly on electrons due to their small mass) and it pushes particles independently of the sign of their charge in the direction opposed to the intensity gradient. A more general expression of Eq. 6 for the average part of the momentum is:

$$d_t(p) = f_p \equiv -\frac{e^2}{2m_e\langle\gamma\rangle}\nabla\left(\vec{A}^2\right) \tag{9}$$

The averaging is done over a laser period. The average relativistic factor is given by:

$$\langle\gamma\rangle = \sqrt{1 + \frac{\langle\vec{p}\rangle^2}{m_e^2c^2} + \frac{e^2\left\langle\vec{A}^2\right\rangle}{m_e^2c^2}}$$

It takes into account the average momentum $\langle\vec{p}\rangle$ and the oscillating momentum $\vec{p}_{osc} = e\vec{A}$. The ponderomotive force in Eq. 9 can be presented in a more simple expression, noting that in $\langle\gamma\rangle$, only the oscillating part depends on coordinates, while the average momentum $\langle\vec{p}\rangle$ is a function of time. The ponderomotive force can therefore be written:

$$f_p = -m_e c^2 \nabla\langle\gamma\rangle .$$

The importance of this latest expression comes from the fact that the ponderomotive force is presented as the gradient of the ponderomotive potential, $U_p = m_e c^2\left(\langle\gamma\rangle - 1\right)$. As for all potential forces, the work of the ponderomotive force does not depend on the particle trajectory but only on the starting and on the ending points. It gives another interpretation of the fact that a free charged particle cannot gain (or lose) kinetic energy in the laser pulse. It is the Lawson-Woodward theorem. Even if this theorem is verified in the ideal case described above, several possibilities exist for the electron to gain energy in the wave.

2.3.3 Nonlinear Thomson scattering

In Thomson scattering, an electron that is initially at rest may acquire relativistic velocities in the fields of high-intensity light and emit radiation at high harmonics of the light frequency. If the electron already possesses a relativistic energy before it encounters the high-intensity laser, there is an additional Doppler-shift of the scattered light.

For lasers with low intensities (a<<1), an electron that is initially at rest undergoes a small amplitude, transverse oscillation at the laser frequency. The Thomson scattering spectrum consists of a single frequency in all directions and the radiation pattern is the same as that from a dipole. If a is increased to a few tens of percent, the electron's oscillation frequency begins to deviate from the laser frequency (see Lau et al., 2003). If a~1, the Lorentz force associated with the laser's magnetic field becomes significant, and the electron acquires an oscillation along the laser propagation direction, in addition to the transverse oscillation (see Section 2.3.2). The electron also acquires an average drift velocity along k. For a>>1, the axial oscillation greatly exceeds the transverse oscillation, and the electron orbital period is much greater than the laser optical period. A non-linear scattering spectrum is therefore obtained.

The tutorial given on this regime in (Lau et al. 2003) is based on the simple, classical model of a single electron interacting with an infinite plane wave. While highly idealized, it suggests that the brightest x-ray source is achieved by head-on collisions of a relativistic electron beam with an intense laser with a>1. Such a configuration, broadly known as laser synchrotron, strongly resembles the conventional synchrotron/FEL.

3. Description of existing FEL compact schemes

3.1 LWFA coupled with a magnetic wiggler

A few groups have proposed to use electron packets, accelerated in the LWFA or bubble regime up to energies around 1 GeV, and inject them into an undulator to obtain the micro-bunching effect of free electron lasers (Grüner et al., 2007). This would result in a XUV or soft X-ray source, in the 1 keV range; an important hurdle is the effect of space charge, that induces a large expansion of the accelerated electron bunch within the wiggler, and might limit the amplification.

Laser wakefield acceleration of electrons is indeed increasingly considered as a potential compact substitute of conventional accelerator technology, at least for extreme UV free electron lasers (Grüner et al., 2007; Schlenvoigt et al., 2008; Nakajima 2008). The main problems of this scheme are the stringent requirements on the mono-energeticity of LWFA electrons, and the stability of electron bunches at very high laser accelerated electron energies.

Gruner at al. (2007) studied the possibility to develop table-tops XFELs. They first studied a proof-of-principle scenario at relatively low electron energies, where space-charge effects play a dominant role leading to a linear energy chirp. The situation of the proposed table-top X-FEL (TT-XFEL) is different. To reach a wavelength of $\lambda = 0.25$ nm, an electron energy of 1.74 GeV is needed in case of a period of $\lambda_0 = 5$ mm. Space charge effects are much weaker here. For this less demanding situation they have confirmed with four different simulation codes that above 1 GeV, with the same parameters as in the extreme case given above, Coulomb-explosion leads to a projected energy chirp of below 0.3% and a bunch elongation with a factor below 1.1. Experimentally the most demanding constraint is that the electron (slice) energy spread should be as small as 0.1%. For the authors, this goal seems to be within reach.

Without the effect of wakefields GENESIS simulations have shown that this TT-XFEL scenario with an undulator length of only 5 m yields 8×10^{11} photons/bunch within ~ 4 fs and 0.2% bandwidth, a divergence of 10 μrad, and a beam size of 20μm. However, the wakefields become the dominant degrading effect as the required undulator length is larger. But since there is no initial space-charge-induced energy chirp, one must find another method for compensating the wakefield-induced energy variation. A suitable method for compensating the wakefields for the TT-XFEL could be tapering, i.e., varying the undulator period along the undulator. Due to the fact that the undulator parameter K is smaller than unity, tapering via K, i.e., by gap variation, could only be used as fine-tuning.

Schlenvoigt et al. (2008) have demonstrated the first successful combination of a laser-plasma wakefield accelerator, producing 55–75 MeV electron bunches, with an undulator to generate visible synchrotron radiation. By demonstrating the wavelength scaling with energy, and narrow-bandwidth spectra, they showed the potential for ultracompact and versatile laser-based radiation sources from the infrared to X-ray energies. In their set-up, Schlenvoigt et al. focus the light from a 5-TW laser pulse into a 2-mm-wide gas jet. The interaction of the laser with the jet produces a beam of electrons with a peak energy of between 55–75 MeV. Directing this beam into a 1-m-long undulator — which consists of a series of alternating magnets — causes its electrons to wiggle in the transverse direction, producing light at the red end of the visible spectrum (with wavelength in the range of 950–550 nm). The authors therefore provide the first demonstration of the production of resonant-like synchrotron radiation from a laser-generated electron beam. The results of several runs of their experiment show that the emission wavelength scales with beam energy just as theory predicts, suggesting that the generation of much shorter wavelengths by this approach should be relatively straightforward. By extending the length of the undulator to 3 m, and feeding it with a more energetic beam — such as the 1-GeV, 30-pC beams recently demonstrated in a 3-cm capillary laser-plasma accelerator (Leemans et al., 2006) — it should soon be possible to reach 3 nm in the soft-X-ray range, at a peak brilliance comparable to that of even the largest modern synchrotron radiation sources.

For Nakajima et al. (Nakajima et al. 2008), once the feasibility of a laser-driven soft-X-ray source is achieved, the next step will be to extend this approach to the more ambitious task of constructing an FEL. This process will require the generation of electron beams of extremely high current and small emittance and energy spread, and the construction of a precisely engineered undulator exceeding a hundred meters in length.

3.2 Laser wiggler concepts

A laser-undulator free electron laser has therefore been repeatedly proposed [Sprangle et al. 1992, Dobiasch et al. 1983, Gea-Banacloche 1987, Sprangle et al. 2009], but has remained so far inapplicable. Many authors have indeed shown that the action of a laser field, counter-propagating with respect to the electrons, is basically equivalent to that of the magnetic field in a wiggler, and thus can be used to obtain an FEL effect (Sprangle 2009b ; Dobiasch et al., 1983; Gea-Banachloche et al, 1987 ; Gallardo et al., 1988). However, the strength parameter K depends strongly on the undulator period. Replacing the magnet period of few cm, by a laser wavelength around 1 μm, results in practice in a very low value of K, and hence in low gain, so that a very high number N of oscillations is required. Since the energy dispersion of the electrons has to be smaller than 1/2N for the Compton free electron laser effect to be effective in the small signal regime, this scheme would require an absolutely outstanding quality of mono-energeticity, together with an equivalent requirement on the constancy of the laser intensity along the interaction region, and an outstanding emittance – all constraints well beyond the present or foreseeable state of the art. Due to these major issues, the scheme was never seriously considered, up to a recent proposal by Petrillo et al. (Petrillo et al. 2008), who suggest to use a mono-energetic electron beam predicted to arise from LFWA with i) a huge electron peak current, ii) a quasi mono-energetic distribution, iii) an emittance three times smaller than the current state of the art. By shining a high energy, few picosecond, and monomode CO2 laser, onto the bunch, Petrillo et al. succeed to meet the conditions outlined above, and predict a coherent emission at 1.4 nm, up to saturation, and have dubbed the process an "All Optical Free Electron Laser". They studied the generation of low emittance high current monoenergetic beams from plasma waves driven by ultrashort laser pulses, in view of achieving beam brightness of interest for free-electron laser (FEL) applications. The aim is to show the feasibility of generating nC charged beams carrying peak currents much higher than those attainable with photoinjectors, together with comparable emittances and energy spread, compatibly with typical FEL requirements. They identified two regimes: the first is based on a laser wakefield acceleration plasma driving scheme in a gas jet modulated in areas of different densities with sharp density gradients. The second regime is the so-called bubble regime, leaving a full electron-free zone behind the driving laser pulse: with this technique peak currents in excess of 100 kA are achievable. They have focused on the first regime, because it seems more promising in terms of beam emittance. Simulations carried out using VORPAL show, in fact, that in the first regime, using a properly density modulated gas jet, it is possible to generate beams at energies of about 30 MeV with peak currents of 20 kA, slice transverse emittances as low as 0.3 mm mrad, and energy spread around 0.4%. These beams achieve very high brightness, definitely above the ultimate performances of photoinjectors, therefore opening a new range of opportunities for FEL applications. The system constituted by the electron beam under the effect of the electromagnetic undulator has been named AOFEL (for all optical free-electron laser).

In a recent article (Sprangle et al. 2009a), Sprangle et al. use the well-known GENESIS code to confirm the scenario, but emphasize the huge technical challenges, with electron energy spread and laser constancy both required of the order of 0.01%.

To summarize, both schemes presented in this Section are extremely interesting in the soft X-ray range around 1 keV. Both require however in a stringent way extremely challenging parameters of mono-energeticity and emittance of the laser-accelerated electron bunches and seem extremely challenging in view of present day electron and laser technologies. To the best of our knowledge, no compact scheme for an XFEL has yet been proposed, robust enough to be adapted to realistic conditions of relativistic electron bunches.

4. Raman XFEL

In this section, a new scheme for an X-ray free electron laser is described: it is based on a Raman process occurring during the interaction between a moderately relativistic bunch of free electrons, and twin intense short pulse lasers interfering to form a transverse standing wave along the electron trajectories. In the high intensity regime of the Kapitza-Dirac effect, the laser ponderomotive potential forces the electrons into a lateral oscillatory motion, resulting in a Raman scattering process. This triggers a parametric process, resulting in the amplification of the Stokes component of the Raman-scattered photons. Experimental operating parameters and implementations are discussed.

4.1 Principle and geometry: Relativistic electrons in a high intensity optical lattice

A totally new approach to create an X-ray laser was recently proposed by Ph. Balcou (Balcou 2010), based on the interaction between a bunch of moderately relativistic electrons, and an optical lattice created by two interfering laser beams. The key issue is to be able to create artificially a new degree of freedom for the electrons, in which case a Raman scattering process can be expected. If one sets up a configuration to maintain the Raman scattering over a certain distance, then a stimulated effect will be switched, leading to exponential amplification of the Raman scattered light, in the extreme UV or X-ray range depending on the Doppler shift.

Let us first consider the typical setup of 90° Thomson scattering of a laser off a relativistic electron bunch, (see Schoenlein et al., 2000). A short pulse, energetic laser impinges at 90°

onto a relativistic electron bunch. Photons and electrons are focused and superpose onto a focal spot of few tens of µm FWHM, thus scattering of X-ray photons along the electron direction. This is a spontaneous scattering process, and the alignment of the X-ray photons along the electron direction is a pure relativistic kinematic effect.

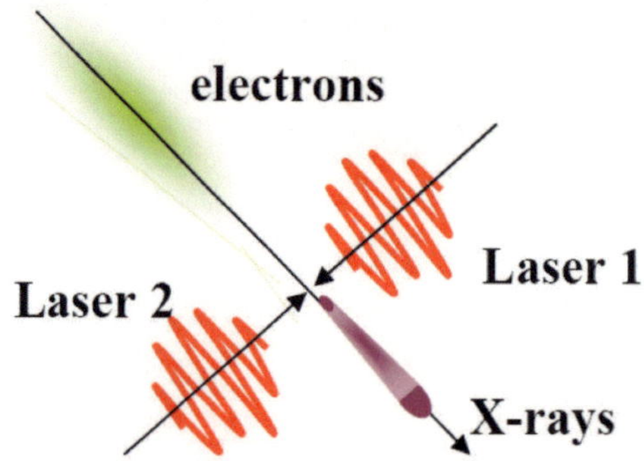

Fig. 8. Proposed 2-beams, 90° scattering.

We now propose to split the incident laser into two identical parts, and send these twin beams in opposite directions, perpendicularly to the electron beam (Fig. 8). In the superposition region, the two beams interfere to form a transverse standing wave. The relativistic electrons interact therefore with an optical lattice, giving rise to an effect known as the strong field Kapitza-Dirac (KD) effect (see section 2 for more details on this effect).

4.2 Electron dynamics: Beam trapping and collective electron modes

Fig. 9 illustrates the proposed scheme, showing the typical behavior of relativistic electrons injected into the transverse optical lattice, in the high intensity regime of the KD effect. The light lattice induces a series of parallel potential wells, aligned along the main electron direction. If the transverse kinetic energy of incident electrons is high, they skip through the potential wells (high emittance / low intensity limit); otherwise, they get trapped into one well, and start oscillating with a characteristic frequency (low emittance/high intensity limit):

$$\Omega = \frac{\sqrt{2}eE_0}{\gamma mc}$$

where E_0 is the laser field of each incident beam, γ the Lorentz factor of electrons of mass m.

This transverse oscillation modifies the photon scattering, splitting the Thomson peak into a doublet of Raman modes. This is illustrated in figure 10, showing the result of an exact numerical calculation of the scattering spectrum of a 10 MeV electron, chosen at random in a 1 µm normalized emittance bunch focused over 30 µm rms, and injected into the transverse light lattice. Individual electrons may oscillate randomly, resulting in a spontaneous Raman scattering.

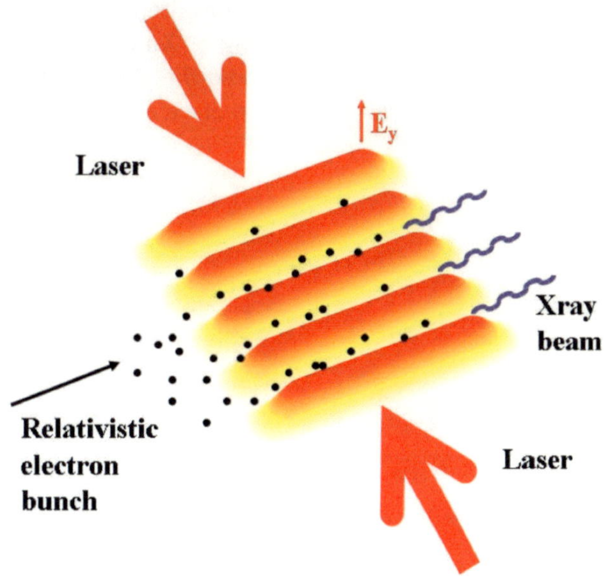

Fig. 9. Typical behavior of relativistic electrons injected into the transverse optical lattice.

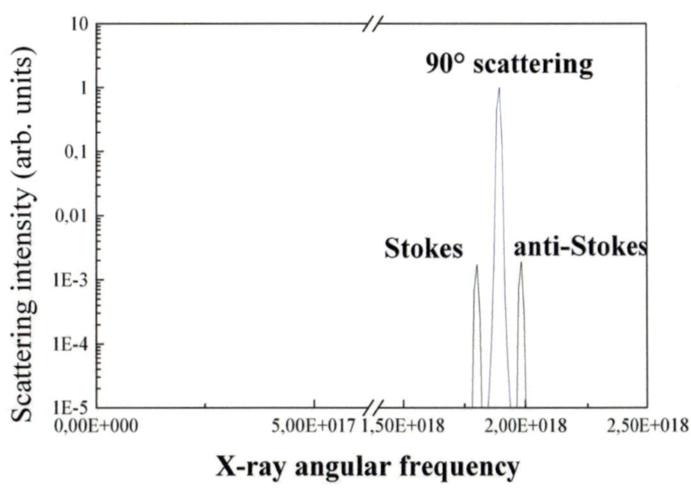

Fig. 10. Result of an exact numerical calculation of the scattering spectrum of a 10 MeV electron, chosen at random in a 1 μm normalized emittance bunch focused over 30 μm rms, and injected into the transverse light lattice.

If a collective electron oscillatory motion is induced, then one obtains a new kind of oscillatory plasma wave, trapped in the light potential, as illustrated in Figure 11. Excitation of this plasma wave will then lead to a coherent scattering process. The excitation mechanism is currently studied, and can be shown to be similar to the excitation mechanism of standard Free Electron Lasers, by means of a Lorentz force inducing a time-dependent, and tranversely-dependent longitunidal bunching.

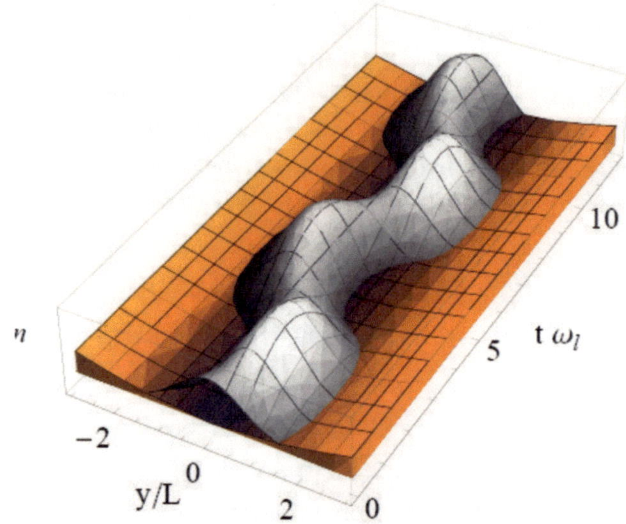

Fig. 11. Oscillatory plasma wave, trapped in the light potential.

The injection of a low emittance, relativistic electron bunch into a transverse light lattice (Fig. 8) in the high intensity regime of the Kapitza-Dirac effect, results therefore in successive phenomena:

1. a trapping of the electron bunch along the stationary wave axis, in the light potential (high intensity optical molasses), as illustrated in Fig. 9;
2. a formation of a new kind of oscillating plasma waves in the light trap (Fig 11) ;
3. a Raman scattering effect of the laser off the oscillating plasma wave (Fig. 10) ;
4. an increased excitation of the plasma wave from the beating between the Raman scattered beam and the laser beams.

These steps are the classical elements required for the growth of a well-known instability, namely, Stimulated Raman Scattering, but with unique characteristics, since the inner degree of freedom is dominated by the transverse oscillations in the light potential, instead of the oscillations in plasma density (which are however taken into account). Whether or not this scattering is efficient enough to yield an actual X-ray laser is the major issue. The two relevant questions are, a) what can be a typical gain length of the instability ?, and b) is it possible to induce and control an interaction over several gain lengths?

Question b) is a very well known issue in plasma-based soft X-ray lasers. In transient collisional schemes, the gain lifetime at any point in the plasma is well below the time needed for a photon to go through the active medium. It is therefore useless to irradiate simultaneously the line focus; on the contrary, irradiation has to be made to follow in time the amplified photons. In (Balcou 2010), it was proposed to use a special optical configuration, called an inhomogeneous wave, in which the pulse energy front is tilted at 45° from the phase fronts, thus irradiating the electron bunch synchronously with its advance. It might also be possible to use a much simpler method, based on the well-known Grazing Incidence Geometry, known as GRIP.

Exploring question a) requires to get estimates for the Raman gain g and gain length. Two separate theoretical models have been developed in that goal. The first model mimics the Compton FEL theory by describing the evolution of the electron bunching, which, in contrast to standard FELs, is not purely longitudinal but also transverse, and oscillating (Fig. 11). The electron density is assumed to be constant, so that the model can be described as a Kinetic Frozen Density model (KFD). It shows that the Stokes mode will undergo exponential amplification, while the anti-Stokes mode is absorbed. The Raman gain is given by:

$$g = \alpha \sqrt{\frac{e^3 n_e E_0}{\varepsilon_0 m^2 c^3 \omega_1}},$$

where n_e is the electron number density, ω_1 the angular frequency of the X-rays in the Stokes mode, and E_0 the external laser field amplitude, and α a constant of order unity.

A second model has been recently proposed by Prof. Vladimir Tikhonchuk and Dr Igor Andriyash (Andriyash et al. 2011), based on a fluid description of excitation of the electron plasma wave within the bunch. This approach is derived from well-known models to describe Raman instabilities in plasmas. It has the advantage to treat in a unified way the plasma oscillations within the electron bunch, and the collective electron motion in the light potential, for a given electron temperature in the bunch rest frame. This plasma fluid model results in similar gain estimates. The important advantage of this second model is to show that the gain depends strongly on the transverse electron temperature.

4.3 Prospective experimental implementation

Let us see how these two formulas from independent models give estimates of gain lengths in a test case. We consider a medium energy electron beam from Laser Wakefield acceleration in the bubble regime, with the parameters given in the recent article by Davoine et al. (2009): 60 MeV, normalized emittance of 1 mm.mrad, peak current of 25 kA. With a laser intensity of 1.2×10^{18} W/cm^2, the KFD model predicts a gain length of 42 µm, for 43 keV X-ray photons. In identical conditions, the fluid plasma model predicts a similar gain length. The two models agree essentially to predict coherent amplification with very short gain lengths, in typical conditions of electron wakefield acceleration.

Assuming that amplification will saturate for a gain.length product g.L between 6 and 10, that the twin lasers are focused onto a vertical spot of 5 µm FWHM, neglecting for the time the initial latency (time required to start the X-ray wave amplification from noise), and considering typical parameters of Ti:Sa lasers, one obtains estimates of laser energy for the optical undulator of the order of 1 to few Joules – readily available with present-day laser technology.

As a result, analytical estimates show that one can expect to operate a Raman X-ray free electron laser with present-day laser technologies.

It is anticipated that the advantages and drawbacks of a Raman X-ray free electron laser, as compared to the Grüner et al. (2007) or Petrillo et al. (2008) schemes, would be:

- It requires far less stringent parameters in terms of mono-energeticity $\delta\gamma/\gamma$ of the electron bunch. A low value of $\delta\gamma/\gamma$ remains favourable, to avoid spreading the gain over a large inhomogeneous bandwidth, but a value higher than 1% should no longer be a killer. Considering the ongoing progress in Laser Wakefield Acceleration techniques, the interest of coupling LWFA and Raman XFEL approaches appears obvious.
- It requires existing laser technologies. Transverse irradiation also allows active optical elements to control finely the intensity and phase distributions of the twin beams along the interaction region.
- Electrons are trapped and guided transversely, at least in one dimension. Incidentally, a 2-dimensional trap is also possible.
- If a set of working parameters is found, then the whole setup may be upscaled to higher X-ray photons energies by scaling up the laser energy by γ, the electron energy by γ as well, while the output X-ray photon energy scales as γ^2. Coherent X-ray generation beyond few tens of keV seems therefore possible. Moreover, increasing the electron energies to values up to hundreds of MeV or even 1 GeV is possible; therefore the current developments of multi-PetaWatt lasers, and of large projects like ELI (Extreme Light Infrastructure), could make it possible to reach lasing in the γ-ray range in the long term.

The scheme may however also present some drawbacks, as :

- It requires a delicate optical setup, with perfectly controlled twin laser undulator beams;
- It is bound to yield a spectrally rather broad X-ray emission, with a $\delta\omega_1/\omega_1$ of the order of Ω/ω_0 (around 1% expected) ;
- The photon number per pulse is bound to be smaller than with large scale XFELs, due to the larger impact of the recoil phenomenon on the electron energies.

However, this list of advantages and drawbacks is still currently being debated; all these aspects must be studied extensively, with a thorough theoretical investigation, and first experimental tests.

5. Conclusion

Due to the strong demand for bright XFEL facilities and the cost of existing systems, the development of compact XFEL schemes is a very attractive field at the crossroads of particle acceleration physics, optics and plasma physics. Significant progresses have been accomplished in the last few years and new schemes involving the coupling of laser wakefield accelerated electrons with either magnetic or optical wigglers have been proposed. Even if some progresses are being made for each of these schemes, they have very stringent requirements on the accelerated electron beams. Another very promising scheme using counter-propagating lasers to create a ponderomotive potential array in which accelerated electrons will propagate is also under study. Modelling of Raman gain through Particle In Cell simulations is now underway and new kinetic theory results are being compared with 2D PIC simulations in the rest frame of the energetic electron beam and encouraging results have been obtained so far (Andriyash et al., 2012; Andriyash et al., in preparation). A proposition of parameters for an all-optical compact XFEL will first be tested using Particle-In-Cell simulations. The original Kinetic Frozen Density model of (Balcou 2010) is also being improved, especially since it was pointed out that the full effect of the magnetic field of the X-ray wave should be better taken into account (Zholents and Zolotorev 2011; Balcou 2011). These efforts will lead to an experimental proposal in order to test the obtained theoretical and numerical results.

6. Acknowledgements

The authors would like to thank Prof. V. T. Tikhonchuk and Dr. I. Andriyash for fruitful discussions.

7. References

Amiranoff F. et al. (1995), Phys. Rev. Lett., 74 p5220–5223.

Andreev N. E. et al. (1992), JETP Lett., 55 p571.

Andriyash I., Balcou Ph. and Tikhonchuk V.T. (2011), Eur. Phys. J. D, 65, 533.

Andriyash I., Balcou Ph., d'Humières E. and Tikhonchuk V.T. (2012) "Scattering of relativistic electron beam by two counter-propagating laser pulses: a new approach to Raman X-ray amplification", accepted by EPJ – WoC

Andriyash I., d'Humières E., Balcou Ph. and Tikhonchuk V.T. « Particle-In-Cell simulations of Raman X-ray amplification in optical free electron lasers » , in preparation.

Antonsen T. M. et Mora P. (1992), Phys. Rev. Lett., 69 p2204.

Arkhiezer A. I. et Polovin R. V. (1956), Sov. Phys. JEPT, 3 p696–705.

Bacci A. et al. (2006), Phys. Rev. ST Accel. Beams 9, 060704.

Balcou Ph. (2010), Eur. Phys. J. D 59, 525–537.

Balcou Ph. (2011), Eur. Phys. J. D 62, 459-459.

Bloembergen N. et al. (1999), Rev. Mod. Phys. 71, S283.

Bucksbaum Ph. Et al. (1988), Phys. Rev. Lett. 61, 1182.

Carroll F. (2003), Journ. Cell. Biochem. 90, 502–508.

Clayton C. E. et al. (1994). Phys. Plasmas., 1 p1753.

Davoine X. et al. (2009), Phys. Rev. Lett. 102, 065001.

Dobiasch P., Meystre P., Scully M.O. (1983), IEEE J. Quantum Electron. 19, 1812.

Emma P. and LCLS commissioning team (2009), Proceedings of the PAC conference.

Emma P. et al. (2010), Nature Photonics 4, 641 - 647.

Everett M.J. et al. (1994), Nature, 368 p527–529.

Faure J. et al. (2004), Nature, 431 p541.

Faure J. et al. (2006), Nature 444, 737-739.

Fedorov V.M. et al. (1988), App. Phys. Lett. 53, 353.

Freimund D.L., Aflatooni K., Batelaan H. (2001), Nature 413, Issue 6852, 142].

Freund H.P. and Antonsen T.M. (1996), Principles of free-electron lasers, Chapman & Hall.

Fritzler S. et al. (2004), Phys. Rev. Lett., 92 p165006.

Gahn C. et al. (1999), Phys. Rev. Lett., 83 p4772.

Gallardo J.C., et al. (1988), IEEE J. Quantum. Electron. 24, 1557.

Gea-Banacloche J. et al. (1987), IEEE J. Quantum. Electron. 23, 1558.

Geddes C. G. R. et al. (2004), Nature, 431 p538.

Gibbon P. (2005) "Short pulse laser interactions with matter. An Introduciton" Imperial College Press, London.
Grüner F. et al. (2007), APB 86, 431.
Hafz N.A.M. et al. (2008) Nature Phot. 2, 571-577.
Hartemann (2000), Astrophys. Journ. Supp. Series 127, 347.
Hidding B. et al. (2006), Phys. Rev. Lett., 96 p105004.
Hosokai T. et al. (2003), Phys. Rev. E, 67 p036407.
Jaeglé P. (2006), Coherent sources of XUV radiation (Springer Verlag, Berlin).
Joshi C., et al. (1981), Phys. Rev. Lett., 60 p1298.
Kapitza P.L., Dirac P.A.M. (1933), Proc. Cambridge Philos. Soc. 29, 297.
Katsouleas T. and Mori W.B. (1988), Phys. Rev. Lett., 61 p90–93.
Kibble T.W.B. (1968), Phys. Rev. 150, 1060.
Kitagawa Y. et al. (1992), Phys. Rev. Lett., 68 p48.
Lau et al. (2003), Phys. Plasmas, Vol. 10, No. 5.
Leemans W.P. et al. (2004), Phys. of Plasmas, 11 p2899–2906.
Leemans, W. et al. (2006), Nature phys. 2, 696.
Liu J.S. et al. (2011), Phys. Rev. Lett. 107, 035001.
Loew G.A. and Talman A. (1983), SLAC-PUB-3221, Sec. 2(Sep.).
Lu W. et al. (2007), Phys. Rev. ST-AB 10, 061301.
Lundh O. et al. (2011), Nature Physics 7, 219.
Malka G., Lefebvre E., and Miquel J.L. (1997) Phys. Rev. Lett. 78, 3314-3317.
Malka V. et al. (2001), Physics of Plasmas, 8 p2605–2608.
Malka V. et al. (2002), Science, 298 p1596–1600.
Mangles S. et al. (2004), Nature, 431 p535.
Miura E. et al. (2005), Appl. Phys. Lett., 86 p25150.
Nakajima K. (2008), Nature Phys. 4, 92.
Petrillo V., Serafini L., and P. Tomassini (2008), Phys. Rev. Sel. Top. Acc. Beams 11, 070703.
Pollock B.B. et al. (2011), Phys. Rev. Lett. 107, 045001.
Pukhov A. and Meyer-ter-Vehn J. (2002), Appl. Phys. B, 74 p355.
Rosenzweig J.B. (1988), Phys. Rev. A, 38 p3634.
Rousse A. et al. (2001), Rev. Mod. Phys. 73, 17.
Schlenvoigt H.P. et al. (2008), Nature Phys. 4, 130.
Schoenlein R.W. et al (1996), Science, 274, 236.
Schoenlein R.W. et al (2000), Science, 287, 2237.
Schoenlein R.W. et al. (2000), Appl. Phys. B, 71, 1
Sprangle P., Esarey E., Krall J. et Joyce G. (1992), Phys. Rev. Lett., 69 p2200.
Sprangle P., Hafizi B., Peñano J.R. (2009), Phys. Rev. Sel. Top. Acc. Beams 12, 050702.
Sprangle P. (2009), JAP 50, 2652 12, 050702.
Strickland D. and Mourou G. (1985), Opt. Comm., 56 p219–221.
Tajima T. and Dawson J.M. (1979) "Laser electron accelerator" Phys. Rev. Lett. 43, 267-270.
Tsung F.S. et al. (2004), Phys. Rev. Lett., 93 p185002.
Ultrafast Phenomena proceedings (1992- 2002), Springer Ser. Chem. Phys. Vol. 55, 60, 62, 63, and 66 (Springer, Berlin).

Wilson P.B. (2008), Rev. Acc. Science Tech., 1, 7.
Winick H. et al. (1981), Physics Today 34, 50.
Zewail A.H. (2000), J. Phys. Chem. A 104, 5660.
Zholents A. and Zolotorev M. (2011), Eur. Phys. J. D 62, 457 (2011).

5

Laser-Driven Table-Top X-Ray FEL

Kazuhisa Nakajima[1,2,3] et al.*
[1]*Shanghai Institute of Optics and Fine Mechanics,*
Chinese Academy of Sciences, Shanghai,
[2]*High Energy Accelerator Research Organization, Tsukuba,*
[3]*Shanghai Jiao Tong University, Shanghai,*
[1,3]*China*
[2]*Japan*

1. Introduction

Synchrotron radiation sources nowadays benefit a wide range of fundamental sciences - from physics and chemistry to material science and life sciences as a result of a dramatic increase in the brilliance of photons emitted by relativistic electrons when bent in the magnetic field of synchrotron accelerators. A trend will tend toward the X-ray free electron laser (FEL) that will produce high-intensity ultrashort coherent X-ray radiations with unprecedented brilliance as kilometer-scale linear accelerator-based FELs are being commissioned to explore new research area that is inaccessible to date, for instance femtosecond dynamic process of chemical reactions, materials and biomolecules at the atomic level (Gerstner, 2011). Such large-scale tool could be built on a table top if high-quality electrons with small energy spread and divergence are accelerated up to the GeV range in a centimetre-scale length (Grüner et al., 2007; Nakajima et al., 1996; Nakajima, 2008, 2011). It is prospectively conceived that a compact source producing high-energy high-quality electron beams from laser plasma accelerators (LPAs) will provide an essential tool for many applications, such as THz and X-ray synchrotron radiation sources and a unique medical therapy as well as inherent high-energy accelerators for fundamental sciences (Malka et al., 2008).

The present achievements of the laser wakefield accelerator performance on the beam properties such as GeV-class energy (Leemans et al., 2006; Clayton et al., 2010; Lu et al., 2011), 1%-level energy spread (Kameshima et al., 2008; Rechatin et al., 2009), a few mm-

---

*Aihua Deng[1], Hitoshi Yoshitama[4], Nasr A. M. Hafz[3], Haiyang Lu[1], Baifei Shen[1], Jiansheng Liu[1], Ruxin Li[1] and Zhizhan Xu[1]
[1]*Shanghai Institute of Optics and Fine Mechanics, Chinese Academy of Sciences, Shanghai,*
[2]*High Energy Accelerator Research Organization, Tsukuba,*
[3]*Shanghai Jiao Tong University, Shanghai,*
[4]*Hiroshima Unversity, Higashi Hiroshima,*
[1,3]*China*
[2,4]*Japan*

mrad emittance (Karsch et al., 2007), 1-fs-level bunch with a 3-4 kA peak current (Lundh et al., 2011), and good stability and controllability (Hafz et al., 2008; Osterhoff et al., 2008) of the beam production allow us to downsize a large-scale X-ray synchrotron radiation source and FEL to a table-top scale including laser drivers and radiation shields. The undulator radiation from laser-plasma accelerated electron beams has been first demonstrated at the wavelength of $\lambda_{\mathrm{rad}} = 740$ nm and the estimated peak brilliance of the order of 6.5×10^{16} photons/s/mrad²/mm² /0.1% bandwidth driven by the electron beam from a 2-mm-gas jet with the beam energy $E_b = 64$ MeV, the relative energy spread $\Delta E / E_b = 5.5\%$ (FWHM) and total charge $Q_b = 28$ pC, which is produced by a 5 TW 85 fs laser pulse at the plasma density $n_p = 2 \times 10^{19}$ cm^{-3} (Schlenvoigt, 2008). The soft X-ray undulator radiation has been also successfully demonstrated at the wavelength $\lambda_{\mathrm{rad}} = 18$ nm and the estimated peak brilliance of the order of 1.3×10^{17} photons/s/mrad²/mm²/0.1% bandwidth radiated by electrons with $E_b = 207$ MeV, $\Delta E / E_b = 6\%$ (FWHM) and $Q_b = 30$ pC from a 15-mm-hydrogen-fill gas cell driven by a 20 TW 37 fs laser pulse at $n_p = 8 \times 10^{18}$ cm^{-3} (Fuchs, 2009). With extremely small energy spread and peak current high enough to generate self-amplified spontaneous emission so-called SASE (Bonifacio, 1984), a photon flux of the undulator radiation can be amplified by several orders of magnitude to levels of brilliance comparable to current large-scale X-ray FELs (Nakajima, 2008).

Here we consider feasibility of a compact hard X-ray FEL capable of reaching a wavelength of $\lambda_X = 0.1$ nm, which requires the electron beam energy of the multi-GeV range in case of a modest undulator period of the order of a few centimeters. One of prominent features of laser-plasma accelerators is to produce 1-fs-level bunch duration, which is unreachable by means of the conventional accelerator technologies. The X-ray FELs rely on SASE, where the coherent radiation builds up in a single pass from the spontaneous (incoherent) undulator radiation. In an undulator the radiation field interacts with electrons snaking their way when overtaking them so that electrons are resonantly modulated into small groups (micro-bunches) separated by a radiation wavelength and emit coherent radiation with a wavelength equal to the micro-bunch period length. This process requires an extremely high-current beam with small energy spread and emittance in addition to a long precisely manufactured undulator. Therefore the conventional accelerator-based FELs need a long section of the multi-stage bunch compressor called as a 'chicane' that compresses a bunch from an initial bunch length of a few picoseconds to the order of 100 fs to increase the current density of the electron beam up to the order of kilo-ampere level before injected to the undulator, whereas the laser-plasma accelerator based FELs would have no need of any bunch compressor. Although the present LPAs need further improvements in the beam properties such as energy, current, qualities and operating stability, the beam current of 100 kA level (i.e. 100 pC electron charge within 1 fs bunch duration) allows a drastic reduction to the undulator length of several meters for reaching the saturation of the FEL amplification. In addition to inherently compact laser and plasma accelerator, a whole FEL system will be operational on the table-top scale. The realization of laser-driven compact table-top X-ray FELs will benefit science and industry over a broad range by providing new tools enabling the leading-edge research in small facilities, such as universities and hospitals.

2. Laser-plasma accelerators

2.1 Laser wakefields in the linear regime

In underdense plasma an ultraintense laser pulse excites a large-amplitude plasma wave with frequency

$$\omega_p = \sqrt{\frac{4\pi e^2 n_p}{m}} \tag{1}$$

and electric field of the order of

$$E_0 = \frac{mc\omega_p}{e} \simeq 96[\mathrm{GV/m}]\sqrt{\frac{n_p}{10^{18}[\mathrm{cm}^{-3}]}} \tag{2}$$

for the electron rest energy mc^2 and plasma density n_p due to the ponderomotive force expelling plasma electrons out of the laser pulse and the space charge force of immovable plasma ions restoring expelled electrons on the back of the ion column remaining behind the laser pulse. Since the phase velocity of the plasma wave is approximately equal to the group velocity of the laser pulse

$$\frac{v_p}{c} \simeq \sqrt{1 - \frac{\omega_p^2}{\omega_L^2}} \sim 1 \tag{3}$$

for the laser frequency ω_L and the accelerating field of ~ 100 GV/m for the plasma density ~ 10^{18} cm^{-3}, electrons trapped into the plasma wave are likely to be accelerated up to ~ 1 GeV energy in a 1-cm plasma. In the linear and quasilinear regime with the normalized laser intensity

$$a_0 = \left(\frac{2e^2\lambda_L^2 I_L}{\pi m^2 c^5}\right)^{1/2} \simeq 0.855 \times 10^{-9} I_L^{1/2}\left[\mathrm{W/cm^2}\right]\lambda_L[\mu\mathrm{m}] \sim 1 \tag{4}$$

where I_L is the laser intensity and $\lambda_L = 2\pi c/\omega_L$ is the laser wavelength, the wake potential Φ is obtained from a simple-harmonic equation (Esarey et al., 1996)

$$\frac{\partial^2 \Phi}{\partial \zeta^2} + k_p^2 \Phi = \frac{1}{2} k_p^2 mc^2 a^2(r,\zeta) \tag{5}$$

where $\zeta = z - v_p t$, $k_p = \omega_p/c$ and $a(r,\zeta) \equiv eA(r,\zeta)/mc^2$ is the normalized vector potential of the laser pulse. The wake potential is given by

$$\Phi(r,\zeta) = -\frac{m_e c^2 k_p}{2}\int_\zeta^\infty d\zeta' \sin k_p(\zeta - \zeta') a^2(r,\zeta') \tag{6}$$

Thus the axial and radial electric fields are calculated by

$$eE_z = -\frac{\partial \Phi}{\partial z}, \text{ and } eE_r = -\frac{\partial \Phi}{\partial r} \tag{7}$$

Considering a temporally Gaussian laser pulse with $1/e$ half-width σ_L, of which the ponderomotive potential is given by

$$a^2(r,\zeta) = U(r)\exp\left(-\frac{\zeta^2}{\sigma_L^2}\right) \tag{8}$$

the wake potential is

$$\Phi(r,\zeta) = -\frac{\sqrt{\pi} mc^2 k_p \sigma_L}{4} U(r)\exp\left(-\frac{k_p^2\sigma_L^2}{4}\right)\left[C(\zeta)\sin k_p\zeta + S(\zeta)\cos k_p\zeta\right] \tag{9}$$

where $C(\zeta)$ and $S(\zeta)$ are defined as

$$C(\zeta) = 1 - \Re\left[\operatorname{erf}\left(\frac{\zeta}{\sigma_L} - i\frac{k_p\sigma_L}{2}\right)\right], \tag{10}$$

and

$$S(\zeta) = \Im\left[\operatorname{erf}\left(\frac{\zeta}{\sigma_L} - i\frac{k_p\sigma_L}{2}\right)\right], \tag{11}$$

respectively, and

$$\operatorname{erf}(z) = \frac{2}{\sqrt{\pi}}\int_0^z e^{-z'^2} dz' \tag{12}$$

is the complex error function. Using Eq. (7), the axial and radial electric fields are

$$eE_z(r,\zeta) = \frac{\sqrt{\pi} mc^2 k_p^2 \sigma_z}{4} U(r)\exp\left(-\frac{k_p^2\sigma_L^2}{4}\right)\left[C(\zeta)\cos k_p\zeta - S(\zeta)\sin k_p\zeta\right], \tag{13}$$

and

$$eE_r(r,\zeta) = \frac{\sqrt{\pi} mc^2 k_p \sigma_L}{4} \frac{\partial U(r)}{\partial r}\exp\left(-\frac{k_p^2\sigma_L^2}{4}\right)\left[C(\zeta)\sin k_p\zeta + S(\zeta)\cos k_p\zeta\right], \tag{14}$$

respectively. Behind the laser pulse at $\zeta \ll -\sigma_L$, taking into account $C(\zeta) \to 2$ and $S(\zeta) \to 0$, the wakefields are approximately given by

$$eE_z(r,\zeta) \simeq \frac{\sqrt{\pi} m_e c^2 k_p^2 \sigma_L}{2} U(r)\exp\left(-\frac{k_p^2\sigma_L^2}{4}\right)\cos k_p\zeta, \tag{15}$$

and

$$eE_r(r,\zeta) \simeq \frac{\sqrt{\pi} m_e c^2 k_p \sigma_L}{2} \frac{\partial U(r)}{\partial r} \exp\left(-\frac{k_p^2 \sigma_L^2}{4}\right) \sin k_p \zeta \tag{16}$$

For $k_p \sigma_L = \sqrt{2}$, the maximum amplitude of the axial field becomes

$$|E_z|_{\max} = \sqrt{\frac{\pi}{2}} U(0) e^{-\frac{1}{2}} \left(\frac{mc\omega_p}{e}\right) \simeq 0.76 E_0 U(0) . \tag{17}$$

As shown in Eq. (15) and (16), the radial electric field shifts a phase by $\pi/2$ with respect to the axial field and radial dependence of the wakefields is determined by the radial component $U(r)$ of the ponderomotive potential. For a Gaussian laser pulse with linear polarization, the ponderomotive potential is given by

$$a^2(r,z,t) = \frac{a_0^2}{2} \frac{r_L^2}{w^2} \exp\left[-\frac{2r^2}{w^2} - \frac{(z - v_p t)^2}{\sigma_L^2}\right], \tag{18}$$

where r_L is the laser spot radius at focus, $w = r_L \sqrt{1 + z^2/z_R^2}$ is the spot radius at z and $z_R = \pi r_L^2/\lambda_L$ is the Rayleigh length for the laser wavelength λ_L. Assuming that the laser pulse propagates the plasma at a constant spot size $w = r_L$ in a matched plasma waveguide, the ponderomotive potential of the Gaussian mode is

$$a^2(r,\zeta) = \frac{a_0^2}{2} \exp\left(-\frac{2r^2}{r_L^2} - \frac{\zeta^2}{\sigma_L^2}\right) \tag{19}$$

Thus the normalized radial potential is defined as

$$U(r) = \frac{a_0^2}{2} \exp\left(-\frac{2r^2}{r_L^2}\right). \tag{20}$$

For the Gaussian mode, the axial and radial fields Eq. (15) and (16) are

$$eE_z(r,\zeta) = \frac{\sqrt{\pi}}{4} a_0^2 mc^2 k_p^2 \sigma_L \exp\left(-\frac{2r^2}{r_L^2} - \frac{k_p^2 \sigma_L^2}{4}\right) \cos k_p \zeta , \tag{21}$$

and

$$eE_r(r,\zeta) = -\sqrt{\pi} a_0^2 mc^2 k_p \sigma_L \frac{r}{r_L^2} \exp\left(-\frac{2r^2}{r_L^2} - \frac{k_p^2 \sigma_L^2}{4}\right) \sin k_p \zeta . \tag{22}$$

For a Gaussian pulse, the maximum accelerating field is

$$E_{z\max} = \frac{1}{2}\sqrt{\frac{\pi}{2}}e^{-\frac{1}{2}}a_0^2 E_0 \simeq 0.38 a_0^2 E_0 \text{ at } r = 0 \text{ for } k_p\sigma_L = \sqrt{2}, \tag{23}$$

and the maximum radial field is

$$E_{r\max} = \sqrt{\frac{\pi}{2}}\frac{e^{-1}}{k_p r_L}a_0^2 E_0 \simeq \frac{0.46}{k_p r_L}a_0^2 E_0 \text{ at } r = \frac{r_L}{2} \text{ for } k_p\sigma_L = \sqrt{2}. \tag{24}$$

Consider the radial potential profile described by super-Gaussian functions as

$$U(r) = \frac{a_0^2}{2}\exp\left[-2\left(\frac{r}{r_L}\right)^n\right], \tag{25}$$

where $n \geq 2$ (Sverto, 1998). A Gaussian profile corresponds to $n = 2$. Substituting Eq. (25) into Eq. (15) and (16), the axial and radial wakefieds for a super-Gaussian potential are

$$eE_z(r,\zeta) = \frac{\sqrt{\pi}}{4}a_0^2 m_e c^2 k_p^2 \sigma_z \exp\left[-2\left(\frac{r}{r_L}\right)^n - \frac{k_p^2\sigma_z^2}{4}\right]\cos k_p\zeta, \tag{26}$$

and

$$eE_r(r,\zeta) = -\frac{n}{2}\sqrt{\pi}a_0^2 m_e c^2 k_p \sigma_z \frac{r^{n-1}}{r_L^n}\exp\left[-2\left(\frac{r}{r_L}\right)^n - \frac{k_p^2\sigma_z^2}{4}\right]\sin k_p\zeta. \tag{27}$$

The maximum accelerating field $E_{z\max}$ at $r = 0$ is given by Eq. (23) and the maximum radial field is

$$E_{r\max} = \sqrt{\frac{\pi}{2}}n\left(\frac{n-1}{2n}\right)^{\frac{n-1}{n}}\frac{e^{-\frac{3}{2}+\frac{1}{n}}}{k_p r_L}a_0^2 E_0 \text{ at } \frac{r}{r_L} = \left(\frac{n-1}{2n}\right)^{1/n} \text{ for } k_p\sigma_L = \sqrt{2}. \tag{28}$$

The peak laser power with the normalized vector potential a_0 is calculated as

$$P_L = \left(\frac{1}{2}\right)^{2/n}\Gamma\left(\frac{n+2}{n}\right)\frac{(k_L r_L a_0)^2}{8}\left(\frac{m^2c^5}{e^2}\right) \simeq 8.7[\text{GW}]\left(\frac{1}{2}\right)^{2/n}\Gamma\left(\frac{n+2}{n}\right)\frac{(k_L r_L a_0)^2}{8}, \tag{29}$$

where $k_L = 2\pi/\lambda_L$ is the laser wave number and $\Gamma(z)$ is the Gamma function. For a Gaussian pulse $n = 2$, the peak power is calculated as

$$P_L = \frac{(k_L r_L a_0)^2}{16}\left(\frac{m^2c^5}{e^2}\right) \simeq 0.544[\text{GW}](k_L r_L a_0)^2, \tag{30}$$

and for a super-Gaussian pulse $n = 4$, the peak power is

$$P_L = \sqrt{\frac{\pi}{2}} \frac{(k_L r_L a_0)^2}{16} \left(\frac{m^2 c^5}{e^2} \right) \simeq 0.681[\text{GW}](k_L r_L a_0)^2 . \tag{31}$$

2.2 Electron acceleration in the quasilinear wakefield

In Eq. (21), for a given plasma density, the maximum field is $E_{z\max} \approx 0.38 a_0^2 E_0$ at the resonant condition $k_p \sigma_L = \sqrt{2}$, while for a given pulse duration, the maximum field is $E_{z\max} \approx 0.33 a_0^2 E_0$ at the resonant condition $k_p \sigma_L = 2$. Changing both plasma density and laser pulse duration, the optimum condition turns out $k_p \sigma_L = 1$ i. e. the FWHM pulse length

$$c\tau_L = 2\sqrt{\ln 2}\,\sigma_L \approx 0.265\lambda_p , \tag{32}$$

for which the maximum field is

$$E_{z\max} \approx 0.35 a_0^2 E_0 \simeq 10.6[\text{GV/m}] a_0^2 \left(\frac{n_p}{10^{17}\left[\text{cm}^{-3}\right]} \right)^{1/2} . \tag{33}$$

In this condition, the laser pulse length is shorter enough than a half plasma wavelength so that a transverse field at the tail of the laser pulse is negligible in the accelerating phase of the first wakefield. The net accelerating field E_z that accelerates the bunch containing the charge $Q_b = eN_b$, where N_b is the number of electrons in the bunch, is determined by the beam loading that means the energy absorbed per unit length,

$$Q_b E_z = \frac{mc^2 k_p^2 \sigma_r^2}{4r_e} \frac{E_{z\max}^2}{E_0^2} \left(1 - \frac{E_z^2}{E_{z\max}^2} \right) . \tag{34}$$

where $r_e = e^2 / mc^2$ is the classical electron radius and

$$1 - \frac{E_z^2}{E_{z\max}^2} \equiv \eta_b \tag{35}$$

is the beam loading efficiency that is the fraction of the plasma wave energy absorbed by particles of the bunch with the rms radius σ_r. In the beam-loaded field $E_z = \sqrt{1-\eta_b} E_{z\max}$, the loaded charge is given by

$$Q_b \simeq \frac{e}{4k_L r_e} \frac{\eta_b k_p^2 \sigma_r^2}{1-\eta_b} \frac{E_z}{E_0} \left(\frac{n_c}{n_e} \right)^{1/2} \approx 232[\text{pC}] \frac{\eta_b k_p^2 \sigma_r^2}{1-\eta_b} \frac{E_z}{E_0} \left(\frac{n_p}{10^{17}\left[\text{cm}^{-3}\right]} \right)^{-1/2} , \tag{36}$$

where $n_c = m\omega_L^2 / 4\pi e^2 = \pi / \left(r_e \lambda_L^2 \right)$ is the critical plasma density and $E_z / E_0 \simeq 0.35 a_0^2 \sqrt{1-\eta_b}$ for $k_p \sigma_L = 1$. Since the loaded charge depends on the accelerating field and the bunch radius, it will be determined by considering the required accelerating gradient and the transverse beam dynamics.

The electron linac with 10 GeV-class beam energy can be composed of a high-quality beam injection stage with beam energy of the order of 100 MeV in a mm-scale length and a high-gradient acceleration stage with meter-scale length. Ideally, the stage length L_{stage} is limited by the pump depletion length L_{pd} for which the total field energy is equal to half the initial laser energy. For a Gaussian laser pulse with the pulse length $k_p\sigma_L = 1$, the pump depletion length is given by

$$k_p L_{\text{pd}} \simeq \frac{8}{\sqrt{\pi} a_0^2 k_p \sigma_L} \frac{\omega_L^2}{\omega_p^2} \exp\left(\frac{k_p^2 \sigma_L^2}{2}\right) \approx \frac{7.4}{a_0^2} \frac{n_c}{n_e} . \tag{37}$$

In laser wakefield accelerators, accelerated electrons eventually overrun the acceleration phase to the deceleration phase, of which the velocity is roughly equal to the group velocity of the laser pulse. In the linear wakefield regime, the dephasing length L_{dp} where the electrons undergo both focusing and acceleration is approximately given by

$$k_p L_{\text{dp}} \simeq \pi \frac{\omega_L^2}{\omega_p^2} = \pi \frac{n_c}{n_p} \tag{38}$$

In the condition for the dephasing length less than the pump depletion length $L_{\text{dp}} \le L_{\text{pd}}$, the normalized vector potential should be $a_0 \le 1.5$. Setting $a_0 = \sqrt{2}$, the maximum accelerating field is $E_{z\max} \approx 0.7 E_0$ for $k_p\sigma_L = 1$. Assuming the beam-loaded efficiency $\eta_b = 0.5$, the net accelerating field becomes

$$E_z \approx \frac{E_{z\max}}{\sqrt{2}} \simeq 0.5 E_0 \approx 15[\text{GV/m}]\left(\frac{n_p}{10^{17}\left[\text{cm}^{-3}\right]}\right)^{1/2} . \tag{39}$$

With the acceleration stage length approximately equal to the dephasing length

$$L_{\text{stage}} \sim L_{\text{dp}} = \frac{\lambda_p}{2} \frac{n_c}{n_p} = \frac{\lambda_L}{2}\left(\frac{n_c}{n_p}\right)^{3/2} \approx 0.92[\text{m}]\left(\frac{0.8[\mu\text{m}]}{\lambda_L}\right)^2 \left(\frac{10^{17}\left[\text{cm}^{-3}\right]}{n_p}\right)^{3/2} , \tag{40}$$

the energy gain in the acceleration stage is given by

$$W_{\text{acc}} = E_z L_{\text{stage}} = \pi m c^2 \frac{E_z}{E_0} \frac{n_c}{n_p} \approx 28[\text{GeV}] \frac{E_z}{E_0} \frac{10^{17}\left[\text{cm}^{-3}\right]}{n_p} \left(\frac{0.8[\mu\text{m}]}{\lambda_L}\right)^2 . \tag{41}$$

Here we assume the accelerating field E_z keeps constant over the whole stage length $k_p L_{\text{stage}} \sim \pi(n_c / n_p)$. In fact the 2D particle-in-cell simulation shows that the laser pulse undergoes self-focusing at the entrance of the plasma channel and propagates over the stage length with the energy depletion, leading the average amplitude to be $\langle a \rangle / a_0 \approx 1$ (Nakajima et al., 2011). The plasma density will be determined by setting the required beam energy E_b for the FEL injector linac as

$$n_p = \frac{\pi mc^2}{E_b}\frac{E_z}{E_0} n_c \approx 2.8\times 10^{17}\left[\mathrm{cm}^{-3}\right]\frac{E_z}{E_0}\frac{10[\mathrm{GeV}]}{E_b}\left(\frac{0.8[\mu\mathrm{m}]}{\lambda_L}\right)^2 . \tag{42}$$

The required accelerator length is given by

$$L_{\mathrm{stage}} = \frac{\lambda_L}{2}\left(\frac{E_b}{\pi mc^2}\right)^{3/2}\left(\frac{E_z}{E_0}\right)^{-3/2} \approx 0.2[\mathrm{m}]\left(\frac{E_z}{E_0}\right)^{-3/2}\frac{0.8[\mu\mathrm{m}]}{\lambda_L}\left(\frac{E_b}{10[\mathrm{GeV}]}\right)^{3/2} \tag{43}$$

In the operation of the staged LPA, self-focusing of the drive laser and self-injection of plasma electrons should be suppressed to prevent the beam quality from deterioration as much as possible. These requirements can be accomplished by the LPA operation in the quasilinear regime, where the laser spot size is bounded by conditions for avoiding bubble formation,

$$\frac{k_p^2 r_L^2}{4} > \frac{a_0^2}{\sqrt{1+a_0^2/2}} , \tag{44}$$

and strong self-focusing,

$$\frac{P_L}{P_c} = \frac{\left(k_p r_L a_0\right)^2}{32} \le 1 , \tag{45}$$

where $P_c = 2\left(m^2c^5/e^2\right)\omega_L^2/\omega_p^2 \simeq 17\left(n_c/n_p\right)[\mathrm{GW}]$ is the critical power for relativistic self-focusing. These conditions put bounds to the spot size

$$2.4 \le k_p r_L \le 4 \text{ for } a_0 = \sqrt{2} . \tag{46}$$

For a given spot radius

$$r_L = \frac{\lambda_L}{2\pi} k_p r_L \left(\frac{n_c}{n_e}\right)^{1/2} \approx 17[\mu\mathrm{m}] k_p r_L \left(\frac{10^{17}\left[\mathrm{cm}^{-3}\right]}{n_p}\right)^{1/2} , \tag{47}$$

the peak laser power P_L becomes

$$P_L = \frac{\left(k_p r_L a_0\right)^2}{32} P_c = \left(\frac{k_p r_L a_0}{4}\right)^2 \frac{m^2c^5}{e^2}\frac{n_c}{n_p} \approx 9.5[\mathrm{TW}]\left(k_p r_L a_0\right)^2 \left(\frac{0.8[\mu\mathrm{m}]}{\lambda_L}\right)^2 \frac{10^{17}\left[\mathrm{cm}^{-3}\right]}{n_p} . \tag{48}$$

With the FWHM pulse duration τ_L given by

$$\tau_L = 2\sqrt{\ln 2}\frac{\sigma_L}{c} = \frac{\sqrt{\ln 2}}{\pi}\frac{\lambda_L}{c} k_p \sigma_L \left(\frac{n_c}{n_p}\right)^{1/2} \simeq 93[\mathrm{fs}] k_p \sigma_L \left(\frac{10^{17}\left[\mathrm{cm}^{-3}\right]}{n_p}\right)^{1/2} , \tag{49}$$

the required laser pulse energy U_L is calculated as

$$U_L = P\tau_L = \frac{\sqrt{\ln 2}}{16\pi}\frac{\lambda_L}{c}\left(\frac{m^2c^5}{e^2}\right)\left(k_p r_L a_0\right)^2 k_p\sigma_L\left(\frac{n_c}{n_e}\right)^{3/2}$$
$$\approx 0.89[\mathrm{J}]\left(k_p r_L a_0\right)^2 k_p\sigma_L\left(\frac{0.8[\mu\mathrm{m}]}{\lambda_L}\right)^2\left(\frac{10^{17}\left[\mathrm{cm}^{-3}\right]}{n_p}\right)^{3/2}. \tag{50}$$

3. Beam dynamics in laser-plasma accelerators

3.1 Betatron oscillation

Beams that undergo strong transverse focusing forces $F_\perp = -mc^2K^2x$ in plasma waves exhibit the betatron oscillation, where x is the transverse amplitude of the betatron oscillation. From the axial and radial fields, Eqs. (21) and (22), driven by the Gaussian pulse , the focusing constant K is given by

$$K^2 \simeq \frac{4k_p^2}{\left(k_p r_L\right)^2}\frac{E_z}{E_0}\langle\sin\Psi\rangle, \tag{51}$$

where $\langle\sin\Psi\rangle$ is set to be the average value over the dephasing phase $0 \le \Psi \le \Psi_{\max}$, where $\Psi_{\max}$ is the maximum dephasing phase at the acceleration distance L_{stage}, i.e. $\Psi_{\max} = (\pi/2)\left(L_{\text{stage}}/L_{dp}\right)$. The envelope equation of the rms beam radius σ_r is given by

$$\frac{d^2\sigma_r}{dz^2} + \frac{K^2}{\gamma}\sigma_r - \frac{\varepsilon_{n0}^2}{\gamma^2\sigma_r^3} = 0, \tag{52}$$

where ε_{n0} is the initial normalized emittance (Schachter, 2011). Assuming the beam energy γ is constant, this equation is rewritten as

$$\frac{d^2\sigma_r^2}{dz^2} + \kappa^2\sigma_r^2 = C, \tag{53}$$

where $\kappa = 2K/\sqrt{\gamma}$ is the focusing strength and $C = 2\sigma_{r0}' + \kappa^2\sigma_{r0}^2/2 + 2\varepsilon_{n0}^2/\gamma^2\sigma_{r0}^2$ is the constant given by the initial conditions $\sigma_{r0}' = \left(d\sigma_r/dz\right)_{z=0}$ and $\sigma_{r0} = \sigma_r(0)$. Thus the beam envelope is obtained as

$$\sigma_r^2(z) = \frac{C}{\kappa^2} + \frac{1}{\kappa}\sqrt{\frac{C^2}{\kappa^2} - \frac{4\varepsilon_{n0}^2}{\gamma^2}}\sin\left(\kappa z + \phi_0\right), \tag{54}$$

where

$$\tan\phi_0 = \frac{\sigma_{r0}^2 - C/\kappa^2}{2\sigma_{r0}\sigma_{r0}'/\kappa} \tag{55}$$

The beam envelope oscillates around the equilibrium radius $\bar{\sigma}_r = \sqrt{C}/\kappa$ with the wavelength $2\pi/\kappa = \pi/k_\beta$, where $\lambda_\beta = 2\pi/k_\beta$ is the betatron wavelength. For the condition $C/\kappa = 2\varepsilon_{n0}/\gamma$ that leads to $\sigma_{r0}^2 = 2\varepsilon_{n0}/\kappa\gamma$ with $\sigma_{r0}' = 0$, the beam propagates at the matched beam radius

$$\sigma_{rM}^2 = \frac{2\varepsilon_{n0}}{\kappa\gamma} = \frac{\varepsilon_{n0}}{k_\beta \gamma} = \frac{\varepsilon_{n0}}{K\sqrt{\gamma}} \approx \frac{r_L \varepsilon_{n0}}{2\sqrt{\gamma}} \left(\frac{E_z}{E_0} \langle \sin \Psi \rangle \right)^{-1/2} \tag{56}$$

Consider the betatron oscillation in the wakefields given by Eqs. (26) and (27), driven by the super-Gaussian pulse. The focusing force is written as

$$F_r = -2n \frac{r^{n-1}}{r_L^n} \frac{E_z}{E_0} \sin k_p \zeta \,, \tag{57}$$

where E_z is the peak amplitude of the accelerating field. For $r \ll r_L$, the equation of betatron oscillation is given by

$$\frac{d^2 x}{dz^2} + \frac{K^2}{\gamma} x^{n-1} = 0 \,, \tag{58}$$

where the focusing constant K^2 is

$$K^2 = \frac{2n}{r_L^n} \frac{E_z}{E_0} \langle \sin k_p \zeta \rangle \tag{59}$$

The envelope equation of the rms beam radius σ_r is given by

$$\frac{d^2 \sigma_r}{dz^2} + \frac{K^2}{\gamma} \sigma_r^{n-1} - \frac{\varepsilon_{n0}^2}{\gamma^2 \sigma_r^3} = 0 \tag{60}$$

Assuming the beam energy γ is constant, Eq. (60) leads to

$$\frac{d^2 \sigma_r^2}{dz^2} + \kappa^2 \sigma_r^n = C \,, \tag{61}$$

where

$$\kappa^2 \equiv \frac{2K^2}{\gamma} \frac{n+2}{n} = \frac{4(n+2)}{\gamma r_L^n} \frac{E_z}{E_0} \langle \sin k_p \zeta \rangle \tag{62}$$

and

$$C = 2\left(\frac{d\sigma_r}{dz} \right)_{z=0}^2 + \frac{4K^2}{n\gamma} \sigma_{r0}^n + \frac{2\varepsilon_{n0}^2}{\gamma^2 \sigma_{r0}^2} \tag{63}$$

With $(d\sigma_r/dz)_{z=0} = 0$, the equilibrium radius is obtained from setting $d^2\sigma_r^2/dz^2 = 0$ as

$$\sigma_{rM}^2 = \left(\frac{\varepsilon_{n0}}{K\sqrt{\gamma}}\right)^{4/n+2} = r_L^{\frac{2n}{n+2}}\left[\frac{\varepsilon_{n0}^2}{2n\gamma}\left(\frac{E_z}{E_0}\right)^{-1}\left\langle \sin k_p \zeta \right\rangle^{-1}\right]^{\frac{2}{n+2}} \tag{64}$$

3.2 Betatron radiation and radiative damping

The synchrotron radiation causes the energy loss of beams and affects the energy spread and transverse emittance via the radiation reaction force. The motion of an electron traveling along z-axis in the accelerating force eE_z and the radial force eE_r from the plasma wave evolves according to

$$\frac{du_x}{cdt} = -K^2 x + \frac{F_x^{\mathrm{RAD}}}{mc^2}, \qquad \frac{du_z}{cdt} = k_p \frac{E_z}{E_0} + \frac{F_z^{\mathrm{RAD}}}{mc^2}, \tag{65}$$

where $\mathbf{F}^{\mathrm{RAD}}$ is the radiation reaction force and $\mathbf{u} = \mathbf{p}/m_e c$ is the normalized electron momentum. The classical radiation reaction force (Jackson, 1999) is given by

$$\frac{\mathbf{F}^{\mathrm{rad}}}{mc\tau_R} = \frac{d}{dt}\left(\gamma \frac{d\mathbf{u}}{dt}\right) + \gamma \mathbf{u}\left[\left(\frac{d\gamma}{dt}\right)^2 - \left(\frac{d\mathbf{u}}{dt}\right)^2\right], \tag{66}$$

where $\gamma = \left(1+u^2\right)^{1/2}$ is the relativistic Lorentz factor of the electron and

$$\tau_R = \frac{2r_e}{3c} \simeq 6.26 \times 10^{-24}\,[\mathrm{s}] \tag{67}$$

Since the scale length of the radiation reaction $c\tau_R$ is much smaller than that of the betatron motion, assumming that the radiation reaction force is a perturbation and $u_z >> u_x$. the equations of motion Eqs. (65) are approximately written as

$$\frac{du_x}{dt} \simeq -cK^2 x - c^2 \tau_R K^2 u_x \left(1 + K^2 \gamma x^2\right), \frac{du_z}{dt} \simeq \omega_p \frac{E_z}{E_0} - c^2 \tau_R K^4 \gamma^2 x^2, \frac{dx}{dt} = \frac{cu_x}{\gamma} \simeq c\frac{u_x}{u_z} \tag{68}$$

Finally the particle dynamics is obtained from the following coupled equations,

$$\frac{d^2 x}{dt^2} + \left(\frac{\omega_p}{\gamma}\frac{E_z}{E_0} + \tau_R c^2 K^2\right)\frac{dx}{dt} + \frac{c^2 K^2}{\gamma} x = 0, \tag{69}$$

and

$$\frac{d\gamma}{dt} = \omega_p \frac{E_z}{E_0} - \tau_R c^2 K^4 \gamma^2 x^2. \tag{70}$$

The particle orbit and the energy are obtained from the coupled equations, Eqs. (69) and (70), describing the single particle dynamics, which can be solved numerically for specified focusing and accelerating fields.

The radiative damping rate is defined by the ratio of the radiated power

$$P_s \simeq \frac{2e^2\gamma^2}{3m^2 c^3} F_\perp^2 \tag{71}$$

to the electron energy as

$$\nu_\gamma = \frac{P_s}{\gamma m c^2} \simeq \frac{\tau_R \gamma}{m^2 c^2} F_\perp^2 \tag{72}$$

For the betatron oscillation of a matched beam in the plasma wave, the damping rate is given by

$$\nu_\gamma \simeq \frac{1}{2}\gamma\tau_R c^2 K^4 \left\langle x^2 \right\rangle \simeq \gamma\tau_R c^2 K^4 \sigma_{r0}^2 \approx 16\gamma\tau_R \omega_p^2 \frac{\left(k_p\sigma_{r0}\right)^2}{\left(k_p r_L\right)^4}\left(\frac{E_z}{E_0}\right)^2 \langle \sin\Psi \rangle^2 . \tag{73}$$

where $\left\langle x^2 \right\rangle = 2\sigma_{x0}^2$ is an average over the beam particles. Assuming the damping time is slow compared to the betatron oscillation

$$\frac{\nu_\gamma}{\omega_\beta} = \frac{\nu_\gamma}{ck_\beta} = \nu_\gamma \frac{\sqrt{\gamma}}{cK} = \tau_R \gamma^{\frac{3}{2}} c K^3 \sigma_{r0}^2 = \tau_R \omega_p \gamma^{\frac{3}{2}} \left(\frac{K}{k_p}\right)^3 \left(k_p \sigma_{r0}\right)^2 \ll 1 , \tag{74}$$

and $\omega_p E_z/E_0 \gg \gamma\tau_R c^2 K^2$ the analytical expression for the mean energy and the relative energy spread are obtained by solving Eqs. (69) and (70) with the initial energy γ_0 and the initial energy spread $\sigma_{\gamma 0}$ (Michel et al., 2006),

$$\gamma = \gamma_0 + \omega_p t \frac{E_z}{E_0} + \frac{2}{5}\gamma_0^2 \frac{\nu_\gamma}{\omega_p}\frac{E_0}{E_z}\left[1 - \left(1 + \frac{\omega_p t}{\gamma_0}\frac{E_z}{E_0}\right)^{5/2}\right], \tag{75}$$

and

$$\frac{\sigma_\gamma^2}{\gamma^2} = \frac{\sigma_{\gamma 0}^2}{\gamma_0^2}\left(1 + \frac{E_z}{E_0}\frac{\omega_p t}{\gamma_0}\right)^{-2}\left\{1 + \frac{4}{25}\left(\frac{\nu_\gamma \gamma_0}{\omega_p}\right)^2 \left(\frac{\sigma_{\gamma 0}}{\gamma_0}\right)^{-2}\left(\frac{E_z}{E_0}\right)^{-2}\left[1 - \left(1 + \frac{E_z}{E_0}\frac{\omega_p t}{\gamma_0}\right)^{5/2}\right]^2\right\}. \tag{76}$$

For $\left(\omega_p t / \gamma_0\right)\left(E_z / E_0\right) << 1$, Eq. (76) is approximated to

$$\frac{\sigma_\gamma^2}{\gamma^2} \simeq \left(1 - 2\frac{E_z}{E_0}\frac{\omega_p t}{\gamma_0}\right)\left(\frac{\sigma_{\gamma 0}^2}{\gamma_0^2} + \nu_\gamma^2 t^2\right). \tag{77}$$

Initially the energy spread decreases linearly with time due to acceleration and for later times, $t^2 > \sigma_{\gamma 0}^2 / \left(\nu_\gamma^2 \gamma_0^2\right)$, the energy spread increases due to the radiative effects. The total energy loss due to the synchrotron radiation is given by

$$\Delta\gamma_{\mathrm{RAD}} \simeq \frac{2}{5}\gamma_0^{-\frac{1}{2}}\gamma_f^{\frac{5}{2}}\frac{v_\gamma}{\omega_p}\left(\frac{E_z}{E_0}\right)^{-1} \simeq \frac{32}{5}\tau_R\omega_p\gamma_0^{\frac{1}{2}}\gamma_f^{\frac{5}{2}}\frac{E_z}{E_0}\frac{\left(k_p\sigma_{r0}\right)^2}{\left(k_p r_L\right)^4}\langle\sin\Psi\rangle^2$$
$$\approx 7.2\times10^{-10}\gamma_0^{\frac{1}{2}}\gamma_f^{\frac{5}{2}}\langle\sin\Psi\rangle^2\frac{E_z}{E_0}\frac{\left(k_p\sigma_{r0}\right)^2}{\left(k_p r_L\right)^4}\left(\frac{n_p}{10^{17}\left[\mathrm{cm}^{-3}\right]}\right)^{1/2}, \tag{78}$$

where γ_f is the final beam energy and

$$\tau_R\omega_p = \frac{2r_e}{3}k_p \approx 1.12\times10^{-10}\left(\frac{n_e}{10^{17}\left[\mathrm{cm}^{-3}\right]}\right)^{1/2}$$

The energy spread at γ_f becomes

$$\frac{\sigma_\gamma^2}{\gamma^2} \simeq \frac{\sigma_{\gamma0}^2}{\gamma_0^2}\left(\frac{\gamma_f}{\gamma_0}\right)^{-2} + \frac{1024}{25}\tau_R^2\omega_p^2\gamma_0\gamma_f^3\frac{\left(k_p\sigma_{r0}\right)^4}{\left(k_p r_L\right)^8}\left(\frac{E_z}{E_0}\right)^2\langle\sin\Psi\rangle^4. \tag{79}$$

Assuming the first term that means an adiabatic decrease of the energy spread is neglected in comparison with a radiative increase given by the second term, the energy spread leads to

$$\left(\frac{\sigma_\gamma}{\gamma_f}\right)_{\mathrm{RAD}} \simeq \frac{32}{5}\tau_R\omega_p\gamma_0^{\frac{1}{2}}\gamma_f^{\frac{3}{2}}\frac{E_z}{E_0}\frac{\left(k_p\sigma_{r0}\right)^2}{\left(k_p r_L\right)^4}\langle\sin\Psi\rangle^2$$
$$\approx 7.2\times10^{-10}\gamma_0^{\frac{1}{2}}\gamma_f^{\frac{3}{2}}\langle\sin\Psi\rangle^2\frac{E_z}{E_0}\frac{\left(k_p\sigma_{r0}\right)^2}{\left(k_p r_L\right)^4}\left(\frac{n_p}{10^{17}\left[\mathrm{cm}^{-3}\right]}\right)^{1/2}. \tag{80}$$

4. Design considerations on a laser-plasma X-ray FEL

4.1 A design example for 6 GeV LPA-driven X-ray FEL

4.1.1 Requirements for emittance and energy spread

The SASE FEL driven by an electron beam with energy γ requires the transverse normalized emittance

$$\varepsilon_n < \gamma\lambda_X, \tag{81}$$

where λ_X is a FEL wavelength of radiation from the undulator with period λ_u,

$$\lambda_X = \frac{\lambda_u}{2\gamma^2}\left(1+\frac{K_u^2}{2}\right), \tag{82}$$

and K_u is the undulator parameter with the magnetic field strength on the undulator axis B_u,

$$K_u = 0.934\lambda_u[\text{cm}]B_u[\text{T}]. \tag{83}$$

For lasing a hard X-ray region $\lambda_X \approx 0.1$ nm (the photon energy $E_{\text{photon}} = 12.4$ keV) from the undulator of $\lambda_u = 2$ cm with the magnetic field of $B_u = 0.5$ T ($K_u = 0.934$) at the beam energy $E_b = 6$ GeV ($\gamma = 1.2 \times 10^4$), the normalized emittance should be $\varepsilon_n < 1.2\ \mu\text{m rad}$. In addition , it is essential for SASE FELs to inject electron beams with a very high peak current of the order of 100 kA. This requirement imposes a charge of ~ 200 pC on the LPA design in the case of accelerated bunch length of 2 fs. Eq. (36) determines the operating plasma density to be $n_p \sim 10^{17}$ cm^{-3} for $k_p\sigma_r = 1$ and $\eta_b \approx 0.5$. The matched beam condition Eq. (56) for a given initial normalized emittance ε_{n0} imposes an allowable linear focusing strength,

$$\frac{K}{k_p} = \frac{2}{k_p r_L}\left(\frac{E_z}{E_0}\right)^{1/2} \sqrt{\langle \sin\Psi \rangle} \le \frac{k_p \varepsilon_{n0}}{k_p^2 \sigma_{r0}^2 \sqrt{\gamma_0}} \approx 4.25\times 10^{-3} \frac{1}{k_p^2\sigma_{r0}^2} \frac{\varepsilon_{n0}}{1[\mu\text{m}]} \left(\frac{100[\text{MeV}]}{E_i}\right)^{1/2} \left(\frac{n_p}{10^{17}[\text{cm}^{-3}]}\right)^{1/2}, \tag{84}$$

where E_i is the injection energy. For $k_p\sigma_{r0} = 1$, $E_i = 100$ MeV and $n_p \sim 10^{17}$ cm^{-3} , a given emittance $\varepsilon_{n0} \approx 1\mu\text{m rad}$ leads to $K/k_p \le 5\times 10^{-3}$, which limits the injection phase to be

$$0 \le \Psi_0 \le \sin^{-1}\left[6.25\times 10^{-6}\left(k_p r_L\right)^2 \left(\frac{E_z}{E_0}\right)^{-1}\right] \tag{85}$$

For $k_p r_L = 3$ and $E_z/E_0 = 0.5$, the injection phase angle should be $0 \le \Psi_0 \le 0.006°$. In the case of the mismatched beam with a finite energy spread, different particles will undergo betatron oscillations at different frequencies, $\omega_\beta \propto \gamma^{-1/2}$, which will lead to decoherence that is a slippage of the particles with respect to each other, and then to emittance growth until the emittance reaches the matched value. This emittance growth rate (Michel et al., 2006) is given by

$$\nu_\varepsilon = \frac{\omega_\beta \sigma_\gamma}{\sqrt{8}\gamma} \simeq \frac{Kc}{\sqrt{8\gamma}} \frac{\sigma_\gamma}{\gamma}. \tag{86}$$

Without the betatron radiation, the evolution of the transverse emittance can be approximately calculated from

$$\varepsilon_n^2 \approx \gamma_0 K^2 \sigma_{r0}^4 \left(1 - e^{-8\nu_\varepsilon^2 t^2}\right). \tag{87}$$

For $8\nu_\varepsilon^2 t^2 \ll 1$,

$$\varepsilon_n \approx 2\sqrt{2}\sigma_{r0}^2 K \gamma_0^{1/2} \nu_\varepsilon t = k_p^2\sigma_{r0}^2 \left(\frac{K}{k_p}\right)^2 \left(\frac{\gamma_0}{\gamma}\right)^{1/2} \frac{\sigma_\gamma}{\gamma} ct = \frac{4}{k_p}\left(\frac{k_p\sigma_{r0}}{k_p r_L}\right)^2 \left(\frac{\gamma_0}{\gamma}\right)^{1/2} \frac{E_z}{E_0}\frac{\sigma_\gamma}{\gamma} k_p z \sin\Psi, \tag{88}$$

where z is the acceleration distance and $\Psi = k_p\left(z - v_p t\right) + \Psi_0$ is the aceleration phase. Considering the energy spread,

$$\frac{\sigma_\gamma}{\gamma} = \left(\frac{\sigma_\gamma}{\gamma}\right)_0 \frac{\gamma_0}{\gamma}, \text{ and } \gamma = \gamma_0 + k_p z \frac{E_z}{E_0},$$

the normalized emittance becomes

$$\varepsilon_n \approx \frac{4\gamma_0}{k_p}\left(\frac{k_p\sigma_{r0}}{k_p r_L}\right)^2 \left(\frac{\sigma_\gamma}{\gamma}\right)_0 \left(\frac{\gamma_0}{\gamma}\right)^{1/2}\left(1 - \frac{\gamma_0}{\gamma}\right)\sin\Psi . \tag{89}$$

Setting $\varepsilon_n \le \varepsilon_{n0}$ at $z = L_{\text{stage}}$ and $\gamma = \gamma_f$, the initial relative energy spread is limited to

$$\left(\frac{\sigma_\gamma}{\gamma}\right)_{0G} \le \frac{k_p\varepsilon_{n0}}{4\gamma_0 \sin\Psi_{\max}}\left(\frac{k_p r_L}{k_p\sigma_{r0}}\right)^2\left(\frac{\gamma_f}{\gamma_0}\right)^{1/2}$$

$$\approx \frac{0.8\times 10^{-4}}{\sin\Psi_{\max}}\left(\frac{k_p r_L}{k_p\sigma_{r0}}\right)^2\left(\frac{E_b}{E_i}\right)^{1/2}\frac{\varepsilon_{n0}}{1[\mu\text{m}]}\left(\frac{100[\text{MeV}]}{E_i}\right)^{1/2}\left(\frac{n_p}{10^{17}\left[\text{cm}^{-3}\right]}\right)^{1/2} . \tag{90}$$

For $k_p\sigma_{r0} = 1$, $k_p r_L = 3$, $E_b = 6$ GeV and $E_i = 100$ MeV, the allowable energy spread scales as

$$\left(\frac{\sigma_\gamma}{\gamma}\right)_{0G} \le 5.5\times 10^{-3}\frac{\varepsilon_{n0}}{1[\mu\text{m}]}\left(\frac{n_p}{10^{17}\left[\text{cm}^{-3}\right]}\right)^{1/2} . \tag{91}$$

For the super-Gaussian drive pulse, as the betatron frequency is dependent on the oscillation amplitude as

$$\omega_\beta = ck_\beta = \frac{cK}{\sqrt{\gamma}}\sigma_r^{\frac{n-2}{2}}, \tag{92}$$

with the matched beam radius Eq. (64), the normalized emittance is

$$\varepsilon_{nSG} \simeq \sigma_{r0}^n K^2\left(\frac{\gamma_0}{\gamma}\right)^{1/2}\frac{\sigma_\gamma}{\gamma}ct = \frac{2n\gamma_0}{k_p}\left(\frac{k_p\sigma_{r0}}{k_p r_L}\right)^n\left(\frac{\sigma_\gamma}{\gamma}\right)_0\left(\frac{\gamma_0}{\gamma}\right)^{1/2}\left(1-\frac{\gamma_0}{\gamma}\right)\sin\Psi . \tag{93}$$

For $n = 2$, Eq. (93) becomes the Gaussian pulse case given by Eq. (89). The allowable energy spread for the super-Gaussian pulse is given by

$$\left(\frac{\sigma_\gamma}{\gamma}\right)_{0SG} \le \frac{k_p\varepsilon_{n0}}{2n\gamma_0}\left(\frac{k_p r_L}{k_p\sigma_{r0}}\right)^n\left(\frac{\gamma_f}{\gamma_0}\right)^{1/2} \approx \frac{1.52\times 10^{-4}}{n}\left(\frac{k_p r_L}{k_p\sigma_{r0}}\right)^n\left(\frac{E_b}{E_i}\right)^{1/2}\frac{\varepsilon_{n0}}{1[\mu\text{m}]}\left(\frac{100[\text{MeV}]}{E_i}\right)^{1/2}\left(\frac{n_p}{10^{17}\left[\text{cm}^{-3}\right]}\right)^{1/2} \tag{94}$$

For $k_p\sigma_{r0}=1$, $k_p r_L=3$, $E_b=6$ GeV, $E_i=100$ MeV and $n=4$ (6),

$$\left(\frac{\sigma_\gamma}{\gamma}\right)_{\mathrm{OSG}} \le 2\,(13)\times 10^{-2}\frac{\varepsilon_{n0}}{1[\mu\mathrm{m}]}\left(\frac{n_p}{10^{17}\left[\mathrm{cm}^{-3}\right]}\right)^{1/2} \tag{95}$$

The super-Gaussian wakefields mitigate the emittance growth due to mismatching of the injected beam.

In order to reach the X-ray wavelength $\lambda_X=0.1$ nm using the undulator with $\lambda_u=2$ cm, the maximum beam energy E_b is set to be 6 GeV. The accelerator stage length becomes $L_{\mathrm{stage}}=0.4\ \mathrm{m}\approx 0.43L_{\mathrm{dp}}$ where the accelerating field is $E_z=15\ \mathrm{GV/m}$ at the operating plasma density $n_p=1\times 10^{17}\ \mathrm{cm}^{-3}$. Setting the injection phase $\Psi_0\simeq 0^\circ$, the acceleration distance corresponds to the dephasing phase

$$0^\circ \le \Psi \le \frac{\pi}{2}\frac{L_{\mathrm{stage}}}{L_{dp}} \approx 39^\circ\left(\frac{\lambda_L}{0.8[\mu\mathrm{m}]}\right)^2\left(\frac{n_p}{10^{17}\left[\mathrm{cm}^{-3}\right]}\right)^{3/2}. \tag{96}$$

Scanning the electron beam energy in the range from 1.2 GeV to 6 GeV can allow the FEL to cover the X-ray wavelengths $0.1\ \mathrm{nm}\le\lambda_X\le 2.5\ \mathrm{nm}$ ($496\ \mathrm{eV}\le E_{\mathrm{photon}}\le 12.4\ \mathrm{keV}$).

The FEL operation is characterized by the FEL Pierce parameter (Bonifacio, 1984),

$$\rho_{\mathrm{FEL}}=\frac{1}{2\gamma}\left[\frac{I_b}{I_A}\left(\frac{\lambda_u K_u A_u}{2\pi\sigma_r}\right)^2\right]^{1/3}, \tag{97}$$

where I_b is the beam current, $I_A=17$ kA is the Alfven current, σ_r is the r.m.s transverse size of the electron bunch, and the coupling factor is $A_u=1$ for a helical undulator and $A_u=J_0(\xi)-J_1(\xi)$ for a planar undulator, where $\xi=K_u^2/4\left(1+K_u^2/2\right)$ and J_0 and J_1 are the Bessel functions of the first kind. For $\lambda_u=2$ cm, $K_u=0.934$ and $A_u\approx 0.92$, assuming the average beta function in the undulator $\overline{\beta}_u=1$ m, the FEL parameter can be obtained as

$$\rho_{\mathrm{FEL}}\simeq 0.0032\left(\frac{I_b}{100[\mathrm{kA}]}\right)^{1/3}\left(\frac{E_b}{6[\mathrm{GeV}]}\right)^{-2/3}\left(\frac{\varepsilon_n}{1[\mu\mathrm{m}]}\right)^{-1/3}. \tag{98}$$

The gain length L_{gain} that is the *e*-folding length of the exponential amplification of the radiation power is

$$L_{\mathrm{gain}}=\frac{\lambda_u}{2\pi\sqrt{3}\rho_{\mathrm{FEL}}} \tag{99}$$

The saturation length is set to be $L_{\mathrm{sat}}\sim(10-15)L_{\mathrm{gain}}\approx\lambda_u/\rho_{\mathrm{FEL}}$, at which the saturation power is approximately $P_{\mathrm{sat}}\sim\rho_{\mathrm{FEL}}I_bE_b$. Accordingly the spectral band width is

$1/N_u \sim \rho_{\text{FEL}}$, where N_u is the number of undulator periods. Assuming a bunch duration of $\tau_b \approx 2$ fs, the beam current scales as

$$I_b = \frac{Q_b}{\tau_b} \approx 58[\text{kA}]\left(\frac{n_p}{10^{17}\left[\text{cm}^{-3}\right]}\right)^{-1/2}, \tag{100}$$

for $k_p\sigma_r = 1$ and $E_z/E_0 = 0.5$. With $L_{\text{stage}} = 0.4$ m, the beam energy scales as

$$E_b \approx 6[\text{GeV}]\left(\frac{n_p}{10^{17}\left[\text{cm}^{-3}\right]}\right)^{1/2}. \tag{101}$$

The required emittance for SASE FEL is

$$\varepsilon_n \simeq \gamma\lambda_X = \frac{\lambda_u}{2\gamma}\left(1+\frac{K_u^2}{2}\right) \approx 1.2[\mu\text{m}]\left(\frac{n_p}{10^{17}\left[\text{cm}^{-3}\right]}\right)^{-1/2}, \tag{102}$$

for $\lambda_u = 2$ cm, $K_u = 0.934$. Accordingly the FEL parameter scales as

$$\rho_{\text{FEL}} \simeq 0.0025\left(\frac{n_p}{10^{17}\left[\text{cm}^{-3}\right]}\right)^{-1/3}, \tag{103}$$

and the relative energy spread requirement for the SASE FEL is given as

$$\frac{\sigma_\gamma}{\gamma} \le \rho_{\text{FEL}}. \tag{104}$$

4.1.2 Numerical studies of betatron radiation effects

According to Eq. (80), we estimate the energy spread growth due to the betatron radiation from the electron beam accelerated in the wakefields from the injection energy $E_i = 100$ MeV ($\gamma_i \simeq 200$) to $E_b = 6$ GeV ($\gamma_f = 1.2\times10^4$) for $E_z/E_0 = 0.5$, $k_p\sigma_{r0} = 1$, $k_p r_L = 3$ and $\langle\sin\Psi\rangle = 1/2$ as

$$\left(\frac{\sigma_\gamma}{\gamma_f}\right)_{\text{RAD}} \approx 2\times10^{-5}\left(\frac{n_p}{10^{17}\left[\text{cm}^{-3}\right]}\right)^{1/2}, \tag{105}$$

which is much smaller than ρ_{FEL}. For avoiding the normalized emittance growth due to the betatron oscillation, setting the initial energy spread $(\sigma_\gamma/\gamma)_0 \approx 1$ % adiabatically decreases to be $(\sigma_\gamma/\gamma)_f \approx 0.02$ % $< \rho_{\text{FEL}}$ after accelerated up to 6 GeV.

The degradation of the energy spread and the emittance due to betatron radiation effects is investigated by solving the coupled equations, Eqs. (69) and (70), describing the single particle dynamics. We have solved them numerically for the case of the aforementioned 6 GeV LPA. Using the numerical results for a set of test particles that can be solved for the initial conditions corresponding to the initial energy, energy spread and transverse emittance, an estimate of the underlying beam parameter can be calculated as an ensemble average over test particles; for example, the mean energy is given by

$$\langle\gamma\rangle = \sum_i \gamma_i / N_p , \tag{106}$$

where γ_i is the energy of the i th particle and N_p is the number of test particles, and the energy spread is defined as

$$\sigma_\gamma^2 = \langle\gamma^2\rangle - \langle\gamma\rangle^2 . \tag{107}$$

The normalized transverse emittance is calculated as

$$\varepsilon_{nx}^2 = \langle(x-\langle x\rangle)^2\rangle\langle(u_x-\langle u_x\rangle)^2\rangle - \langle(x-\langle x\rangle)(u_x-\langle u_x\rangle)\rangle^2 , \tag{108}$$

where $u_x = \gamma dx / cdt$, with averaging over the ensemble of particles. The single particle equations of motion, Eqs. (69) and (70), are integrated numerically using the Runge-Kutta algorithm. We have carried out numerical calculations for an ensemble of $N_p = 10^4$ particles for the parameters of the 6 GeV LPA operated at $n_p = 1\times10^{17}$ cm^{-3} as shown in Table 1.

The results of our numerical calculations are shown in Fig. 1 together with the analytical estimates for $\langle\gamma\rangle$ and $\sigma_\gamma / \langle\gamma\rangle$, calculated from Eqs. (75) and (76). The numerical calculations are in excellent agreement with the analytical expressions. The relative energy spread in the beginning decreases linearly in time due to the linear increase of the zeroth order mean energy. For the numerical calculations the final value is $(\sigma_\gamma / \langle\gamma\rangle)_{\text{numerical}} = 1.7\times10^{-4}$ at the end of the stage, while the analytical estimate calculates $(\sigma_\gamma / \langle\gamma\rangle)_{\text{analytical}} = 1.7\times10^{-4}$. The $\langle\gamma\rangle$ increases almost linearly in time and reaches a final value of 1.2×10^4 for both the analytical and the numerical calculations after the LPA stage, which corresponds to an electron energy of 6 GeV. Since the radiative effects are negligibly small as estimated from Eq. (105), the emittance is well conserved over the LPA stage, where the matched beam is injected. In Fig. 1, an analytical estimate of the emittance evolution is given by

$$\begin{aligned}\Omega(\gamma) = \Omega_0 \exp[-(\frac{\gamma}{\gamma_{\text{ld}}} - \frac{\gamma_0}{\gamma_{\text{ld}}}) - e^{\frac{\gamma_0}{\gamma_{ld}}}[(\frac{\gamma}{\gamma_{\text{nd}}})^{\frac{3}{2}} - (\frac{\gamma_0}{\gamma_{\text{nd}}})^{\frac{3}{2}}]] \\ \{1 + \frac{1}{4\alpha\sqrt{\chi\gamma_0}}(\frac{2\gamma_0}{\gamma_{\text{ld}}} + \frac{3}{\chi}(\frac{\gamma_0}{\gamma_{\text{nd}}})^{\frac{3}{2}})(1 - e^{-4\alpha\chi^{\frac{1}{2}}(\sqrt{\gamma}-\sqrt{\gamma_0})})\}\end{aligned} , \tag{109}$$

where $\Omega(\gamma) = k_p\varepsilon_{nx}$ is the dimensionless normalized emittance at γ , $\Omega_0 = \Omega(\gamma_0)$, $\chi \equiv E_z/E_0$, $K/k_p = \alpha\chi^{1/2}$, γ_{ld} is the linear damping energy, at which the emittance damps to $\Omega/\Omega_0 \sim 1/e$,

$$\gamma_{\rm ld} = \frac{2}{\tau_R \omega_p \alpha^2} = \left(\frac{2\sqrt{\pi}}{3} \alpha^2 r_e^{\frac{3}{2}} \sqrt{n_p} \right)^{-1} \approx 1.8 \times 10^{10} \frac{E_z}{E_0} \left(\frac{K}{k_p} \right)^{-2} \left(\frac{n_p}{10^{17} \left[\mathrm{cm}^{-3} \right]} \right)^{-1/2}, \quad (110)$$

and $\gamma_{\rm nd}$ is the nonlinear damping energy, given by

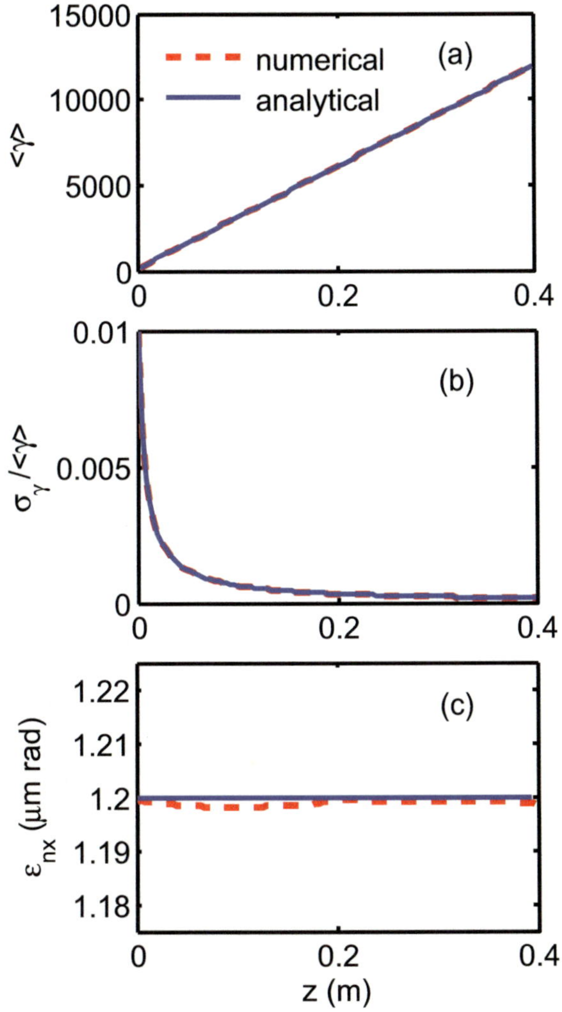

Fig. 1. The results of the beam dynamics calculations for the 6 GeV LPA at the operating plasma density $n_p = 1 \times 10^{17}$ cm^{-3}: (a) mean energy, (b) relative energy spread and (c) normalized transverse emittance over the stage length 0.4 m for the initial energy 100 MeV, initial energy spread of 1%, initial normalized emittance 1.2 μm rad, and constant accelerating field $E_z = 15$ GV/m . The dashed curves show the numerical calculations, while the solid curves show the analytical expressions Eqs. (75) and (76) for mean energy and relative energy spread, and Eq. (109) for normalized emittance, respectively.

$$\gamma_{\rm nd} \simeq \frac{1}{\alpha^2}\left(\frac{1}{2\tau_R\omega_p\chi^{1/2}\Omega_0}\right)^{2/3} = \frac{1}{\alpha^2\chi^{1/3}}\left(\frac{3}{16\pi r_e^2 n_p \varepsilon_{n0}}\right)^{2/3} \approx 3.84\times10^6\left(\frac{E_z}{E_0}\right)^{-1/3}\left(\frac{K}{k_p}\right)^{-2}\left(\frac{n_p}{10^{18}\left[\text{cm}^{-3}\right]}\frac{\varepsilon_{n0}}{1[\mu\text{m}]}\right)^{-2/3}. \quad (111)$$

For the given parameters $\chi = 0.5$, $\alpha = 0.37$ and $\varepsilon_{n0} = 1.2\ \mu\text{m rad}$ at $n_p = 10^{17}\ \text{cm}^{-3}$, the linear damping energy is $\gamma_{\rm ld} \sim 1.3\times10^{11}$ (66 PeV), while the nonlinear damping energy is $\gamma_{\rm nd} \sim 1.5\times10^8$ (75 TeV).

4.1.3 Attainable peak brilliance of the laser-driven X-ray FEL

In the saturation regime, the photons flux of X-ray radiation is

$$N_{\rm photon} \approx \rho_{\rm FEL}\frac{E_b}{E_{\rm photon}}\frac{I_b}{e} \approx 4.4\times10^{26}\left[\text{s}^{-1}\right] \quad (112)$$

for $E_b = 6\ \text{GeV}$, $E_{\rm phton} = 12.4\ \text{keV}$ ($\lambda_X = 0.1\ \text{nm}$) , $\rho_{\rm FEL} = 0.0025$ and $I_b = 58\ \text{kA}$. Thus the peak brilliance can be obtained as

$$B_{\rm peak} = \frac{N\gamma^2}{4\pi^2\varepsilon_n^2}\left(\frac{\Delta\lambda}{\lambda}\right)^{-1} \approx 4.4\times10^{32}\left[\text{photons/s mm}^2\ \text{mrad}^2\ 0.1\%\text{BW}\right]. \quad (113)$$

This peak brilliance is comparable to large-scale X-ray FELs based on the conventional linacs (Ackermann et al., 2007).

The parameters of the required drive laser pulse can be obtained from Eq. (47)-(50) for $a_0 = 1.4$, $k_p\sigma_L = 1$, $k_p r_L = 3$ and $n_p = 1\times10^{17}\ \text{cm}^{-3}$ as follows: the laser spot radius is $r_L = 51\ \mu\text{m}$, the peak laser power is $P_L = 171\ \text{TW}$, the pulse duration is $\tau_L = 93\ \text{fs}$ and the laser pulse energy is $U_L \approx 16\ \text{J}$ at the laser wavelength $\lambda_L = 0.8\ \mu\text{m}$. It is necessary for laser-plasma accelerators to propagate ultraintense laser pulses with peak power of the order of 200 TW over the single stage distance of the order of 0.4 m at repetition rate of 10 Hz. Stably propagating laser pulses through the plasma channel with a parabolic density profile

$$n(r) = n_p + \Delta n\frac{r^2}{r_{ch}^2}, \quad (114)$$

where Δn is the density depth at the channel radius r_{ch}, requires that its spot radius r_L should be equal to the matched radius

$$r_{LM} = \left(\frac{r_{ch}^2}{\pi r_e \Delta n}\right)^{1/4}. \quad (115)$$

For this condition, the density depth of plasma channel is given by

$$\frac{\Delta n}{n_p} = \frac{\Delta n_c}{n_p}\left(\frac{r_{ch}}{r_L}\right)^2 = \frac{4}{k_p^2 r_L^2}\left(\frac{r_{ch}}{r_L}\right)^2, \tag{116}$$

where $\Delta n_c = 1/\left(\pi r_e r_L^2\right)$ is the critical channel depth (Sprangle, 1992).

The design parameters of a table-top X-ray FEL driven by the 6 GeV LPA are summarized for a wavelength of $\lambda_X = 0.1$ nm in Table 1.

Electron beam parameters	
Beam energy E_b	6 GeV
Peak beam current I_b	58 kA
Energy spread (rms) σ_γ/γ	0.02%
Pulse duration τ_b	2 fs
Normalized emittance ε_n	1.2 mm mrad
Laser-plasma accelerator parameters	
Plasma density n_p	1×10^{17} cm^{-3}
Plasma wavelength λ_p	106 μm
Channel depth at r_L $\Delta n_c/n_p$	0.44
Accelerating field E_z	15 GV/m
Injection beam energy E_i	100 MeV
Stage length L_{stage}	40 cm
Charge per bunch Q_b	116 pC
Laser wavelength λ_L	0.8 μm
Normalized laser intensity a_0	1.4
Laser pulse duration τ_L	93 fs
Laser spot radius r_L	51 μm
Laser peak power P_L	171 TW
Laser pulse energy U_L	16 J
Undulator parameters	
Undulator period λ_u	2 cm
Undulator parameter K_u	0.934
FEL parameter ρ_{FEL}	0.0025
Gain length L_{gain}	73.5 cm
Saturation length L_{sat}	8 m
X-ray parameters	
Wavelength λ_X	0.1 nm ($E_{phton} = 12.4$ keV)
Photon flux N_{photon}	4.4×10^{26} s^{-1}
Peak brilliance B_{peak}	4.4×10^{32} [photons/s mm^2 mrad^2 0.1%BW]

Table 1. The design parameters of a table-top X-ray FEL driven by a laser-plasma-based accelerator.

4.2 A high-quality electron beam injector

4.2.1 Self-injection in the bubble regime

Most of LPA experiments that successfully demonstrated the production of quasi-monoenergetic electron beams with narrow energy spread have been elucidated in terms of self-injection and acceleration mechanism in the bubble regime (Kostyukov et al. 2004; Lu et al., 2006). In these experiments, electrons are self-injected into a nonlinear wake, often referred to as a bubble, i.e. a cavity void of plasma electrons consisting of a spherical ion column surrounded with a narrow electron sheath, formed behind the laser pulse instead of a periodic plasma wave in the linear regime. Plasma electrons radially expelled by the radiation pressure of the laser form a sheath with thickness of the order of the plasma skin depth c / ω_p outside the ion sphere remaining unshielded behind the laser pulse moving at relativistic velocity so that the cavity shape should be determined by balancing the Lorentz force of the ion sphere exerted on the electron sheath with the ponderomotive force of the laser pulse. This estimates the bubble radius R_B matched to the laser spot radius r_L, approximately as

$$k_p R_B \approx k_p r_L \approx 2\sqrt{a_0} , \qquad (117)$$

for which a spherical shape of the bubble is created. This condition is reformulated as

$$a_0 \simeq 2\left(\frac{P}{P_c}\right)^{1/3} , \qquad (118)$$

where $P_c \simeq 17(\omega_0^2 / \omega_p^2)$ GW is the critical power for the relativistic self-focusing (Lu et al., 2006). The electromagnetic fields inside the bubble is obtained from the wake potential of the ion sphere moving at the velocity v_B as

$$\frac{eE_z}{mc\omega_p} = -\frac{1}{2}k_p\zeta \text{ , and } \frac{eE_r}{mc\omega_p} = -\frac{1}{2}k_p r , \qquad (119)$$

where $\zeta = z - v_B t$ is the coordinate in the moving frame of the bubble and r the radial coordinate with respect to the laser propagation axis (Kostyukov et al. 2004). One can see that the maximum accelerating field is given by $eE_{z\max} = mc^2 k_p^2 R_B / 2$ at the back of the bubble and the focusing force is acting on an electron inside the bubble. Assuming the bubble phase velocity is given by

$$v_B \approx v_g - v_{\text{etch}} \approx c\left[1 - \left(\frac{1}{2} + 1\right)\frac{\omega_p^2}{\omega_L^2}\right] , \qquad (120)$$

where $v_{\text{etch}} \approx c\omega_p^2 / \omega_L^2$ is the velocity at which the laser front etches back due to the local pump depletion, the dephasing length leads to

$$L_{\text{dp}} \approx \frac{c}{c - v_B} R_B \approx \frac{2}{3}\frac{\omega_L^2}{\omega_p^2} R_B \approx \frac{2}{3}\frac{n_c}{n_p} R_B . \qquad (121)$$

Hence the electron injected at the back of the bubble can be accelerated up to the energy

$$W \approx \frac{1}{2} eE_{z\max} L_{dp} \approx \frac{1}{6} mc^2 \left(k_p R_B\right)^2 \left(\frac{\omega_L}{\omega_p}\right)^2 \approx \frac{2}{3} mc^2 a_0 \frac{n_c}{n_p}. \tag{122}$$

Using the matched bubble radius condition, the energy gain is approximately given by

$$W \approx mc^2 \left(\frac{P}{P_r}\right)^{1/3} \left(\frac{n_c}{n_p}\right)^{2/3}, \tag{123}$$

where $P_r = m^2c^5 / e^2 \approx 8.72$ GW (Lu et al., 2007).

The 2D and 3D particle-in-cell simulations confirm that quasi-monoenergetic electron beams are produced due to self-injection of plasma electrons at the back of the bubble from the electron sheath outside the ion sphere as the laser intensity increases to the injection threshold. As expelled electrons flowing the sheath are initially decelerated backward in a front half of the bubble and then accelerated in a back half of it toward the propagation axis by the accelerating and focusing forces of the bubble ions, their trajectories concentrate at the back of the bubble to form a strong local density peak in the electron sheath and a spiky accelerating field. Eventually the electron is trapped into the bubble when its velocity reaches the group velocity v_g of the laser pulse. The trapping cross section (Kostyukov et al. 2004)

$$\sigma_{\text{trapping}} \simeq \frac{2\pi}{k_p^3 d} \left(\ln \frac{k_p R_B}{\sqrt{8}}\right)^{-1}, \tag{124}$$

with the sheath width d imposes $k_p R_B \approx 2\sqrt{a_0} \geq \sqrt{8}$, i.e. $a_0 \geq 2$ for the matched bubble radius. Once an electron bunch is trapped in the bubble, loading of trapped electrons reduces the wakefield amplitude below the trapping threshold and stops further injection. Consequently the trapped electrons undergo acceleration and bunching process within a separatrix on the phase space of the bubble wakefield. This is a simple scenario for producing high-quality monoenergetic electron beams in the bubble regime.

However, in most of laser-plasma experiments aforementioned conditions and scenarios are not always fulfilled. In the experiment for the plasma density $n_p = (1-2) \times 10^{19}$ cm^{-3}, observation of the self-injection threshold on the normalized laser intensity gives $a_{0\text{th}} \approx 3.2$ after accounting for self-focusing and self-compression that occur during laser pulse propagation in the plasma. In terms of the laser peak power

$$\frac{P}{P_c} = \frac{\pi^2}{8} \frac{a_0^2 r_L^2}{\lambda_p^2}, \tag{125}$$

the self-injection threshold for the power $(P / P_c)_{th} \approx 12.6$ as the laser spot size reduces to the plasma wavelength due to the relativistic self-focusing (Mangles et. al, 2007). In the experiment at $n_p = (3-5)\times 10^{18}$ cm^{-3}, the self-injection threshold is $(P/P_c)_{\text{th}} = 3$, corresponding to $a_{0\text{th}} = 1.6$ (Froula et al., 2009).

4.2.2 A design example of self-injection bubble-regime LPA

We study the production of high-quality electron beams by means of the particle-in-cell (PIC) simulations for the self-injection. We have confirmed the qualities of accelerated electron beams with the r.m.s. energy spread less than 1%, the normalized transverse emittance of the order of a few π mm mrad and the r.m.s. bunch duration of the order of 1 fs. These parameters can satisfy the criteria of the electron beam injector that are required for X-ray FELs.

The self-injection electron beam production has been investigated by the 2D PIC simulation code VORPAL (Nieter et al., 2004), using the 2D moving window, of which the size is $83.5 \times 120\ \mu\text{m}^2$ and the number of simulation cells is 1472×320, assuming H^+ immobile ions and 4 electrons per simulation cells. The laser pulse of the wavelength $\lambda_L = 0.8\ \mu$m is focused on a spot size $r_L = 20\ \mu$m, so that the peak normalized vector potential becomes $a_0 = 3$. Initially the laser pulse is located at $z = 0$ mm and after approximately $t = 1.7$ ps, it moves at the focal point $z = 0.5$ mm distant from the plasma edge. The transverse electron density forms a parabolic radial profile, given by Eq. (114). In this simulation, we set the on-axis plasma density and the density depth to be $n_p = 2\times 10^{18}$ cm^{-3} and $\Delta n = 0.3 n_p$, respectively. The longitudinal electron density and H^+ ion density of the plasma increase along with laser propagating axis from $z = 0$ mm to $z = 0.5$ mm and are constant over the rest of the simulation distance. The simulation has been carried out for the FWHM pulse duration $\tau_L = 27, 35, 38,$ and 40 fs. For each pulse duration, two bunches are trapped and accelerated as follows: the higher energy bunch with narrow energy spread is trapped and accelerated in the first bucket of the wake, while the lower energy bunch with large energy spread is trapped to the second bucket of the wake. Figure 2 shows the energy spectrum of accelerated electron bunch for $\tau_L = 38$ fs at $z = 2.9$ mm, where the first bunch reaches the maximum energy and the minimum energy spread.

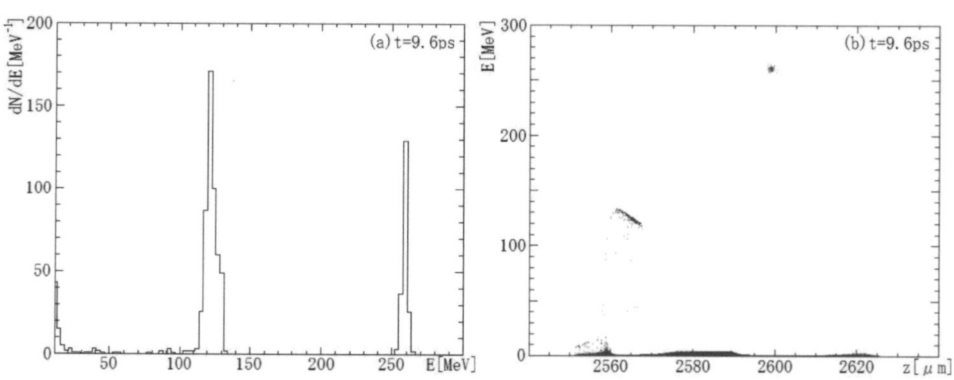

Fig. 2. The 2D PIC simulation results of the self-injected laser wakefield acceleration in the plasma channel for $a_0 = 3$, $\tau_L = 38$ fs, $r_L = 20\ \mu$m, $n_p = 2\times 10^{18}$ cm^{-3} and $\Delta n_c = 0.3 n_p$: (a) the energy spectrum and (b) the phase space distribution of accelerated electrons at $t = 9.6$ ps.

The beam parameters such as the bunch energy, the energy spread, the charge, the normalized emittance and the bunch length of the first bucket are investigated as a function of the laser pulse duration, when the bunch energy reaches the maximum value, for which the bunch has travelled approximately the dephasing length. For the optimum pulse duration, we obtains the best beam parameters characterized by the energy $E_b = 283$ MeV, the r.m.s. relative energy spread $\sigma_E / E_b = 0.5$ %, the normalized emittance $\varepsilon_n = 2.2\pi$mm mrad, the r.m.s. bunch length $\sigma_b = 0.38$ μm (1.3 fs) and the peak beam current $I_b = 15$ kA, when a drive laser pulse with the peak power $P_L = 120$ TW ($a_0 = 3$) and pulse duration $\tau_L = 38$ fs is focused at the spot radius $r_L = 20$ μm into the entrance of the plasma channel with the channel depth $\Delta n / n_p = 0.3$. These electron beam parameters satisfy requirements for the table-top soft X-ray FEL capable of generating a 10 GW-class saturation power at the radiation wavelength of 13.5 nm (92 eV photon energy) using a 1.1-m long undulator with 5-mm period and the 1-Tesla magnetic field that give the undulator parameter $K_u = 0.465$ (Nakajima, 2011).

In practical applications, high-quality beams from laser-plasma injectors are transported and injected to the undulator or the next LPA stage through a beam transport system. We consider the compact beam transport system for focusing the above-mentioned accelerated electron beam into the next accelerator stage or the miniature undulator for the soft X-ray FEL. The design has been studied using TRACE3D (Crandall & Rusthoi, 1997), which is an envelope code based on a first-order matrix description of the transport. The focus system consists of four permanent-magnet-based quadrupoles (PMQs), arranged in the defocus-defocus-focus-focus lattice configuration. The simulation results of TRACE3D are shown in Fig. 3, where the electron bunch is transported from the left to the right for the aforementioned beam parameters.

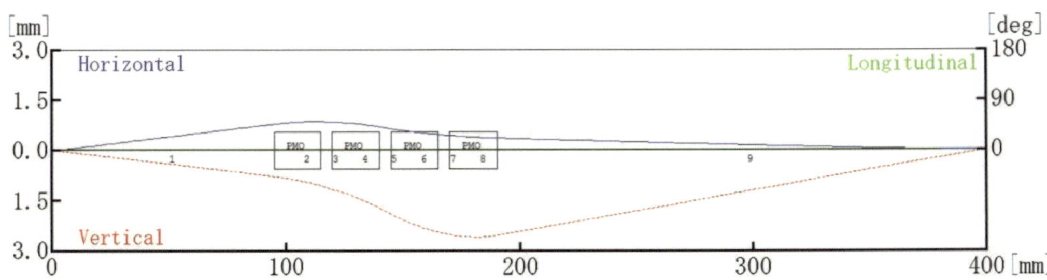

Fig. 3. TRACE3D simulation result of the electron beam transporting through the PMQ focus system for the beam energy $E_b = 283$ MeV, the normalized emittance $\varepsilon_{nx} = \varepsilon_{ny} = 2.2\pi$mm mrad, the relative energy spread $\sigma_E / E_b = 0.5$ %, the bunch duration $\sigma_{\tau b} = 1.3$ fs and the beam current $I_b = 15$ kA.

The field gradient of the two dimensional Halbach-type PMQ (Lim et al., 2005) is given by

$$B' = 2B_r \left(\frac{1}{r_i} - \frac{1}{r_0} \right), \tag{126}$$

where B_r is the tip field strength, r_i is the bore radius and r_0 is the outer radius of the PMQ. With $B_r = 2$ T and $r_i = 3$ mm, and $r_0 = 12$ mm, we can obtain the field gradient

$B' = 500\ \mathrm{T/m}$. In this simulation, we assume the parameters of the PMQs as follows: for PMQ1 and PMQ2, the field gradient $B' = -600\ \mathrm{T/m}$ and for PMQ3 and PMQ4, $B' = 500\ \mathrm{T/m}$. All PMQs have the same permanent-magnet geometry, i.e. $r_i = 3\ \mathrm{mm}$, $r_0 = 12\ \mathrm{mm}$ and the length of 20 mm. Since the electron beam only with the energy $E_b = 283$ MeV from the LPA located at the position $z = 0$ mm focuses on the horizontal and vertical r.m.s. beam sizes of $\sigma_x = 5.6\ \mu\mathrm{m}$ and $\sigma_y = 6.9\ \mu\mathrm{m}$, respectively, at the position $z = 400$ mm, the second bunch with the lower energy $E_b \approx 120$ MeV and the larger energy spread overfocuses on the plane at $z = 400$ mm. Hence, we can tune the beam energy of the focus system so as to discriminate the first bunch with high energy and high qualities from the second bunch with low energy and less qualities.

5. Stability control of laser-plasma acceleration

Many of applications require the stability of the beam parameters as well as their qualities. In particular, the stability issue is crucial for the X-ray FEL relying on the SASE mechanism. The stability of the laser system itself is very important for achieving stable LPAs (Hafz et al., 2008). However, up to date there is no conclusive proposal for stabilizing the production of electron beams from LPAs at the low plasma densities which are relevant to GeV energies. Here we show the effect of laser pulse skewness (asymmetry) on minimizing the electron beam pointing angle in the weakly-nonlinear laser wakefield accelerator operating at the low densities in a gas jet target.

5.1 Experiment for stabilizing electron beam production

5.1.1 Setup and parameters

The experimental setup is described as follows. A laser beam from a titanium sapphire system had, after compressor, the energy of ~ 0.9 J per pulse. The laser pulses are delivered to a target chamber and focused above a 4-mm long supersonic helium gas jet by using a focusing optic having F-number of 22 that focuses the laser pulse on the FWHM spot size $w_0 \approx 23 \pm 1\ \mu\mathrm{m}$ in vacuum. The gas jet stagnation pressure is ~ 1 bar and the laser is focused at the height of a few millimeters above the nozzle, where the gas density is in the range of $10^{17} - 10^{18}\ \mathrm{cm}^{-3}$. Therefore the expected wavelength of the wakefield is in the range $\lambda_p \approx 30 - 100\ \mu\mathrm{m}$. The electron beam pointing angle is detected by using a LANEX Kodak phosphor screen which is located at the distance of 78.5 cm from the gas jet. The LANEX is imaged onto an intensified charge-coupled device located near the interaction chamber in a radiation shielded area (Hafz et al., 2008). In order to obtain temporally-asymmetric laser pulses, the distance between two gratings of the pulse compressor is detuned from its optimum value which produces the shortest (37 fs) and symmetric pulses. The temporal pulse shape is measured by using a spectral phase interferometer for direct electric field reconstruction (SPIDER) device. Of interest is the negative detuning (positive chirp) which produces fast rise time laser pulses. Through negative detuning values from 0 to 250 μm, the laser pulse asymmetry increases and its length increases from 37 fs to 74 fs. In this range, the laser intensity is in the range $7.5 \times 10^{17} - 1.46 \times 10^{18}\ \mathrm{W/cm^2}$, corresponding to the normalized vector potential of the laser pulse in the range $a_0 = 0.59 - 0.83$. Therefore, this experiment is characterized roughly with the parameters $a_0 \leq 1$ and $c\tau_L \leq w_0 < \lambda_p$.

5.1.2 Results

In the following, the reference direction for the electron beam pointing angle is the laser beam direction itself. At first, we set the laser compression to optimum (no detuning), so that the laser intensity is $\sim 1.46 \times 10^{18}$ W/cm^2. The helium gas jet backing pressure is 1 bar and the interaction point is located at 1 mm height to the nozzle. At this height the gas density is $\sim 1 \times 10^{18}$ cm^{-3}. With those interaction conditions, the probability of observing an electron beam is as low as 1% or lower. However, the situation dramatically changes by detuning the compressor grating distance toward negative values. At a detuning distance of -100 μm the electron beam started to appear, however, the beam pointing angle is as large as ± 40 mrad. (The ± sign here means the direction of deflection angle with respect to the laser reference. In what follows we will remove the ± sign for simplicity). By changing the detuning distance to –200 μm the electron beam pointing angle is improved to 25 mrad and to 15 mrad at the detuning of –300 μm. Then the electron beam pointing angle has increased again to 25 mrad by increasing the detuning to –500 μm. In this experiment, the electron beam pointing in the vertical direction is smaller than that in the horizontal one. At laser height of 1.75 mm to the nozzle position, we notice that the electron beam pointing angle is improved to 8.5 and 10 mrad for the detuning distances of –200 and –400 μm. The electron beam pointing angle (horizontally and vertically) has been dramatically reduced to 2 mrad at a laser height of 3.25 mm where the gas density is in the range of 10^{17} cm^{-3}. Each data point is an average of 10 successive shots. From these data we can conclude that a detuning distance of –200 through –250 μm and the height of 3.25 mm are almost the optimum conditions for producing the smallest electron beam pointing angles. It should be noted that for this detuning range the laser intensity is in the range of $7.5 \sim 9 \times 10^{17}$ W/cm^2. More precise scanning for the grating detuning distance at a fixed laser height of 3.25 mm shows an interesting result as illustrated in Fig. 4. For this height and at zero detuning, the electron beam pointing angle is severely large ~100 mrad and the beam generation reproducibility is ~50%. Again, within the grating detuning range from –200 to –300 μm the electron beam pointing angle reaches its minimum value at 2 mrad. In addition, the electron generation reproducibility is almost 100%, and the electron beam charge is ~30 ±10 pC as measured by an integrating current transformer. The data points of Fig. 4 are averages over hundreds of successive laser shots except for those at 0 or positive detuning values where the electron beam production is null or extremely rare.

Finally, we measured the electron beam energy by using a bending dipole magnet (H-shaped) which had a uniform magnetic field intensity of 0.94 Tesla and a longitudinal length of 20 cm (Hafz et al., 2008). The distance from the gas jet to the magnet entrance is ~1.5 m and the LANEX is located at 25 cm from the end of the magnet. Between the gas jet and magnet we installed 1-m long helical undulator with 0.5 T magnetic field and 2.4 cm period for generating a synchrotron radiation. The distance from the gas jet to the undulator is 30 cm, and the inner diameter of the undulator tube is 9 mm. Thus an electron beam from the gas jet must enter the undulator, propagate through it and then enter the dipole magnet region which bends the beam into the LANEX screen. The measured electron beam have a quasi-monoenergetic energy peak at ~ 165 MeV. This article is focused on minimizing the fluctuation of the electron beam pointing angle, thus our results are crucial for on-going world-wide experiments on compact free-electron laser and undulator radiation using intense laser irradiated gas jets as a compact electron beam accelerator (Hafz et al., 2010; Nakajima, 2008).

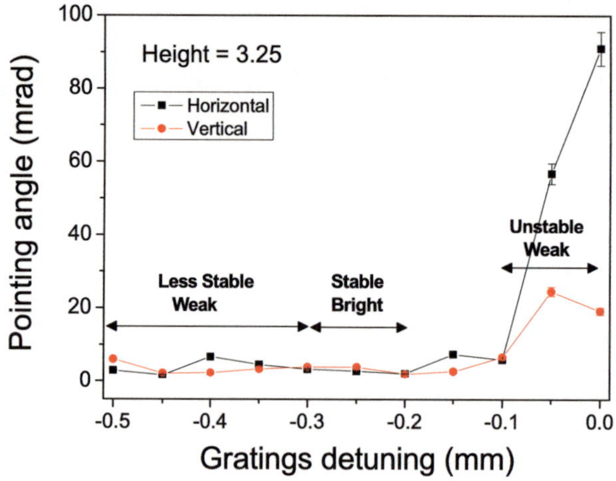

Fig. 4. Electron beam pointing angle versus detuning distance at the optimum height of 3.25 mm above the gas jet nozzle.

6. Conclusion

We have worked out the design considerations of a compact X-ray FEL that can reach the wavelength of 0.1 nm corresponding to the hard X-ray with photon energy of 12 keV. The system consists of a cm-scale 100 MeV-class electron beam injector, a 0.4-m long PMQ-based transport beam line , a 0.4-m long 6 GeV LPA linac and a 8-m long undulator. Including a 100 TW-class table-top laser system and an application space for the coherent X-ray research, main system can be installed within a 10-m long, 2-m wide space. The present considerations are based on the current achievements of laser-plasma accelerators and currently available technologies on drive lasers and undulators, for which we have not assumed new technologies and developments as well as new physics concepts on FEL. In this context, the present design of the hard X-ray FEL would be rather conventional and therefore it may be materialized in a near term at a reasonably low cost, guaranteeing the performance comparable to large-scale X-ray FELs. Harnessing miniature undulators with period of $\lambda_u \sim 5\,\mathrm{mm}$ (Grüner et al., 2007; Eichner et al., 2007) may make the required saturation length shorter by a factor of 3, i.e. a 2.5-m long undulator, and the required electron beam energy becomes approximately half, i.e. $E_b = 3\,\mathrm{GeV}$, for a 0.1 nm X-ray wavelength, assuming the saturation length scales as $L_{\mathrm{sat}} \propto \lambda_u^{5/6}$ with $\gamma \propto \lambda_u^{1/2}$ and the LPA is operated at the same plasma density. This option may turn out a whole system to be on a 3-m long table top under the condition of trading off requirements for further high-quality, high-stability electron beam production from the LPAs.

Another way to build X-ray FELs on a table top is to produce the interaction between an electron beam and a laser pulse via coherent Thomson scattering or Compton scattering

(Bonifacio, 2005; Smetanin & Nakajima, 2004), where intense laser fields are used as an optical undulator or a laser wiggler/undulator that replaces the magnetic undulator field to the laser field with 3~4 orders magnitude shorter wavelength. Combining laser-plasma accelerated electron beams with laser undulators leads to further compact X-ray FELs, which have been recently proposed as all-optical-free electron lasers (Petrillo et al, 2008). However, these options must satisfy harsh requirements in terms of beam current, emittance, energy spread and stability of both laser and electron beams. We would like to expect further research and future progress in this new approach to a compact X-ray FEL using laser-plasma accelerators.

7. Acknowledgment

The work has been supported by the National Natural Science Foundation of China (Project Nos. 10834008, 60921004 and 11175119) and the 973 Program (Project No. 2011CB808104). K. Nakajima is supported by Chinese Academy of Sciences Visiting Professorship for Senior International Scientists.

8. References

Ackermann, W., Asova, G., Ayvazyan, Y. et al., Operation of a free-electron laser from the extreme ultraviolet to the water window. *Nature Photonics*, Vol. 1, No. 6, (June 2007), pp. 336 – 342, ISSN 1749-48852007

Bonifacio, R., Pellegrini, C., Narducci, L.M. (1984). Collective instabilities and high-gain regime in a free electron laser. *Optics Communications*, Vol. 50, No. 6, (July 1984), pp. 373-378

Bonifacio, R. (2005). Quantum SASE FEL with laser wiggler. *Nuclear Instruments and Methods in Physics Research A*, Vol. 546, pp. 634-638

Clayton, C.E., Ralph, J.E., Albert, F., Fonseca, R.A., Glenzer, S.H., Joshi, C., Lu, W., Marsh, K. A., Martins, S. F., Mori, W.B., Pak, A., Tsung, F.S., Pollock, B.B., Ross, J.S., Silva, L.O. & Froula, D. H. (2010). Self-guided laser wakefield acceleration beyond 1 GeV using ionization-induced injection. *Physical Review Letters*, Vol. 105, pp. 105003-1-4

Crandall, K., & Rusthoi, D. (1997). *TRACE 3-D documentation* (Third Edition), Los Alamos National Laboratory Technical Report No. LA-UR-97-886, Los Alamos National Laboratory, Los Alamos, New Mexico 87545

Eichner, T., Grüner, F., Becker, S., Fuchs, M., Habs, D., Weingartner, R., Schramm, U., Backe, H. (2007). Miniature magnetic devices for laser-based, table-top free-electron lasers. *Physical Review Special Topics - Accelerators and Beams*, Vol. 10, pp. 082401-1-9

Esarey, E., Sprangle, P., Krall, J., & Ting, A. (1996). Overview of plasma-based accelerator concepts. *IEEE Transactions on Plasma Science*, Vol. 24, No. 2, (April 1996), pp. 252-288

Froula, D.H., Clayton, C.E., Döppner, T., Marsh, K.A., Barty, C.P.J., Divol, L., Fonseca, R.A., Glenzer, S.H., Joshi, C., Lu, W., Martins, S.F., Michel, P., Mori, W.B., Palastro, J.P., Pollock, B.B., Pak, A., Ralph, J.E., Ross, J.S., Siders, C.W., Silva, L.O., & Wang, T. (2009) Measurements of the critical power for self-injection of electrons in a laser wakefield accelerator, *Physical Review Letters*, Vol. 103, pp. 215006-1-4.

Fuchs, M., Weingartner, R., Popp, A., Major, Zs., Becker, S., Osterhoff, J., Cortrie, I., Benno Zeitler, B., Rainer Hörlein, R., Tsakiris, G.D., Schramm, U., Rowlands-Rees, T.P., M. Hooker, S.M., Habs, D., Krausz, F., Karsch, S., & Grüner, F. (2009). Laser-driven soft-X-ray undulator source, *Nature Phys*. Vol. 5, pp. 826 – 829, ISSN 1745-2473

Gerstner, E. (2011). Free Electron Lasers - X-ray crystallography goes viral. *Nature Physics*, Vol. 7, No. 3, (March 2011), pp. 194-194, ISSN 1745-2473

Grüner, F., Becker, S., Schramm, U., Eichner, T., Fuchs, M., Weingartner, R., Habs, D., Meyer-Ter-Vehn., J., Geissler, M., Ferrario, M., Serafini, L., Van der Geer, B., Backe, H., Lauth W., & Reiche, S. (2007). Design considerations for table-top, laser-based VUV and X-ray free electron lasers. *Applied Physics B*, Vol. 86, pp. 431-435

Hafz, N.A.M., Jeong, T.M., Choi, I.W., Lee, S.K., Pae, K.H., Kulagin, V.K., Sung, J.H., Yu, T.J., Hong, K.H., Hosokai, T., R. Cary, J.R., Ko, D.K., & Lee, J.M. (2008). Stable generation of GeV-class electron beams from self-guided laser–plasma channels. *Nature Photonics*, Vol. 2, No. 9, (September 2008), pp. 571 - 577, ISSN 1749-4885

Hafz, N.A.M., Yu, T.J., Lee, S.K., Jeong, T.M., Sung, J.H., & Lee, J.M. (2010). Controlling the Pointing angle of a relativistic electron beam in a weakly-nonlinear laser wakefield accelerator. *Applied Physics Express*, Vol. 3, pp. 076401-1-3

Jackson, J. D. (1999). *Classical Electrodynamics* (Third Edition), Wiley, New York

Kameshima, T., Hong, W., Sugiyama, K., Wen, X.L., Wu, Y.C., Tang, C.M., Qihua Zhu, Q.H., Gu, Y.Q., Zhang, B.H., Peng, H.S., Kurokawa, S., Chen, Tajima, T., Kumita, T. & Nakajima, K. (2008). 0.56 GeV laser electron acceleration in ablative-capillary-discharge plasma channel. *Applied Physics Express*, Vol. 1, pp. 066001-1-3

Karsch, S., J Osterhoff, J., Popp, A., Rowlands-Rees, T.P., Major, Zs., Z M Fuchs, Z.M., Marx, B., Hörlein, R., K Schmid, K., Veisz, L., Becker, S., Schramm, U., Hidding, B., Pretzler, G., Habs, D., Grüner, F., Kraus, F., & S M Hooker, S.M. (2007). GeV-scale electron acceleration in a gas-filled capillary discharge waveguide. *New Journal of Physics*, Vol. 9, pp. 415-425

Kostyukov, I., Pukhov, A., Kiselev, S. (2004). Phenomenological theory of laser-plasma interaction in "bubble" regime. *Physics of Plasmas*, Vol. 11, pp. 5256-5264

Leemans, W.P., Nagler, B., Gonsalves, A.J., Toth, Cs., Nakamura, K., Geddes, C.G.R., Esarey, E., Schroeder, C.B. & Hooker, S.M. (2006). GeV electron beams from a centimetre-scale accelerator. *Nature Physics*, Vol. 2, pp. 696-699, ISSN 1745-2473

Lim, J.K., Frigola, P., Travish, G., Rosenzweig, J.B., Anderson, S.G., Brown, W.J., Jacob, J.S., Robbins, C.L., & Tremaine, A.M. (2005). Adjustable short focal length permanent-magnet quadrupole based electron beam final focus system. *Physical Review Special Topics - Accelerators and Beams*, Vol. 8, pp. 072401-1-17

Lu, H.Y., Liu, M.W., Wang, W.T., Wang, C., Liu, J.S., Deng, A.H., Xu, J.C., Xia, C.Q., Li, W.T., Zhang, H., Lu, X.M., Wang, C., Wang, J.Z., Liang, X.Y., Leng, Y.X., Shen, B.F., Nakajima, K., Li, R.X. & Xu, Z.Z. (2011). Laser wakefield acceleration of electron beams beyond 1 GeV from an ablative capillary discharge waveguide. *Applied Physics Letters*, Vol. 99, pp. 091502-1-3

Lundh, O., Lim, J., C. Rechatin1, Ammoura, L., Ben-Ismaïl, A., Davoine, X., Gallot, G., Goddet, J-P., Lefebvre, E., Malka, V. & Faure, J. (2011). Few femtosecond, few kiloampere electron bunch produced by a laser–plasma accelerator. *Nature Physics*, Vol. 7, No. 3, (March 2011), pp. 219-222, ISSN 1745-2473

Lu, W., Huang, C., Zhou, M., Mori, W.B. & Katsouleas, T. (2006). Nonlinear theory for relativistic plasmawakefields in the blowout regime. *Physical Review Letters*, Vol. 96, pp. 165002-1-4

Lu, W., Tzoufras, M., Joshi, C., Tsung, F.S., Mori, W.B., Vieira, J., Fonseca, R.A., & L. O. Silva, L.O. (2007). Generating multi-GeV electron bunches using single stage laser wakefield acceleration in a 3D nonlinear regime. *Physical Review Special Topics - Accelerators and Beams*, Vol. 10, pp. 061301-1~12

Malka, V., Faure, J., Gauduel, Y.A., Lefebvre, E., Rousse, A., & Phuoc, K.T. (2008). Principles and applications of compact laser–plasma accelerators. *Nature Physics*, Vol. 4, No. 6, (June 2008), pp. 447-453, ISSN 1745-2473

Mangles, S.P.D., Thomas, A.G.R., Lundh, O., Lindau, F., Kaluzac, M. C., Persson, A., Wahlström, C.-G., Krushelnickd, K., & Najmudin, Z., (2007). On the stability of laser wakefield electron accelerators in the monoenergetic regime, *Physics of Plasmas*, Vol. 14, pp. 056702-1~7.

Michel, P., Schroeder, C.B., Shadwick, B.A., Esarey, E. & Leemans, W.P. (2006). Radiative damping and electron beam dynamics in plasma-based accelerators. *Physical Review E* , Vol. 74, pp. 026501-1-14

Nakajima, K. (2008). Compact X-ray sources-Towards a table-top free-electron laser. *Nature Physics*, Vol. 4, No. 2, (February 2008), pp. 92-93, ISSN 1745-2473

Nakajima, K. (2011). Recent Progress on Laser Plasma Accelerators and Applications for Compact High-Quality Particle Beam and Radiation Sources, *Journal of Materials Science and Engineering A*, Vol. 1, pp. 293-300, ISSN 1934-8959

Nakajima, K., Kando, M., Kawakubo, T., Nakanishi, T., & Ogata, A. (1996). A table-top X-ray FEL based on the laser wakefield accelerator-undulator system. *Nuclear Instruments and Methods in Physics Research*, Vol. A375, pp. 593-596

Nakajima, K., Deng, A., Zhang, X., Shen, B., Liu, J., Li, R., Xu, Z., Ostermayr, T., Petrovics, S., Klier, C., Iqbal, K., Ruhl, H. & Tajima, T. (2011). Operating plasma density issues on large-scale laser-plasma accelerators toward high-energy frontier. *Physical Review Special Topics - Accelerators and Beams*, Vol. 14, pp. 091301-1-12

Nieter, C. & Cary, J.R. (2004) VORPAL: a versatile plasma simulation code. *Journal of Computational Physics*, Vol. 196, No. 2, pp. 448-473

Osterhoff, J., Popp, A., Major, Zs., Marx, B., Rowlands-Rees, T.P., Fuchs, M., Geissler, M., Hörlein, R., Hidding, B., Becker, S., Peralta, E.A., Schramm, U., Grüner, F., Habs, D., Krausz, F., Hooker, S.M., & Karsch, S. (2008). Generation of stable, low-divergence electron beams by laser-wakefield acceleration in a steady-state-flow gas cell. *Physical Review Letters*, Vol. 101, pp. 085002-1-4

Petrillo, V., Serafini, L., & Tomassini1, P. (2008). Ultrahigh brightness electron beams by plasma-based injectors for driving all-optical free-electron lasers. *Physical Review Special Topics - Accelerators and Beams*, Vol. 11, pp. 070703-1-7

Rechatin, C., Faure, J., Ben-Ismail, A., Lim, J., Fitour, R., Specka, A., Videau, H., Tafzi, A., Burgy, F., & Malka, V. (2009). Controlling the phase-space volume of injected electrons in a laser-plasma accelerator. *Physical Review Letters*, Vol. 102, pp. 164801-1-4

Schachter, L. (2011). *Beam-Wave Interaction in Periodic and Quasi-Periodic Structures* (Second Edition), Springer-Verlag, ISBN 978-3-642-19847-2, Berlin

Schlenvoigt, H.-P., Haupt, K., Debus, A., Budde, F., Jackel, O., Pfotenhauer, S., Schwoerer, H., Rohwer, E., Gallacher, J.G., Brunetti, E., Shanks, R.P., Wiggins, S.M., & Jaroszynski, D.A. (2008). A compact synchrotron radiation source driven by a laser-plasma wakefield accelerator. *Nature Physics*, Vol. 4, No. 2, (February 2008), pp. 130-133, ISSN 1745-2473

Sprangle, P., & Esarey, E. (1992). Interaction of ultrahigh laser fields with beams and plasmas. *Physics Fluids B*, Vol.4, No. 7, (July 1992), pp. 2241-2248

Smetanin, I.V., & Nakajima, K. (2004). Quantum effects in laser-beam Compton interaction and stimulated electron positron annihilation in a strong field. *Laser and Particle Beams*, Vol. 22, pp. 479-484

Sverto, O. (1998). *Principles of Lasers* (Fourth Edition), Plenum Press, ISBN 0-306-45748-2, New York

6

Theoretical Analysis on Smith-Purcell Free-Electron Laser

D. Li[1], M.Hangyo[2], Y. Tsunawaki[3], Z. Yang[4], Y. Wei[4], S. Miyamoto[5], M. R. Asakawa[6] and K. Imasaki[1]

[1]Institute for Laser Technology
[2]Osaka University
[3]Osaka Sangyo University
[4]University of Electronic Science and Technology of China
[5]University of Hyogo
[6]Kansai University
[1,2,3,5,6]Japan
[4]China

1. Introduction

The first observation of radiation emitted by an electron passing over a diffraction grating was made long ago by Smith and Purcell [Smith, S. J.], and the idea of using this effect in a free-electron laser has been proposed by numerous authors [Gover, A., Schachter, L., Leavitt, R.P.], and a cavity is usually used to provide feedback in those schemes. A renewed interest in Smith-Purcell radiation has been raised in recent years, since the analysis of the dispersion relation of surface waves for lamellar gratings by Andrews and Brau [Andrews, H.]. They pointed out that the interaction between an initially continuous electron beam and a surface wave on the grating could lead to bunching at the frequency of the wave, and subsequently the periodic electron bunches induce strong radiation at a certain angle, and it is called super-radiant Smith-Purcell radiation. Thus, the Smith-Purcell radiation is recongnized to be a promising alternative in the development of a compact, tunable and high power terahertz device. The terahertz sources, a currently active research area, are of importance in a variety of applications to biophysics, medical and materials science.

It is well known that the Smith-Purell radiation is emitted as an electron passes close to the surface of a periodic metallic grating. The incoherent Smith-Purcell radiation has been analyzed in many ways, such as diffraction theory, integral equation method and induced surface current model [van den Berg, P.M., Walsh, J., Shibata, Y.]. The super-radiant Smith-Purcell radiation is regarded as the result of the periodic electron bunches, which can be generated from a system of pre-bunched beam, or the interaction of initial continuous beam with the enhanced surface wave traveling along the grating. Several theories have been proposed to explain the super-radiant phenomenon and calculate the growth rate of the radiation[Kuma, V., Schächter, L, Andrews. H.]. With particle-in-cell simulations, the incoherent, coherent and super-radiant Smith-Purcell radiation have been demonstrated, and some ideas on improvement of the gorwth rate were proposed [Donohue, J.T.,Li, D.]. Recently, experiments of proof-of-principal are carried out [Andrews, H., Gardelle, J.], and the experimental results are in agreement with the theoretical analysis.

In this Chapter, we introduce the Smith-Purcell free-electron lasers through theoretical analysis and particle-in-cell simulation. We are trying to understand the incoherent and superradiant Smith-Purcell radiation by a complete calculation of a single electron bunch, a train of bunches and a continuous beam, respectively. Some proposals on enhanceing the interaction of electron beam with surface waves are demonstrated. We also explore the characterics of the Smith-Purcell radiation from a grating made of negative-index materials.

2. Theory of grating emission

When an electric charge moves, at constant speed, parallel to a grating, it excites Smith-Purcell radiation and surface wave. The surface wave is not radiant, unless it comes to the ends of the grating. It partially reflects and diffracts there, and the part of diffraction is radiant. In this secetion, we theoretically analyze the two phenomena.

2.1 Smith-Purcell radiation

We consider a two-dimensional scheme of lamellar grating made of perfect conductor embedded in vacuum as shown in Fig. 1. The periodicity of the grating is chosen in the z direction, whereas the grating's rulings are parallel to the y axis. We consider the idealized case that a line charge is supposed to pass over the grating along the z axis. The single

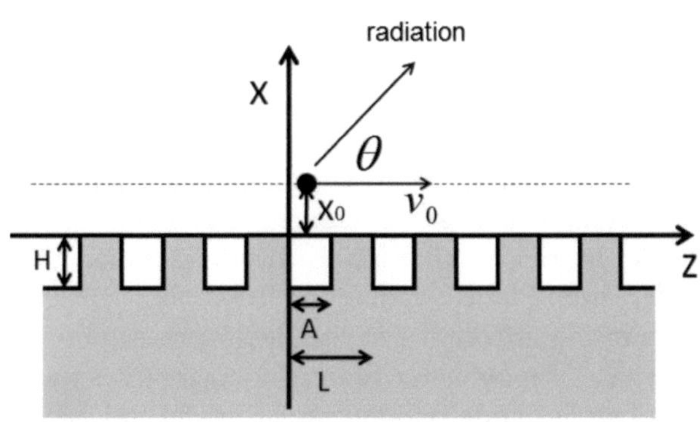

Fig. 1. Cross section of the configuration. Rectangular grating; A=groove width, H=groove depth, L=period.

line charge, with a charge-density distribution q per unit length in the y direction, moves with velocity v_0 along the trajectory $x = x_0$. Both the geometry of the configuration and all field quantities are independent of y direction since it is two dimensional. Thus, the expression for the current density of the line charge can be written as

$$J_z(x,z,t) = qv_0\delta(x - x_0)\delta(z - v_0 t) \tag{2.1}$$

The Fourier transform of the current density is then given by

$$J_z(x,z,\omega)=\int_{-\infty}^{\infty} J_z(x,z,t)e^{j\omega t}dt = q\delta(x-x_0)e^{j\frac{\omega}{v_0}z} \tag{2.2}$$

The current distribution described by Eq. (2.2) excites the z-directed component of the magnetic vector potential which in turn satisfies

$$(\frac{\partial^2}{\partial x^2}+\frac{\partial^2}{\partial z^2}+\frac{\omega^2}{c^2})A_z(x,z,\omega)=-\mu_0 J_z(x,z,\omega) \tag{2.3}$$

Its solution is assumed to have the form $A_z(x,z,\omega)=A_z(x,\omega)e^{j\frac{\omega}{v_0}z}$, and $A_z(x,\omega)$ satisfies

$$(\frac{d^2}{dx^2}-\frac{\omega^2}{v_0^2}+\frac{\omega^2}{c^2})A_z(x,\omega)=-\mu_0 q\delta(x-x_0) \tag{2.4}$$

For $x>x_0$ the solution of this equation is

$$A_z(x>x_0,\omega)=C_+e^{-\alpha_0(x-x_0)},$$

and for $x<x_0$

$$A_z(x<x_0,\omega)=C\_e^{\alpha_0(x-x_0)},$$

where $\alpha_0^2=(\omega/v_0)^2-(\omega/c)^2$. The $A_z(x,\omega)$ should be continuous at $x=x_0$, thus we have $C_+=C_-=C$, whereas its derivative is discontinuous. The discontinuity is determined by the Dirac delta function in Eq. (2.4), therefore by integrating Eq. (2.4) we obtain

$$[\frac{d}{dx}A_z(x)]_{x=x_0^+}-[\frac{d}{dx}A_z(x)]_{x=x_0^-}=-\mu_0 q,$$

hence

$$-\alpha_0 C-\alpha_0 C=-\mu_0 q,$$

Then we get

$$C=\frac{\mu_0 q}{2\alpha_0}.$$

We conclude that the magnetic vector potential reads

$$A_z(x,z,\omega)=\frac{\mu_0 q}{2\alpha_0}e^{-\varepsilon(x)\alpha_0(x-x_0)}e^{j\frac{\omega}{v_0}z},$$

where $\varepsilon(x)=-1$ for $x<x_0$, and $\varepsilon(x)=1$ for $x>x_0$. The magnetic field of y component can be obtained as

$$H_y(x,z,\omega) = -\frac{1}{\mu_0}\frac{\partial A_z(x,z,\omega)}{\partial x} = \frac{q\varepsilon(x)}{2}e^{-\varepsilon(x)\alpha_0(x-x_0)}e^{j\frac{\omega}{v_0}z},$$

and the incident field on the grating surface should be $H_y^i = -\frac{q}{2}e^{\alpha_0(x-x_0)}e^{j\frac{\omega}{v_0}z}$.

Above the grating we expand the reflected fields H_y^r in Floquet series (space harmonics), and the total magnetic field can be written as

$$H_y^T = H_y^i + H_y^r = -\frac{q}{2}e^{\alpha_0(x-x_0)}e^{j\frac{\omega}{v_0}z} + \sum_{p=-\infty}^{\infty} A_p e^{j\alpha_p x}e^{jk_p z}, \tag{2.5}$$

where $k_p = \omega/v_0 + 2\pi p/L$, $\alpha_p^2 = (\omega/c)^2 - k_p^2$, and A_p is the scalar coefficient to be determined. It is clear that, the refracted waves are evanescent for $p \geq 0$, since α_p is imaginary; when p is negative integer, some of the refracted modes radiate in a certain range of frequency to guarantee $\alpha_p^2 > 0$. The frequency range is worked out, and we express it in wavelength

$$\frac{L}{|p|}(\frac{1}{\beta}-1) \leq \lambda \leq \frac{L}{|p|}(\frac{1}{\beta}+1).$$

Then, we can define an angle of emergence θ ,between the radiating wave and the charge moving direction. Thus, from the relation $k_p = (\omega/c)\cos\theta = \omega/v_0 + 2\pi p/L$, it is straightforward to achieve

$$\lambda = \frac{L}{|p|}(\frac{1}{\beta} - \cos\theta), \tag{2.6}$$

which is well known for Smith-Purcell radiation.

The z-directed electric field can be obtained from Eq.(2.5) by using the Maxwell equations, and it reads

$$E_z^T = \frac{j}{\omega\varepsilon_0}(-\frac{q\alpha_0}{2}e^{\alpha_0(x-x_0)}e^{j\frac{\omega}{v_0}z} + \sum_{p=-\infty}^{\infty} jA_p\alpha_p e^{j\alpha_p x}e^{jk_p z}) \tag{2.7}$$

For simplicity without sacrificing the generality, only the lowest mode in the groove, which is the most easily excited mode, is considered. Its electromagnetic fields are given by

$$H_y^g = B(\cos(\frac{\omega}{c}x) + \tan(\frac{\omega}{c}H)\sin(\frac{\omega}{c}x)), \tag{2.8}$$

$$E_z^g = \frac{-j}{c\varepsilon_0}B(\sin(\frac{\omega}{c}x) + \tan(\frac{\omega}{c}H)\cos(\frac{\omega}{c}x)), \tag{2.9}$$

where B is a scalar coefficient to be determined.

At the surface of the grating $(x=0)$ the tangential component of the electric field is continuous. Since the tangential field vanishes on the surface of the conductor, we get

$$E_z^T(x=0)=\begin{cases}E_z^g(x=0) & for 0\le z\le A\\ 0 & for A\le z\le L\end{cases}. \tag{2.10}$$

We multiply by $e^{-jk_r z}$ and integrate over $0<z<L$, we get

$$A_r=\frac{j}{2\alpha_r L}\left(\frac{2B\omega}{c}\tan\left(\frac{\omega}{c}H\right)\psi_1^r-q\zeta\alpha_0 L e^{-\alpha_0 x_0}\delta_{r,0}\right), \tag{2.11}$$

where $\psi_1^r=\int_0^A e^{-jk_r z}dz$. Likewise, the tangential component of the magnetic field must be continuous at the grating surface,

$$H_y^T(x=0)=H_y^g(x=0). \tag{2.12}$$

We integrate over $0<z<A$, and we get

$$B=\frac{1}{A}\left(\sum_{p=-\infty}^{\infty}A_p\psi_2^p-\frac{1}{2}qe^{-\alpha_0 x_0}\psi_2^0\right), \tag{2.13}$$

where $\psi_2^p=\int_0^A e^{jk_p z}dz$. If we change r in Eq. (2.11) by p and substitute it into Eq. (2.13), we get

$$B=-\frac{1}{2}\frac{q\psi_2^0(j+1)L}{e^{\alpha_0 x_0}\left(-j\frac{\omega}{c}\tan\left(\frac{\omega}{c}H\right)\left(\sum_{p=-\infty}^{\infty}\frac{\psi_1^p\psi_2^p}{\alpha_p}\right)+AL\right)}. \tag{2.14}$$

Substitute Eq.(2.14) into Eq.(2.11), we finally get the expression for coefficients of reflected waves,

$$A_r=j\frac{1}{2\alpha_r L}\left(-\frac{q\psi_2^0(1+j)L\frac{\omega}{c}\tan\left(\frac{\omega}{c}H\right)\psi_1^r}{e^{\alpha_0 x_0}\left(-j\frac{\omega}{c}\tan\left(\frac{\omega}{c}H\right)\left(\sum_{p=-\infty}^{\infty}\frac{\psi_1^p\psi_2^p}{\alpha_p}\right)+AL\right)}-q\alpha_0 L e^{-\alpha_0 x_0}\delta_{r,0}\right). \tag{2.15}$$

In order to be independent of influence from the charge, we define

$$A_r^u=\frac{A_r}{qe^{-\alpha_0 x_0}}=j\frac{1}{2\alpha_r L}\left(-\frac{\psi_2^0(1+j)L\frac{\omega}{c}\tan\left(\frac{\omega}{c}H\right)\psi_1^r}{-j\frac{\omega}{c}\tan\left(\frac{\omega}{c}H\right)\left(\sum_{p=-\infty}^{\infty}\frac{\psi_1^p\psi_2^p}{\alpha_p}\right)+AL}-\alpha_0 L\delta_{r,0}\right). \tag{2.16}$$

So, the coefficients of reflected waves can be numerically calculated when relevant parameters are given.

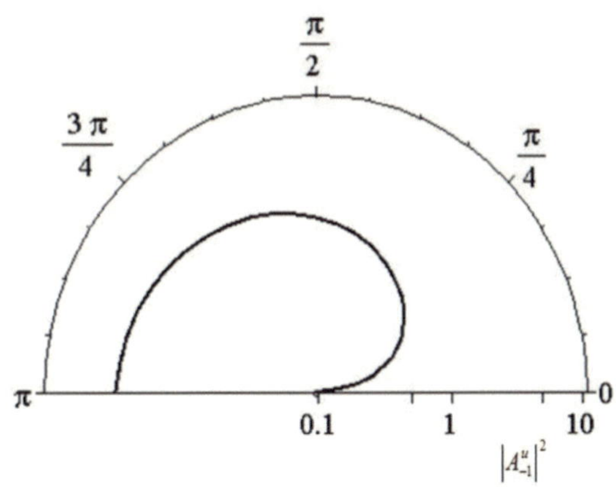

Fig. 2. Radiated intensity $\left|A_{-1}^{u}\right|^{2}$ in the -1st order Smith-Purcell radiation as a function of the radiating angle.

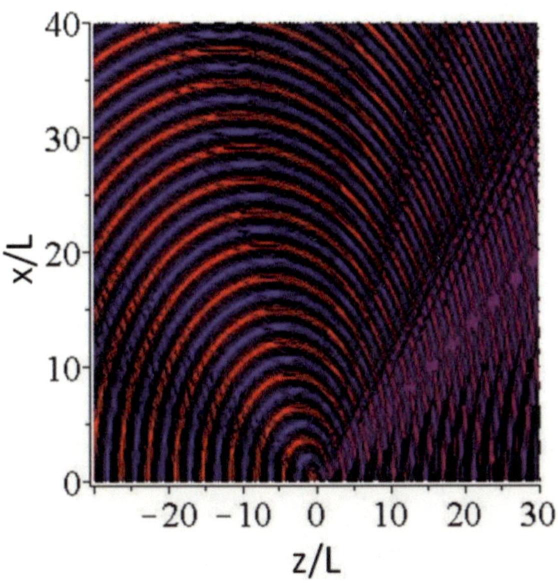

Fig. 3. Contour plot of the magnetic field of -1st order Smith-Purcell radiation H_y^{-1}.

The grating to be calculated is with the parameters, L=173 μ m, A=62 μ m, and H=100 μ m. The electron energy is 40 keV. We focus on the –1st order($p=-1$) wave, which is the lowest mode. Only the radiating waves (Smith-Purcell radiation) are considered, so we substitute Eq.(2.6) into Eq.(2.16), and calculate the dependency of radiated intensity $\left|A_{-1}^{u}\right|^{2}$ on the emission angle θ. The results are given in Fig.2. It is shown that the Smith-Purcell radiation induced by a single line charge is dependent on the radiation angle, and the backward radiation is stronger than the forward radiation. Remember that the radiation wave length and the radiation angle satisfy the Eq. (2.6), so, the wave length of the backward radiation is longer than that of the forward one. The –1st order refracted magnetic field is expressed as

$$H_y^{-1} = A_{-1}^u e^{j\alpha_{-1}x} e^{jk_{-1}z}. \tag{2.17}$$

Based on Eq. (2.17), the space distribution of H_y^{-1} can be calculated, and the contour plots are given in Fig.3.

2.2 Surface wave

Considering the case of abscence of electric charge in Fig. 1, we search the evenescent mode on the surface of the grating . Our analysis focuses on the transverse magnetic wave mode, which has longitudinal electric field. The y-directed component of the magnetic field above the grating ($x > 0$) can be expanded in Floquet series, and it is written as

$$H_y^a = \sum_{p=-\infty}^{\infty} A_p e^{-\alpha_p x} e^{jk_p z}, \tag{2.18}$$

where $k_p = k + 2\pi p/L$, $\alpha_p^2 = k_p^2 - (\omega/c)^2$, and A_p is the scalar coefficient to be determined. The z-directed electric field can be obtained from Eq.(2.18) by using the Maxwell equations, and it reads

$$E_z^a = \frac{-j}{\omega\varepsilon_0} \sum_{p=-\infty}^{\infty} A_p \alpha_p e^{-\alpha_p x} e^{jk_p z} \tag{2.19}$$

As mentioned above, only the lowest mode in the groove ($0 > x > -H$), which is the most easily excited mode, is considered. Its electromagnetic fields are same to Eq.(2.8) and Eq. (2.9), and we write here again

$$H_y^g = B(\cos(\frac{\omega}{c}x) + \tan(\frac{\omega}{c}H)\sin(\frac{\omega}{c}x)) \tag{2.20}$$

$$E_z^g = \frac{-j}{c\varepsilon_0} B(\sin(\frac{\omega}{c}x) + \tan(\frac{\omega}{c}H)\cos(\frac{\omega}{c}x)) \tag{2.21}$$

where B is scalar coefficient to be determined.

At the surface of the grating ($x = 0$) the tangential component of the electric field is continuous. Since the tangential field vanishes on the surface of the conductor, we get

$$E_z^a(x=0) = \begin{cases} E_z^g(x=0) & for 0 \le z \le A \\ 0 & for A \le z \le L \end{cases}$$

We multiply by $e^{-jk_q z}$ and integrate over $0 < z < L$, we get

$$A_q \alpha_q L = \frac{\omega}{c} B \tan(\frac{\omega}{c}H) \psi_1^q \tag{2.22}$$

where $\psi_1^q = \int_0^W e^{-jk_q z} dz$. Likewise, the tangential component of the magnetic field must be continuous at the grating surface,

$$H_y^a(x=0) = H_y^g(x=0) .$$

We integrate over $0 < z < A$, and we get

$$BA = \sum_{p=-\infty}^{\infty} A_p \psi_2^p , \tag{2.23}$$

where $\psi_2^p = \int_0^A e^{jk_p z} dz$. If we change q in Eq. (2.11) by p and substitute it into Eq. (2.23), we get the dispersion equation for the surfave wave

$$1 = \sum_{p=-\infty}^{\infty} \frac{\omega \tan(\frac{\omega}{c} H) \psi_1^p \psi_2^p}{\alpha_p LAc} . \tag{2.24}$$

With the parameters mentioned above, L=173 μ m, A=62 μ m, and H=100 μ m, Eq.(2.24) is calculated and the result is shown in Fig.4. This is a typical dispersion curve for a slow-wave structure. In the region $0 < kL/2\pi < 0.5$, the phase velocity is in the same direction of the group velocity, and it is called tarvelling wave. In the region $0.5 < kL/2\pi < 1$, the phase velocity and the group velocity are in opposite directions, and it is called backward wave. A beam line for 40 keV electron beam is also plotted in Fig. 4. The beam line interacts with the backward wave, thus, the electromagnetic energy flows opposite to the beam moving direction and form the backward wave oscillator, which external feedback is not necessary.

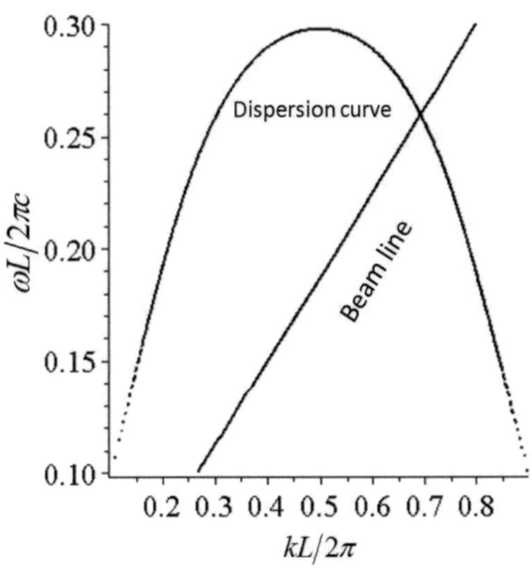

Fig. 4. Dispersion relation of the surface wave of the grating with parameters L=173 μ m, A=62 μ m, and H=100 μ m. The beam line is for 40 keV electron beam.

3. Simulation of grating emission

The particle-in-cell code employed in our simulation, MAGIC, is developed by Mission Research Corporation. It is a finite-difference, time-domain code for simulating processes that involve interactions between space charge and electromagnetic fields.

We consider a two-dimensional model consisting a grating with rectangular form, a sheet electron beam and vacuum region for electron-wave interaction and radiation propagation. The simulation geometry is as shown in Fig. 5, where the grating is set in the center of the bottom of the vacuum box. The simulation uses the Cartesian coordinate system, with the origin being chosen at the center of the grating. The surface of the grating is assumed to consist of a perfect conductor whose grooves are parallel and uniform in the y direction. The electron beam is chosen as the sheet form, with a finite thickness. The beam, a perfect laminar beam produced by the MAGIC algorithm and moving in the *z*-axis, is generated from a cathode, which is located at the left boundary of the simulation box and 34μm above the grating. The vacuum box is bordered with a special region (called *free-space* in MAGIC language), in which incident electromagnetic waves and electrons can be absorbed. The whole simulation area is divided into mesh with rectangle cell of very small size in the region of beam propagation and large size in the remainder. Since it is a two-dimensional simulation, it assumes that all fields and currents are independent of the z coordinate, and it should be noted that the current value mentioned in the paper represents the current per meter in the z direction.

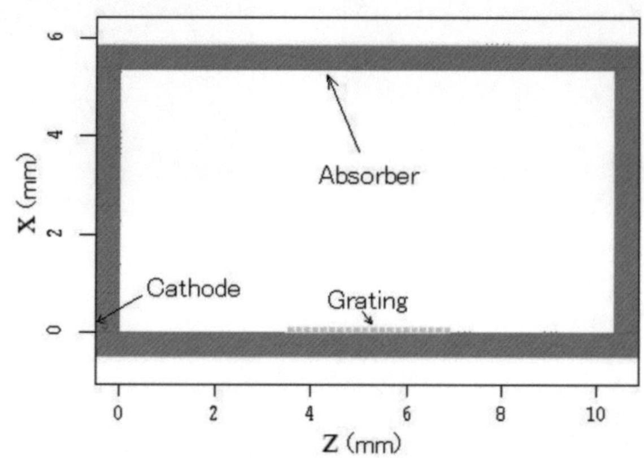

Fig. 5. Geometry used in simulation. (The surface of the grating is assumed to consist of a perfect conductor)

grating period	173 μm
groove width	62 μm
groove depth	100 μm
electron beam energy	40 keV
beam thickness	24 μm
beam-grating distance	34 μm
external magnetic field	2 T

Table 1. Main parameters for simulation

The main parameters chosen in our simulation are summarized in table 1. The electron beam specification will be varied and described in detail in each of the following simulation cases when it is necessary. The external magnetic field (z direction) is only used for the continuous beam simulation, in order to ensure stable beam propagation above the grating.

As to the diagnostics, MAGIC allows us to observe a variety of physical quantities such as electromagnetic fields as functions of time and space, power outflow, and electron phase-space trajectories . We can set the relevant detectors anywhere in the simulation area.

3.1 A single bunch

First of all, we simulate a single electron bunch passing over a grating possessing 20 periods, so as to clearly observe the radiation process. The electron beam current is 480 mA, and the bunch is 0.1 ps long, small compared to the Smith-Purcell radiation wavelength allowed by Eq. (2.6), and consequently the radiation is coherent. The code runs enough time to ensure that all the radiation emitted arrive at the detector.

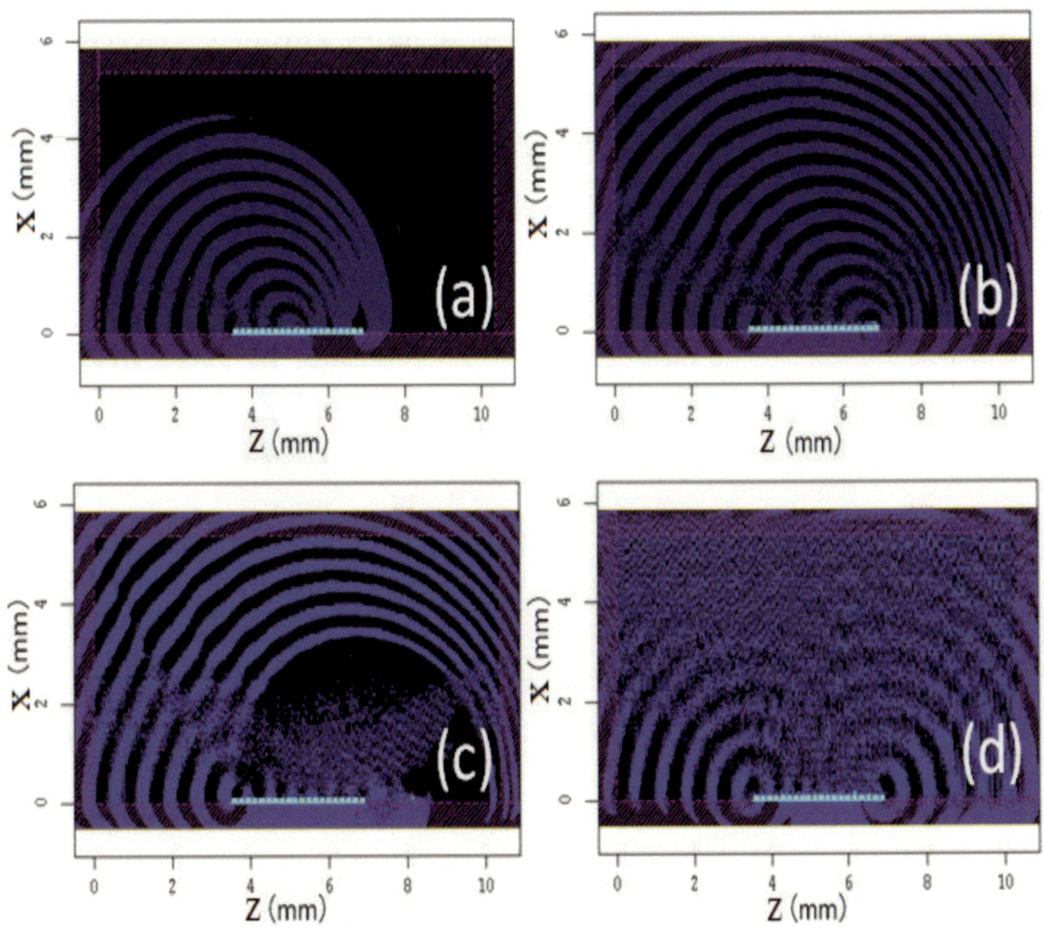

Fig. 6. Contour plots of B_y at (a) t=50.9 ps, (b) t=65.6 ps, (c) t=77 ps, (d) t=148.9 ps

The radiation process can be understood through the contour plots of magnetic field B_y as shown in Fig. 6, where the crescent-shaped wave fronts of the radiation are illustrated. In Fig. 6 (a), we see that the bunch has reached the center of the grating and has covered 10 periods, and 10 crescents clearly appeared in the vacuum box. From this phenomenon we can deduce that the radiation is attributed to Smith-Purcell effect, which has the characteristic that the electron diffracts at every period of grating. Fig. 6(b) shows that the bunch has covered all the periods and has arrived at the end of the grating. From Fig. 6(c) we understand that the Smith-Purcell radiation will no longer be emitted when the bunch moves beyond the grating. But the interesting thing happens in Fig. 6(d), where clear interference pattern appears. It is seen that the pattern is formed by two waves radiated from the two ends of the grating, and we deduce that is the surface wave radiation. The surface wave does not radiate until it reaches the ends of a grating, undergoing partial reflection and partial diffraction. The reflected portion oscillates between the grating's ends, and that is the reason that even the electron bunch disappears from the simulation area the radiation is still emitted.

A set of Bz detectors is placed at the same distance 5.346 mm and various angles from the grating center. One of the temporal behavior observed at 120° is given in Fig. 7, where again we see that the surface wave oscillates after the Smith-Purcell radiation ends. The corresponding FFT of Fig. 7 is as shown in Fig. 8, clearly demonstrating the Smith-Purcell radiation and surface wave signals, respectively. The surface wave frequency is 425 GHz, independent of radiation angle, lower than the allowed Smith-Purcell radiation frequency at any angles. This value may be read from the dispersion relation for the grating, which was developed in last section. The dependence of the Smith-Purcell radiation frequency on the radiation angle is very close to the theoretical calculation from Eq. (2.6), as shown in Fig. 9. Also plotted is the distribution of Bz amplitude in Fig. 9, which shows that the Smith-Purcell radiation is weak at small angle, and it reaches its maximum at about 125° in this simulation. This result is in agreement with Fig.2.. The amplitude of the surface wave is not plotted for clarity, and hereafter, we only concentrate on the Smith-Purcell radiation.

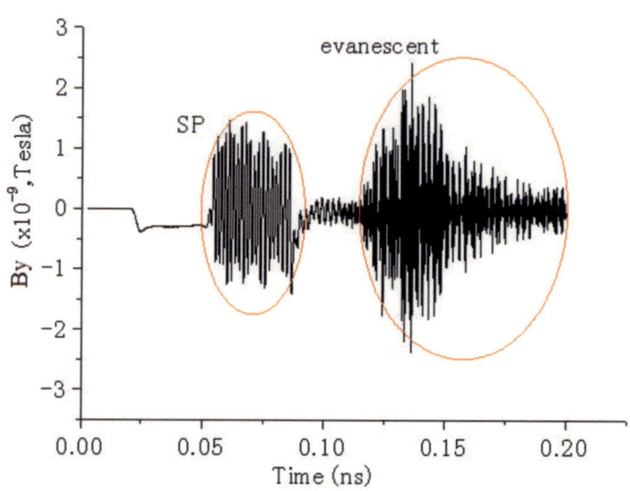

Fig. 7. Time signal of B_y, observed at point 5.346 mm and 120° from the center of the grating.

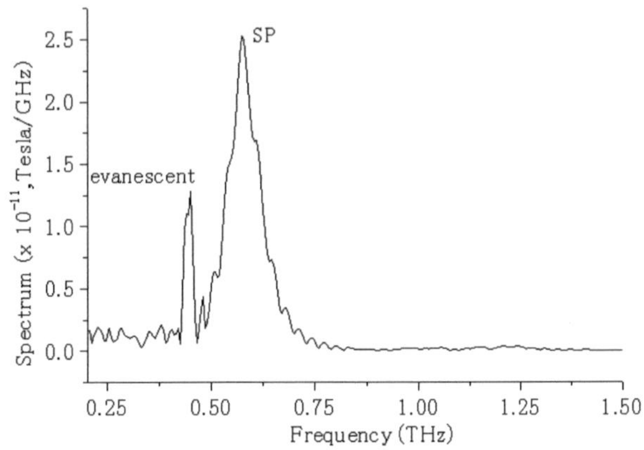

Fig. 8. FFT of time signal corresponding to Fig.7.

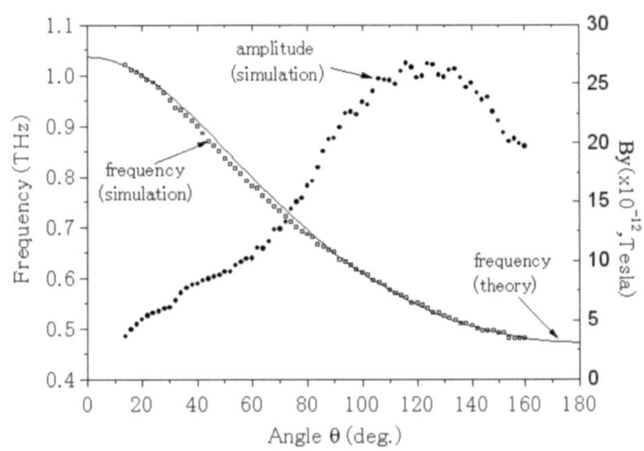

Fig. 9. Radiation frequency and amplitude as a function of angle.

3.2 Train of bunches

When the electron bunches are repeated periodically, the spectral intensity of the radiation is enhanced at the bunching frequency and its harmonics, which is called superradiance. The electron beam can be bunched by some proper devices to form a train of bunches before they are introduced in the grating system. Under certain conditions, a continuous beam can also be bunched by the interaction with the surface wave during its propagation along a grating. In order to clearly demonstrate the properties of superradiant radiation, we generate a train of electron bunches to drive the grating in our simulation.

In this simulation, we also use a grating that has 20 periods. The repeat frequency of bunches is set to be 300 GHz, and the parameters for each single bunch are the same to those mentioned earlier. During the running time, 60 bunches are produced and enter the simulation area. From the FFT of the temporal behavior observed by the Bz detectors we know that the radiation is focused on two frequencies, the second and the third harmonic of bunches' frequency, which falls into the allowed frequency of the first order Smith-Purcell radiation. The results are illustrated in Fig. 10, where two signals are clearly demonstrated.

The dominant radiation is at the second harmonic of bunches' frequency and peaked at the angle 104°, which is the angle satisfying Eq. (2.6) for such a frequency, while the other one radiates at the frequency of the third harmonic, and at the angle of about 40°. Also in Fig. 10, comparing to the second harmonic radiation, the third one is weak, which roughly corresponds to the amplitude distribution shown in Fig. 9. Another evident to prove the fact that the two radiations emit at certain angles can be found in the contour plot of B_y, as shown in Fig. 11. The second harmonic radiation is dominant and clearly observed to radiate at the angle of about 104°, corresponding to what is shown in Fig. 10.

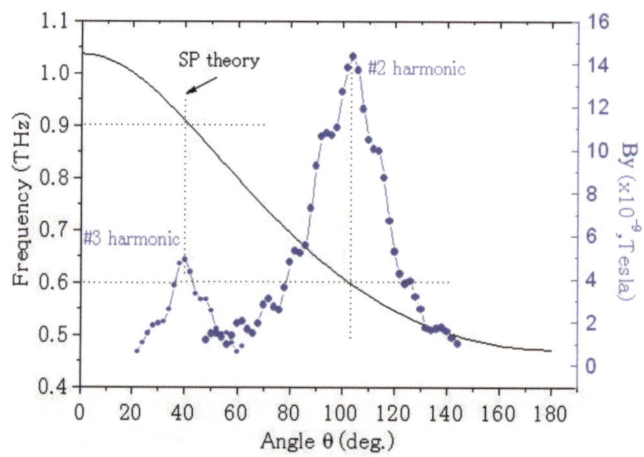

Fig. 10. Amplitude of superradiant radiation as a function of angle. Also shown are the frequency characteristics.

It should be noted that, about the frequency characteristics of superradiant radiation, Andrews and co-workers already predicated in their theoretical analysis[Andrews, H.], and Donohue also discussed it based on the simulation of a continuous beam[Donohue, J.D.]. The results of our pre-bunched beam simulation support their predication.

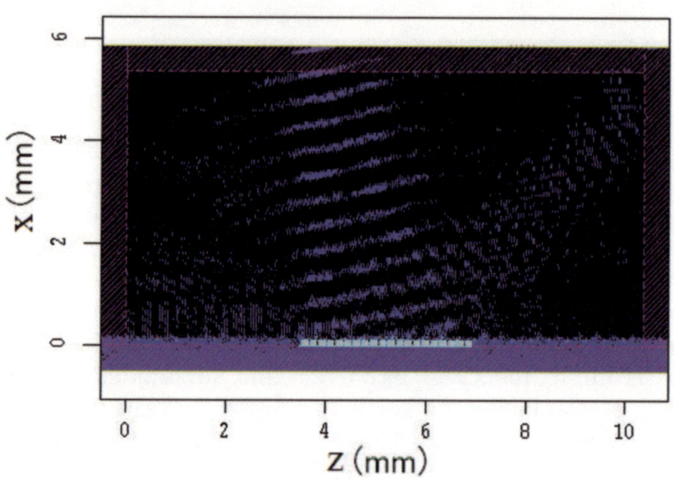

Fig. 11. Contour plot of B_y for superradiant ratiation.

3.3 Continuous beam

The electron beam bunching induced by the surface wave has been addressed by Donohue and Gardelle [Donohue, J.T.]. They also analyzed the relation of growth rate with the electron beam current. Here, we only concentrate on the output power of Smith-Purcell radiation

In this simulation we deal with a grating with 50 periods. The electron beam from the cathode is continuous, not modulated, and an external magnetic field of 2T is introduced to prevent the beam from diverging. We vary the beam current and observe the total power flow out the top plane. From the theoretical analysis in section 2.1, we know that the beam line intersects the dispersion curve at a point representing a backward wave, which means the device operates in the mode of backward-wave oscillator (BWO). Under certain conditions, the device can start to oscillate without external feedback if the beam current exceeds a threshold value beyond gain occurs. During the simulation, we experienced that there surely exists a certain value for the beam current, over which the beam bunching can occur. We read the peak power of Smith-Purcell radiation through the observation of power spectrum, and plot the result in Fig. 12, where we can easily identify two regimes for incoherent and superradiant radiation, respectively. We can also deduce the threshold of beam current.

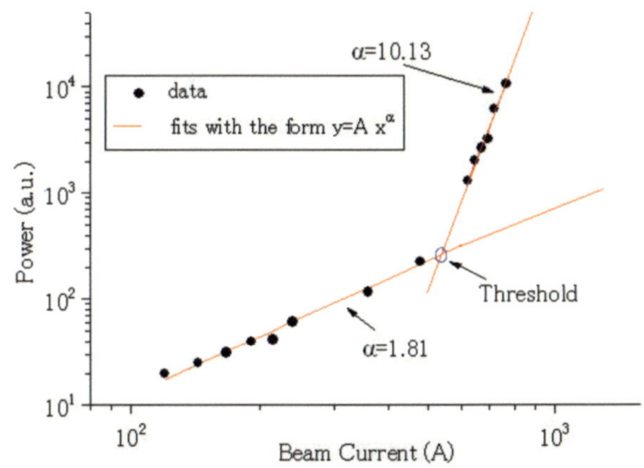

Fig. 12. Output power with respect to the beam current.

The frequency characteristic is given in Fig. 13. It is seen that the frequencies of the incoherent radiation and superradiant radiation are completely different. We found that the frequency of incoherent radiation shows classical Smith-Purcell radiation characteristics, and the 640 GHz shown in Fig. 13 is close to the radiation frequency emitted at 90°, while the superradiant radiation frequency 840 GHz is the second harmonic of the evanescent wave which bunches the beam, independent of radiation angle, and it corresponds to the characteristic discussed earlier. We note that the frequency slightly decreases as the beam current increases, because the space charge reduces the electrons' velocity. The velocity reduction influences the frequency of incoherent and superradiant radiation in different ways: for the case of incoherent radiation, the lower velocity will give rise to lower frequency determined by Eq. 2.6; while for the case of superradiant radiation, the reduction of particle velocity makes the intersection of beam line and dispersion curve shift to smaller frequency (see Fig.4), which means a decrease of the frequency of the surface wave.

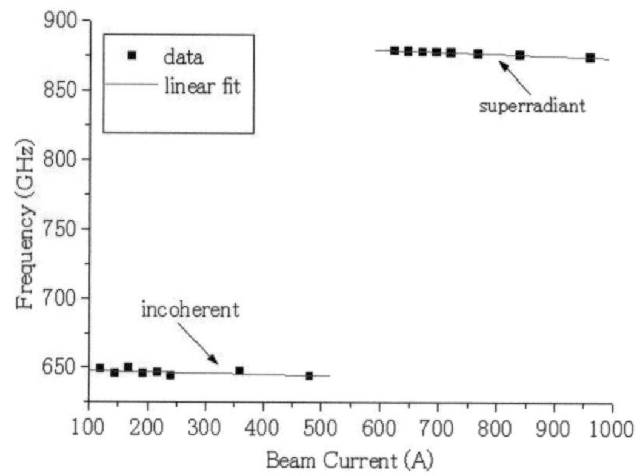

Fig. 13. Frequencies of incoherent and superradiant radiation with respect to beam current.

4. Improvement of Smith-Purcel free-electron laser

In order to improve the efficiency of Smith-Purcell free-electron laser, some new schemes have been worked out. Here we introduce two methods. By those methods, the beam-wave interaction can be enhanced, and then the growth rate could also be improved and, consequently, the start current is expected to be reduced.

4.1 Grating with Bragg reflectors

Based on the fact that the surface wave cannot radiate and it is partially reflected and partially diffracted at the ends of the grating , a scheme of grating with Bragg reflectors is proposed to improve the reflection coefficient, as shown in Fig. 14. For the case of operation point being at the backward-wave region, the Bragg reflector may correspond to the zero harmonic or the -1th harmonic. The reflected zero (or -1st) harmonic increases the entire field when the phase is well matched, and certainly this leads to the increase of the field of zero harmonic and, consequently, the beam-wave interaction would be enhanced. We can tune the lengths of g_1 and g_2 as shown in Fig.14 to optimize the phase-matching. In the following, we demonstrate the scheme of reflecting -1st order harmonic by a two-dimensional particle-in-cell simulation.

The simulations are carried out with using CHIPIC code, which is a finite-difference, time-domain code designed to simulate plasma physics processes. The grating system is assumed to be perfect conductor, and it has uniform rectangular grooves along the y direction, with parameters mentioned above. The main grating is assumed to have 60.5 periods. A sheet electron beam with the thickness of 24 μm propagates along the z direction, and its bottom is over the grating surface by height of 34 μm. It is a perfect beam produced from a small cathode located at the left boundary of the simulation area. The beam wave interaction and radiation propagation occur in the vacuum box, which is enclosed with absorber regions. Since it is a two-dimensional simulation, it is assumed that all fields and currents are independent of the y direction. We have simulated the reflection effect of a surface wave by the end of grating and by Bragg grating with using the method mentioned somewhere[Andrews, H.]. It has been shown that for the frequency of our interest the reflection coefficient can be improved from 0.35 to 0.85 when a Bragg grating is connected with the end of the main grating, and 10 periods of the Bragg grating is enough.

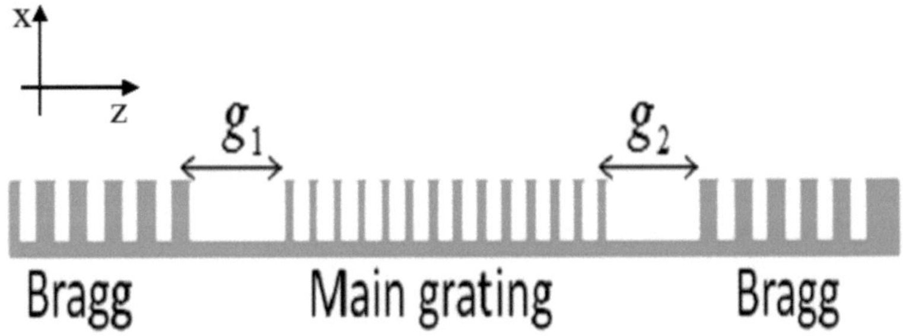

Fig. 14. A scheme of a main grating with Bragg gratings connected at both ends

We firstly determine the wavelength of the -1st order harmonic through simulating the main grating alone, and with using 35 keV electron beam it turns out to be $\lambda_{-1} = 599.2\mu m$, thus, the period of the corresponding Bragg grating should be $d_B^{-1} = \lambda_{-1}/2 = 299.6\mu m$. We simplify choose the groove width as $d_B^{-1}/2 = 149.8\mu m$, and groove depth is 100μm same as that of the main grating. The frequency of the surface wave is 432 GHz, which is a little bit lower than that of the analytical calculation, due to the decrease of the electron's energy induced by the effect of space charge. The procedure of optimization is as follows: since the energy carried by -1st harmonic moves backward (negative z direction), we firstly set the Bragg grating at the upstream end only, and tune g_1 to find the biggest growth rate through observation of the evolution of the y-component magnetic field. The observation point is set 17.3 μm above the grating surface at the center of the main grating; Next, we set another Bragg grating at the downstream end and optimize g_2. The simulation result is shown in Fig. 15, where the evolution of the y-component magnetic field is given.

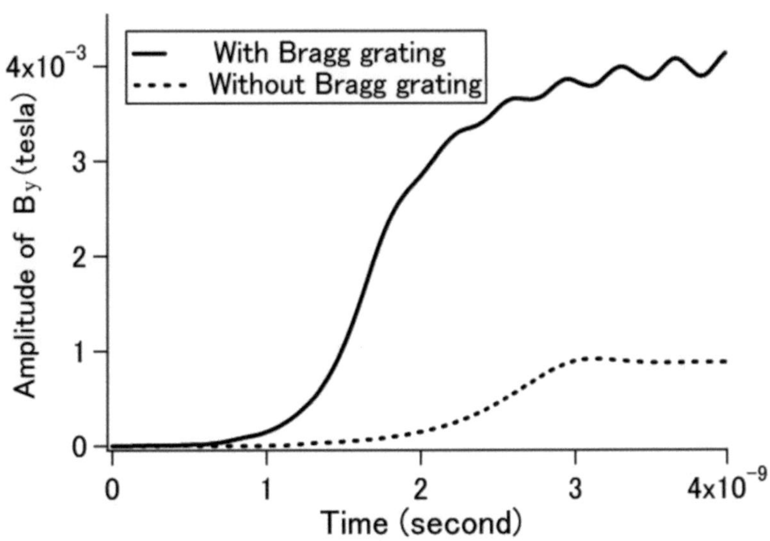

Fig. 15. Evolution of the amplitude of y-component magnetic field.

For comparison, the result obtained without Bragg gratings is also plotted. It is found that for the same current density (1.7×10^6 A/m^2), oscillation starts earlier when Bragg reflectors are used, and saturation occurs sooner. It is found that the growth rate is $\mathrm{Im}(\omega) = 4.12 \times 10^9$ s^{-1} when Bragg reflectors are used and $\mathrm{Im}(\omega) = 2.32 \times 10^9$ s^{-1} when they are not. Bragg reflectors increase the growth rate by a factor of 1.8. By varying the current density it is found that the start current density is 0.9×10^6 A/m^2 when Bragg reflectors are used and 1.6×10^6 A/m^2 when they are not used. Bragg reflectors reduce the start current by a factor of 1.8.

4.1 Grating with sidewall

A sidewall grating is proposed to enhance the coupling of the optical beam with the electron beam. By such a way, the requirements on the electron beam is possible to be relaxed; the growth rate could be improved and consequently the start current could be reduced. It is expected that the optical beam is confined between the two sidewalls to keep a good coupling with the electron beam during the interaction. Furthermore, such a configuration adds no impact on the super-radiant Smith-Purcell emission, which emits over the grating at a certain angle relative to the direction of electron beam propagation, because there is not a top plane above the grating. With the help of three-dimensional particle-in-cell simulations, we compare the general grating (without sidewall) with the sidewall grating and then show the advantages of the latter one.

The simulation models for the general grating and sidewall grating are shown in Fig. 16 (a) and (b), respectively. A cylindrical electron beam is supposed to fly over the grating.

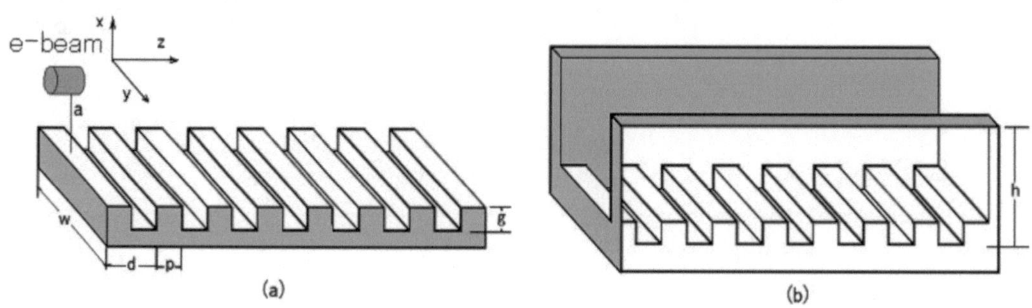

Fig. 16. Schematic of grating model. (a) general grating (b) sidewall grating.

Main parameters are summarized here: period d=2 cm, ridge width p=1 cm, groove depth g=1 cm, grating width w=10 cm, beam hight a=2mm, beam radius r=2.5 mm, beam energy E= 100 keV, wall hight h=14 cm and period number N=46. By these parameters the device operates as a backward wave oscillator, and the synchronous evanescent wave is with the frequency of 4.5 GHz. Details can be found in our previous work[Li, D.]. The grating, assumed to be a perfect conductor, is set in the center of the bottom of a vacuum box bounded by an absorption region. A continuous beam produced from a cathode moves in the z-axis. The simulation area is divided into a mesh with a rectangular cell of very small size in the region of beam propagation and large in the rest. The simulation is performed in the gigahertz region for the convenience to run the code, and we believe the physics applies to the terahertz regime.

The result of the beam-wave interaction is directly reflected by the evolution of the electromagnetic field of the evanescent wave, such as the longitudinal component of electric field Ez. When certain conditions are satisfied, the electric field Ez indicates the processes from spontaneous radiation, exponential growth to saturation. In Fig. 17, the comparisons for the general and sidewall gratings are given. For the case of 0.5 A electron beam, the general grating device cannot reach the saturation even over 500 ns while the sidewall grating device saturates at about 110 ns; for the case of 0.6 A electron beam, the general grating device saturates at around 400 ns while the sidewall grating device saturates at 90 ns. Apparently, the time required to get saturation is dramatically reduced. Furthermore, by the sidewall grating the amplitude of the electric field is also improved.

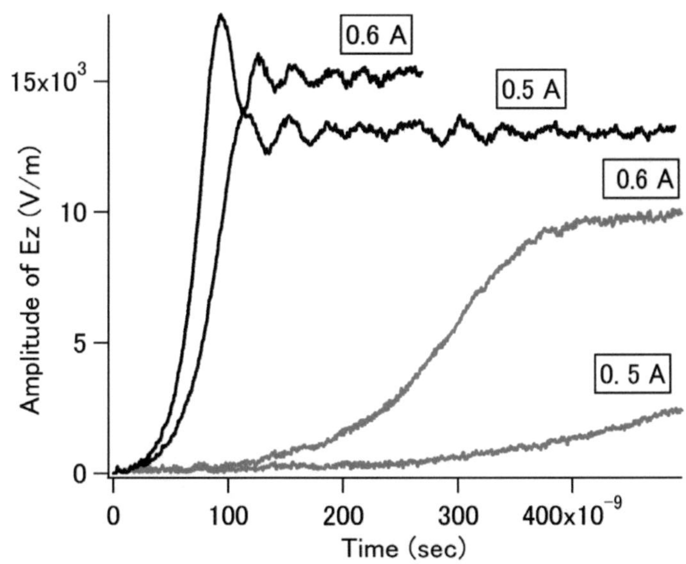

Fig. 17. Evolution of amplitude of electric field Ez. (gray curves for general grating and black curves for sidewall grating.)

5. Grating of negative-index material

There is currently interest on the research of negative-index material, which shows many exotic and remarkable electromagnetic phenomena, such as reversed Cherenkov radiation and reversed Doppler shifts[Agranovich, V. M.]. Recent successes in fabricating these artificial materials [Shelby, R. A.] have initiated an exploration into the use of them to investigate new physics and to develop new applications. It has been demonstrated in theoretical analysis and simulation that enhanced diffraction can be achieved from a grating with negative-index material compared with a grating with positive-index material when a plane-wave is incident [Depine, R. A.]. And this implies a possibility of realizing a high-performance Smith-Purcell free-electron laser.

5.1 Smith-Purcell radiation

We calculate the two-dimensional Smith-Purcell radiation from a grating with a homogeneous, isotropic, and linear material. The grating is with a sinusoidal profile $x = g(z) = 0.5h\cos(\frac{2\pi}{d}z)$, where d is the period and h the amplitude, as shown in Fig.18. The region $x > g(z)$ is vacuous, whereas the medium occupies the region $x < g(z)$, characterized by relative permittivity ε_r and permeability μ_r. If the medium is of the negative-index material, the real part of both permeability and permittivity is negative. And the positive-index material requires that the real part of both permittivity and permeability are positive. The incident wave is supposed to be from a line charge with q coulombs per length passing above the grating with the distance x_0 and velocity v along z-axis. Only considering the TM mode, the y-directed component of magnetic field of the evanescent

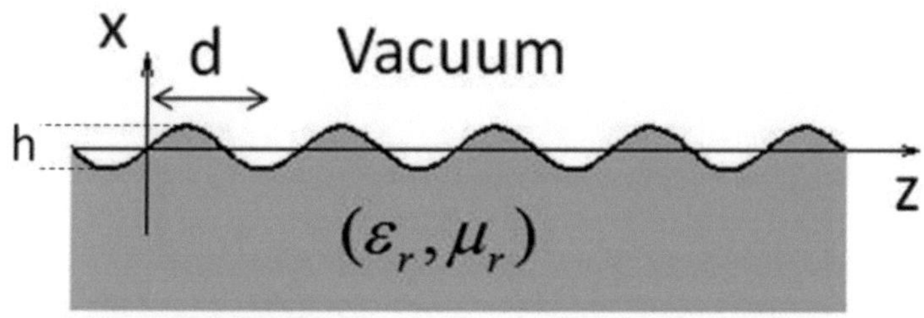

Fig. 18. Schematic diagram of grating

wave from the line charge is given by $\frac{q}{2}e^{\varsigma_0(x-x_0)}e^{j\frac{\omega}{v}z}$ (Hereafter the time part $e^{-j\omega t}$ is omitted), where $\varsigma_0 = \sqrt{\frac{\omega^2}{v^2} - \frac{\omega^2}{c^2}}$, ω the angular frequency, c the light velocity in vacuum. The diffracted and refracted fields by the grating can be represented by Rayleigh expansions, so, the y-directed component of the total magnetic field outside the corrugations can be written as

$$H_y^{(1)} = \frac{q}{2}e^{\varsigma_0(x-x_0)}e^{j\frac{\omega}{v}z} + \sum_{p=-\infty}^{\infty} A_p e^{j\alpha_p x} e^{jk_p z} \quad x > \max g(z) \tag{5.1}$$

$$H_y^{(2)} = \sum_{p=-\infty}^{\infty} B_p e^{-j\beta_p x} e^{jk_p z}, x < \min g(z) \tag{5.2}$$

Here, A_p and B_p are scalar coefficients to be determined and

$$k_p = \frac{\omega}{v} + \frac{2p\pi}{d},\ \alpha_p = \sqrt{\frac{\omega^2}{c^2} - k_p^2},\ \beta_p = \sqrt{\frac{\varepsilon_r\mu_r\omega^2}{c_2} - k_p^2}.$$

For the vacuous half-space the conditions $\mathrm{Re}(\alpha_p) > 0$ and $\mathrm{Im}(\alpha_p) > 0$ are required, whereas the medium half-space requires $\mathrm{Re}(\beta_p) > 0$ for positive-index material, $\mathrm{Re}(\beta_p) < 0$ for negative-index material, and $\mathrm{Im}(\beta_p) > 0$. The corresponding electric fields can be achieved from Maxwell equations. The continuity of the tangential components of the total electric field and magnetic field at the boundary $x = g(z)$ requires $H_y^{(1)} = H_y^{(2)}$ and $\vec{n} \cdot \nabla H_y^{(1)} = \varepsilon_r^{-1} \vec{n} \cdot \nabla H_y^{(2)}$, where $\vec{n}$ is a unit vector normal to the boundary. According to the Rayleigh hypothesis we assume that the expansions in Eqs.(5.1-5.2) can be used in the boundary conditions. We multiply both sides of a boundary condition by $e^{jk_r z}$ and then integrate with respect to z over one period. After some algebraic works, we obtain a system of linear equations as

$$\sum_{p=-\infty}^{\infty} -A_p \cdot U_{r-p}^p + B_p \cdot V_{r-p}^p = \frac{qe^{-\varsigma_0 x_0}}{2} T_r \tag{5.3}$$

$$\sum_{p=-\infty}^{\infty} A_p \cdot \frac{\frac{\omega^2}{c^2} - k_p k_r}{\alpha_p} \cdot U_{r-p}^p + B_p \cdot \frac{\frac{\varepsilon_r \mu_r \omega^2}{c^2} - k_p k_r}{\varepsilon_r \beta_p} \cdot V_{r-p}^p$$

$$= -\frac{jqe^{-\varsigma_0 x_0}}{2} \cdot \frac{\frac{\omega^2}{c^2} - \frac{\omega}{v} \cdot k_r}{\varsigma_0} \cdot T_r \tag{5.4}$$

where

$$U_{r-p}^p = \frac{1}{d} \int_0^d e^{j\alpha_p g(z)} e^{-j\frac{2\pi(r-p)z}{d}} dz$$

$$V_{r-p}^p = \frac{1}{d} \int_0^d e^{-j\beta_p g(z)} e^{-j\frac{2\pi(r-p)z}{d}} dz$$

$$T_r = \frac{1}{d} \int_0^d e^{\varsigma_0 g(z)} e^{-j\frac{2\pi r z}{d}} dz$$

If the grating is made of perfect conductor, the equations can be simplified as

$$\sum_{p=-\infty}^{\infty} A_p \cdot \frac{\frac{\omega^2}{c^2} - k_p k_r}{\alpha_p} \cdot U_{r-p}^p = -\frac{jqe^{-\varsigma_0 x_0}}{2} \cdot \frac{\frac{\omega^2}{c^2} - \frac{\omega}{v} \cdot k_r}{\varsigma_0} \cdot T_r \tag{5.5}$$

As is known, when the integer p is negative, there exist radiating modes, so called Smith-Purcell radiation. The radiation frequency is dependent on the observation angle and electron velocity, and it can be known from the dispersion relation $k_p = \frac{\omega}{c} \cos\theta_p = \frac{\omega}{v} + \frac{2p\pi}{d}$, where θ_p is measured from electron moving direction. The diffraction coefficient A_p can be worked out through solving the equations numerically. In order to be independent of influence from charge q, we define the radiation factor as $R_p = A_p / (\frac{1}{2} qe^{-\varsigma_0 x_0})$.

Some computations are carried out and the radiated flux $|R_{-1}|^2$ in the -1st-order radiating wave as a function of the observation angle θ_{-1} for 35 keV electrons are given in Fig. 19 and Fig. 20. In Fig.19, comparing with the positive-index material ($\varepsilon_r = 5$, $\mu_r = 1$), it is shown that the radiated flux is higher in the region from $\pi/2$ to π for the negative-index material

($\varepsilon_r = -5$, $\mu_r = -1$), which means strong radiation can be obtained. In Fig. 20, we plot the cases of negative-index material and perfect conductor. Comparing with the perfect conductor, the negative-index material shows strong radiation in the region from $\pi/4$ to $3\pi/4$, while outside this region the radiation from the perfect conductor predominates.

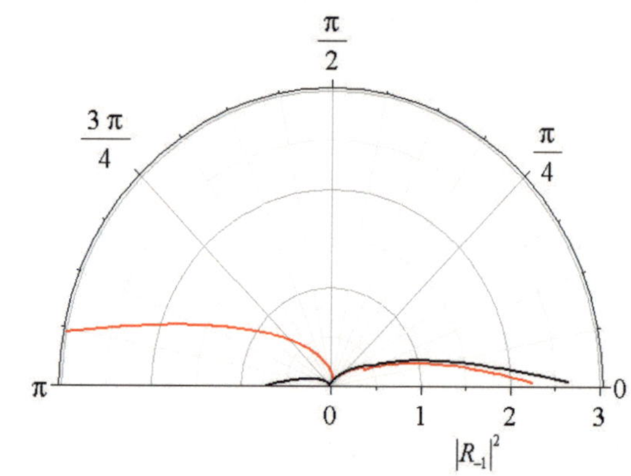

Fig. 19. Radiated flux $|R_{-1}|^2$ as a function of the observation angle θ_{-1} for 35 keV electrons. Grating period d =1 mm, h/d =0.1. (Red line for negative-index material $\varepsilon_r = -5$, $\mu_r = -1$. Black line for positive-index material $\varepsilon_r = 5$, $\mu_r = 1$)

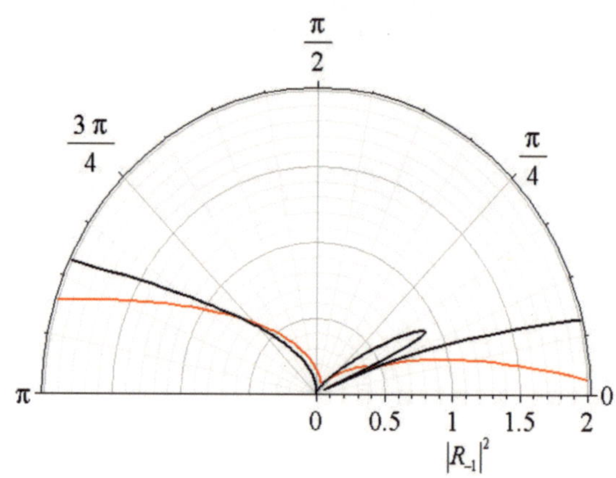

Fig. 20. Radiated flux $|R_{-1}|^2$ as a function of the observation angle θ_{-1} for 35 keV electrons. Grating period d =1 mm, h/d =0.1. (Red line for negative-index material $\varepsilon_r = -5$, $\mu_r = -1$. Black line for perfect conductor)

5.2 Surface wave

We have known that the surface modes of a grating play an important role in the operation of a Smith-Purcell free-electron laser[Andrews, H., Li, D]. The continuous electron beam interacts with the surface mode and is bunched periodically when beam current is beyond so called start current, then the periodic bunches emits in the form of super-radiant Smith-Purcell radiation at a certain angle. Next, we explore the surface mode of a grating with negative-index material.

Considering the case of without incident wave, there are only evanescent wave near the corrugation boundary. Thus, the y-directed component of the total magnetic field outside the corrugations can be written as

$$H_y^{(1)} = \sum_{m=-\infty}^{\infty} C_m e^{-\alpha_m x} e^{jk_m z} \ , \ x > \max g(z) \tag{5.6}$$

$$H_y^{(2)} = \sum_{m=-\infty}^{\infty} D_m e^{\beta_m x} e^{jk_m z} \ , \ x < \min g(z) \tag{5.7}$$

Here, C_m and D_m are scalar coefficients and

$$k_m = k + \frac{2m\pi}{d} \ , \ \alpha_m = \sqrt{k_m^2 - \frac{\omega^2}{c^2}} \ , \ \beta_m = \sqrt{k_m^2 - \frac{\varepsilon_r \mu_r \omega^2}{c^2}} \ .$$

For the vacuous half-space the conditions $\mathrm{Re}(\alpha_m) > 0$ and $\mathrm{Im}(\alpha_m) < 0$ are required, whereas the medium half-space requires $\mathrm{Re}(\beta_m) > 0$, $\mathrm{Im}(\beta_m) > 0$ for negative-index material, and $\mathrm{Im}(\beta_m) < 0$ for positive-index material. Using the same boundary conditions mentioned above, it is straightforward to get a system of linear equations as

$$\sum_{m=-\infty}^{\infty} -C_m \cdot F_{n-m}^m + D_m \cdot G_{n-m}^m = 0 \tag{5.8}$$

$$\sum_{m=-\infty}^{\infty} C_m \cdot \frac{\frac{\omega^2}{c^2} - k_m k_n}{\alpha_m} \cdot F_{n-m}^m + D_m \cdot \frac{\frac{\varepsilon_r \mu_r \omega^2}{c^2} - k_m k_n}{\varepsilon_r \beta_m} \cdot G_{n-m}^m = 0 \tag{5.9}$$

where

$$F_{n-m}^m = \frac{1}{d} \int_0^d e^{-\alpha_m g(z)} e^{-j\frac{2\pi(n-m)z}{d}} dz$$

$$G_{n-m}^m = \frac{1}{d} \int_0^d e^{\beta_m g(z)} e^{-j\frac{2\pi(n-m)z}{d}} dz$$

The dispersion relation for surface mode is obtained by equating to zero the determinant of the coefficients in these equations. Through numerically calculating we get the dispersion curve for negative-index material ($\varepsilon_r = -9$, $\mu_r = -0.1$) and ($\varepsilon_r = -0.1$, $\mu_r = -11$) as is shown in Fig. 21. The 35 keV beam line is also plotted, and the intersection implies the operation point, where the surface wave is synchronous with electron beam. It is also shown that the 35 keV electron beam interacts with the backward-wave, thus, the external feedback system is not necessary. If the beam current is high enough for a device to oscillate, the electron beam would be periodically bunched, and those periodic bunches can emit at the second harmonic in the form of super-radiant Smith-Purcell radiation [Andrews, H., Li, D.].

The permittivity ε_r and permeability μ_r are limited in a narrow region, $(-1<\mu_r<0, -1>\varepsilon_r>1/\mu_r)$ and $(\mu_r<-1, -1<\varepsilon_r<1/\mu_r)$, for the existence of surface wave. However, this limitation can be relaxed by placing a conductor boundary at the bottom of the grating. Such a grating scheme is under research.

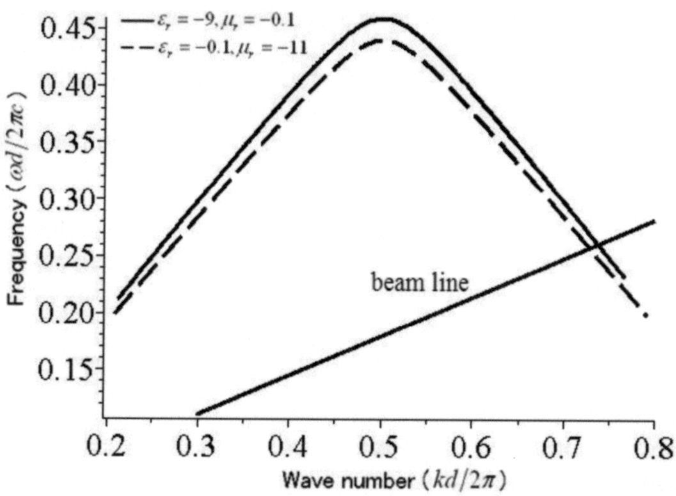

Fig. 21. Dispersion relation of surface waves from the grating of negative-index materials.

6. Conclusion

In conclusion, we theoretically analyzed the the Smith-Purcell radiation and the surface waves induced on the surface of a grating. The evanescent wave of a moving charge is reflected by the periodic structure on the surface of a grating, and the radiating part in the reflected waves forms the Smith-Purcell radiation. The grating also supports surface wave, which plays an important role in the operation of a Smith-Purcell free-electron laser. The surface wave cannot radiate, but it interacts with the electron beam and realize periodic beam bunch, resulting in the generation of super-radiant Smith-Purcell radiation. These phenomena are demonstrated with the help of particle-in-cell simulation. The sidewall grating and Bragg reflection system are proposed to improve the efficiency of Smith-Purcell free-electron laser. Both of them can enhance the beam-wave interaction, improve the growth rate and reduce the start current, which are promissing technologies in the development of Smith-Purcell free—electron lasers. We explored the grating made of negative-index materials, and find that in a certain range of radiation angle the Smith-Purcell radiation is stronger than that from a grating made of metal or positive-index materials. The surface wave from such a grating is also invegistated, which shows possibility in developing a Smith-Purcell free-electron laser.

7. Acknowledgment

This work is supported by a Grant-in-Aid for Scientific Research on Innovative Areas "Electromagnetic Metamaterials" (No. 22109003) from The Ministry of Education, Culture, Sports, Science, and Technology (MEXT), Japan.

8. References

Agranovich, V.M.; Gartstein, Yu. N. (2006). Spatial dispersion and negative refraction of light, *Physics-Uspekhi*, Vol. 49, No. 10, pp.1029-1044.

Andrews, H.; Brau, C.A. (2004). Gain of a Smith-Purcell free-electron laser, *Physical Review Special Topics - Accelerators and Beams*, Vol. 7, pp. 070701-1-070701-7. (2009). Observation of THz evanescent waves in a Smith-Purcell free-electron laser, *Physical Review Special Topics - Accelerators and Beams*, Vol. 12, pp. 080703-1-080703-5.

Chen,H.-T; Padilla,W.J., Zide,J.M.O., Gossard,A.C. & Taylor, A.J. (2006). Active terahertz metamaterial devices, *Nature*, Vol. 444,597-600.

Depine, R. A.; Lakhtakia, A. (2004). Plane-wave diffraction at the periodically corrugated boundary of vacuum and a negative-phase-velocity material, *Physical Review E*, Vol. 69, pp.057602-1-057602-4

Donohue, J.T.; Gardelle, J. (2006). Simulation of Smith-Purcell terahertz radiation using a particle-in-cell code, *Physical Review Special Topics - Accelerators and Beams*, Vol. 9, pp. 060701-1-060701-7. (2005). Simulation of Smith-Purcell radiation using a particle-in-cell code, *Physical Review Special Topics - Accelerators and Beams*, Vol. 8, pp. 060702-1-060702-9.

Gardelle, J.; Modin, P. & Donohue, J.T. (2010). Start Current and Gain Measurements for a Smith-Purcell Free-Electron Laser, *Physical Review Letters*, Vol. 105, No. 26, pp. 224801-1-224801-4. (2009). Observation of coherent Smith-Purcell radiation using an initially continuous flat beam, *Physical Review Special Topics - Accelerators and Beams*, Vol. 12, pp. 110701-1-110701-6.

Gover, A.; Livni Z. (1978). Operation Regimes of Cerenkov-Smith-Purcell Free Electron Lasers and T.W. Amplifiers, *Optics Communications*, Vol. 26, No. 3, pp. 375-380

Kuma, V.; Kim, K. J. (2006). Analysis of Smith-Purcell free-electron lasers, *Physical Review E*, Vol. 73, pp. 026501-1-026501-15.

Leavitt, R.P.; Wortman, D. E. & Morrison C.A. (1979). The orotron-A free-electron laser using the Smith-Purcell effect, *Applied Physics letters*, vol. 35, pp.363-365.

Li, D.; Yang, Z., Imasaki, K. & Park, G. S. (2006). Particle-in-cell simulation of coherent and superradiant Smith-Purcell radiation, *Physical Review Special Topics - Accelerators and Beams*, Vol. 8, pp. 060702-1-060702-9. Vol. 9, pp. 040701-1-040701-6. (2006). Three-dimensional simulation of super-radiant Smith-Purcell radiation, *Applied Physics Letters*, Vol. 88, pp. 201501-1-201501-2. (2011). Improve growth rate of Smith–Purcell free-electron laser by Bragg reflector, *Applied Physics Letters*, Vol. 98, pp. 211503-1-211503- 3

Schächter, L.; Ron, A. (1989). Smith-Purcell free-electron laser, *Phys. Rev. A*, Vol. 40, No.2, pp.876-896.

Shelby ,R.A.; Smith,D.R.& Schultz,S. (2011). Experimental Verification of a Negative Index of Refraction, *Science*, Vol. 292, 77-79.

Shelby, R. A.; Smith, D. R. & Schultz, S. (2001). Experimental erification of anegative index of refraction, *Science*, Vol. 292, pp. 77-79.

Shibata, Y.; Hasebe, S., Ishi, K., Ono, S. & Ikezawa, M. (1998). Coherent Smith-Purcell radiation in the millimeter-wave region from a short-bunch beam of relativistic electrons, *Physical Review E*, Vol. 57, No. 1, pp. 1061-1074

Smith, S. J.; Purcell, E. M. (1953). Visible Light from Localized Surface Charges Moving across a Grating, *Phys. Rev.* , Vol.92, pp. 1069-1069.

van den Berg, P.M. (1973); Smith-Purcell radiation from a line charge moving parallel to a reflection grating, *Journal of the Optical Society of America*, Vol. 63, No. 6, pp.689-698

Walsh, J.; Woods, K. & Yeager, S. (1994). Intensity of Smith-Purcell radiation in the relativistic regim, *Nuclear Instruments and Methods in Physics Research A*, Vol. 341, pp. 277-279

7

Interference Phenomena and Whispering–Gallery Modes of Synchrotron Radiation in a Cylindrical Wave–Guide

Sándor Varró
Wigner Research Centre for Physics
of the Hungarian Academy of Sciences
Hungary

1. Introduction

In 'usual lasers' the light amplification is a consequence of stimulated emission stemming from bound-bound quantum transitions in the active medium (e.g. gas or vapour atoms, molecules, or radiation centers in a solid), on the other hand, in free electron lasers (FELs) the radiation is produced in free-free transitions, taking place in static magnetic fields. In a homogeneous static magnetic field an electron (or other charged particle) moves along an ideal circular (or, in general, helical) trajectory, if the radiation damping is negligible, and emits a 'broad-band radiation' containing all the higher harmonics of its frequency of revolution. The characteristics of this ideally periodic radiation field has been first analysed by Schott [1]. Later several refinements and asymtotic formulae have been derived for describing the quasi-continuous radiation (which in the meantime received the name 'synchrotron radiation') of ultrarelativistic electrons moving along an instantaneous circular orbit (see e.g. Schwinger [2]). A very detailed analysis of this radiation can be found in the book by Sokolov and Ternov [3], which deals with this phenomenon both in the frame of classical electrodynamics and in the frame of relativistic quantum mechanics. By the beginning of the eighties of the last century the synchrotron radiation has become a widely used experimental tool in many branches of science and technology [4].

The idea of using periodic magnetic fields (undulators, wigglers) for properly bending the trajectories of the (ultra)relativistic electrons, and generating coherent radiation is due to Motz [5]. Besides working out the classical theory, he, with his coworkers, made the first experimental demonstrations, too. In these experiments a 100-MeV electron beam from the Stanford linear accelerator passed through the undulator, and besides millimeter waves, light radiated by the beam was also observed [6]. Later Schneider [7] found, by using a phenomenological rate equation approach, that the absorption of radiation around the cyclotron resonance of an electron may go over to negative absorption (stimulated emission), if we take into account the relativistic kinematics of the electrons. His analysis was based on the transitions between the Landau levels of electrons in a homogeneous magnetic field. In 1971 Madey's work [8] gave a new impetus to the research on generation of coherent radiation by free electron beams. He used the Weizsacker-Williams approximation (or in other words, the method of equivalent photons, see e.g. Heitler [9]) to calculate the gain due to the induced emission of radiation into a single electromagnetic mode parallel to the motion of a relativistic electron through a periodic transverse static magnetic field. He has proved that "Finite gain is available from the far-infrared through the visible region raising the possibility of continuously tunable amplifiers and oscillators at these frequencies with the further possibility of partially coherent radiation sources in the ultraviolet and x-ray regions to beyond 10 keV."[8]. In this description the periodic magnetic field is represented by an 'equivalent photon beam' in the nearly comoving frame of the relativistic electrons. By going over to this frame, the macroscopic period (on the order of centimeters) of the magnet is down-converted, due to the Lorentz contraction, to microns or even shorter wavelengths, depending on the velocity of the electron. According to the appropriate Lorentz transformation, this periodic field is essentially equivalent to a zero-mass-shell true radiation field, which is back-scattered on the electron. In this frame the amplification may come from the stimulated Compton scattering, if the radiation is coupled back by mirrors, e.g. in a Fabry–Perot arrangement. Concerning the further development of the theory, it was a remarkable result in Madey's work [8] that, though the quantum kinematics of the Compton process has been used in the analysis, the final gain formula does not contain Planck's constant in the relevant regime of the parameters. In fact, most part of the later analyses of FELs were relying on classical electrodynamics and statistics (see e.g. [10-12]). Really, the first true laser operations [13] in the infrared, at $\lambda \sim 3.4 \mu m$ and in the visible [14] at wavelength $\lambda \sim 0.65 \mu m$ could be satisfactorly described in the framework of the cassical theory. From the conceptual point of view, this has been in good accord with an earlier analysis of Borenstein and Lamb on the "classical laser"[15]. Nevertheless, Madey [8] has already made some estimates concerning the photon statistics of the proposed FEL. Also in a later special issue on free electron lasers Chen and Madey [16] concluded that "Both the experiment data and the simulations indicate that the reduction of phase noise are directly associated with electron bunching. In fact, higher degree of electron bunching leads to less phase fluctuations in the optical field. The conclusion is that our study has once again supported the idea of imparting statistical properties from radiating particles to radiated photons. Hence the generation of phase squeezed light out of the FEL radiation is anticipated, given the reduction of electron phase fluctuations due to the FEL bunching mechanism." A more general analysis of related questions has been given by Tanabe [17] in this same special issue. According to his conclusions "An ideal single-mode amplitude-stabilized gas laser operating well above its threshold could produce coherent state, however, many so-called 'coherent' light may possess only the first-order coherence. Some experimental efforts have been made for spontaneous emission; however, no experimental investigation was made for FEL stimulated emission either in exponential gain regime or in saturated one." In fact, there have already been theoretical results published, according to which the interactions of free electrons with quantized radiation fields lead to non-classical states [18-21], like squeezed coherent states or number-phase minimum-uncertainty states. Both in theory and in practice, for the characterization of the statistical properties of such radiations, the higher-order correlation functions have to be studied, as is done in Hanbury Brown and Twiss type experiments [22-23]. Recent measurements [24-25] on undulator radiation in the x-ray regime have shown that the spontaneous signal produces positive second order (intensity-intensity) correlation, similar to that of thermal radiation. This is set to be the consequence of the chaotic spatial distributions of the electrons. For the FEL a sort of 'phase-squeezed state' has been expected by Chen and Madey [16] earlier, on the other hand, extremely short (attosecond) pulses produced in very high-order harmonic generation processes of atoms or solids [26-27], can certainly be reasonably represented by quantum mechanical phase eigenstates [28]. At present it seems that the higher-order quantum correlation properties of the existing FEL sources [29-31] do not play a significant role in the applications.

We have seen above that he cyclotron and synchrotron radiation in free space has long been thoroughly studied both theoretically and experimentally. The feedback of the radiation, in case of a true laser system is usually secured by a Fabry–Perot type linear resonator arrangement. Besides, already in the sixties of the last century, in the context of cyclotron masers [32] and cavity electrodynamics, more general cavity effects have also been studied [33-34]. In the cyclotron maser the fundamental or low harmonics are generated, on the other hand, in the synchrotron radiation very high harmonics are dominant. The question naturally emerges whether could one somehow feed back these high harmonics in order to have stimulated emission and manage perhaps lasing also. This idea dates back to the principle of operation of the so-called 'halo laser' [35], on the basis of which two-dimensional stimulated emission and planar laser action in the whole 2π angle have been achieved. In this device the laser active dye material in a cylindrical cuvette is pumped from below, and the feedback is secured by partial radial and oblique reflections on the wall of the cuvette. Besides, the whispering-gallery modes of light may have also served for energy storage. The functioning of this system has clearly illustrated that the linear cavity and very small divergence is not a necessary condition to the laser action. We have studied [36-37] a slightly analogous feedback mechanism, where the radiation, emanating tangentially from an electron, is coupled back by a cylindrical mirror. The radiation follows the electron in the form of very high angular momentum whispering gallery modes. The present chapter is devoted to the discussion of the case in which the trajectories of the gyrating ultrarelativistic electrons are completely surrounded by a coaxial mirror, in a wave-guide configuration. We will show that, if the ratio of the the cylinder's radius and radius of the electron's trajectory satisfies certain geometrical resonance conditions, then the resonance practically for all very high harmonics can be achieved. As a general remark, we would like to note that recently there has been a wide-spreading interest in the physics of electromagnetic whispering-gallery modes in nanostructures [38-39]. For instance, in ref [38] it has been demonstrated that e.g. crystalline optical whispering-gallery resonators with very high Q factors can be constructed, and used for studying nonlinear processes, owing to the large field concentration.

In Section 2 the combined initial-value and boundary-value problem of the relevant Maxwell equations are solved exactly on the basis of the suitable Green's function, and the modal structure of the synchrotron radiation in the cylindrical mirror is analysed. The direct radiation reaction will be left out of consideration, however the effect of damping at resonance will be briefly discussed. In Section 3 we prove the existence of a 'broad-band resonance' where an accumulation process is taking place, due to which the intensity of the resulting radiation can be considerably increased. The present analysis aims to show that, if the electron meets with its own radiation field emitted earlier in a tangential narrow cone, and reflected back by the cylindrical mirror, then a constructive interference shows up between this retarded self-field and the actually emitted radiation field. This arrangement could perhaps serve as a basis for constructing a compact „disk-synchrotron-lasers" radiating in a plane. In section 4 we shall briefly summarize our results.

2. Excitation of cylindrical wave–guide modes by an ultrarelativistic electron

In the present section we give an exact solution of Maxwell's equations driven by an ultrarelativistic electron gyrating inside of a perfectly reflecting cylindrical wave-guide.

The transverse position of the electron moving in the homogeneous magnetic field $\vec{e}_z B_0$ has the Cartesian components

$$x(t) = r_0 \cos(\omega_0 t + \varphi_0)\,,\; y(t) = r_0 \sin(\omega_0 t + \varphi_0)\,,\; z(t) = 0\,, \qquad (1)$$

where $r_0 = \upsilon / \omega_0$, $\omega_0 = \omega_c / \gamma$ with $\omega_c = |e| B_0 / mc$ being the cyclotron frequency, υ denotes electron's velocity and $\gamma \equiv (1-\beta^2)^{-1/2}$, $\beta \equiv \upsilon / c$. We assume that the longitudinal component of the electron's velocity is zero, i.e. $\upsilon_z = 0$, and the gyration takes place in the $z = 0$ plane. The charge density ρ and the current density $\vec{j}$, associated to the trajectory of one single electron with initial phase ϕ_0 given by Eq. (1), can be conveniently expressed in cylindrical coordinates

$$\{\rho(r,\varphi,z;t),\; \vec{j}(r,\varphi,z;t)\} = \{1,\; \upsilon\vec{e}_\varphi\} e r^{-1} \delta(r - r_0) \delta[\varphi - (\omega_0 t + \varphi_0)] \delta(z) u(t) \qquad (2)$$

where we have also introduced the unit step function $u(t)$, describing a sharp switching–on of the interaction. Equations (1) and (2) correspond to a highly idealistic situation for many reasons. Because in reality no perfectly homogeneous and stationary magnetic fields can be sustained over large spatio-temporal regions, exactly planar and circular trajectories of charges can never be secured, and, moreover the position of the guiding center cannot be fixed up to an arbitrary accuracy. The production and injection of a perfectly monoenergetic electron beam is impossible, either. The radiation loss necessarily distorts the trajectory, and the reacceleration cannot be solved without causing additional oscillations. In short, the spatio-temporal inhomogenities and spectral imperfections of both the boundary conditions and of the charged particle beams, of course, do not allow to prepare and sustain such an ideal current distribution shown in Eq. (2). The 'fragility' of this external current density may be a subject of a separate study, like the investigation of the possible bunching effect caused by the feedback of the radiation which would result in superradiance and lasing. In the present we shall not consider these important questions, rather we concentrate on the simplest, but exactly solvable part of the radiation problem.

In order to obtain a physically meaningful solutions of Maxwell's equations driven by the densities given by Eq. (2) inside the cylinder, we have to take into account the boundary conditions $[\vec{n} \times \vec{E}]_C = 0$ and $[\vec{n} \cdot \vec{B}]_C = 0$, where the subscript C symbolizes the boundary of the cylinder of radius a (larger than r_0), and $\vec{n}$ is the outward unit normal of it. In words this means that the tangential component of the electric field strength $\vec{E}$ and normal component of the magnetic induction $\vec{B}$ must vanish at the surface of the cylinder (at any vertical position z). Thus $\vec{E}$ and $\vec{B}$ are expanded into a superposition of the so-called cross-sectional vector eigenfunctions [40-41],

$$\vec{E} = \sum_{m\,p} a_{mp} \vec{\nabla}_\perp \Phi_{mp} + \sum_{n\,s} b_{ns} \vec{e}_z \times \vec{\nabla}_\perp \Psi_{ns} + \vec{e}_z \sum_{m\,p} c_{mp} \Phi_{mp} \qquad (3a)$$

$$\vec{B} = \sum_{m\,p} \alpha_{mp}\vec{e}_z \times \vec{\nabla}_\perp \Phi_{mp} + \sum_{n\,s} \beta_{ns}\vec{\nabla}_\perp \Psi_{ns} + \vec{e}_z \sum_{n\,s} \gamma_{ns}\Psi_{ns} \tag{3b}$$

The symbol $\vec{\nabla}_\perp = (\partial / \partial x, \partial / \partial y, 0)$ in Eqs. (3a-b) denotes the transverse nabla vector. In Eqs. (3a,b) Φ_{mp} and Ψ_{ns} are Dirichlet and Neumann eigenfunctions satisfying the scalar Helmholtz equation $(\nabla_\perp^2 + k^2)f = 0$ with eigenvalues k_{mp} and k_{ns}, respectively. The unknown coefficients a_{mp}, b_{ns}, c_{mp} and α_{mp}, β_{ns}, γ_{ns} are to be determined as functions of time (t) and vertical position (z). According to the boundary condition $[\Phi_{mp}]_C = 0$, Φ_{mp} must have the form

$$\Phi_{mp} = J_m(x_{mp}(r / a))\begin{Bmatrix}\sin(m\varphi)\\ \cos(m\varphi)\end{Bmatrix},\ J_m(x_{mp}) = 0 \tag{4a}$$

where x_{mp} is the p-th root of the ordinary Bessel function of first kind J_m of order m. Because of the Neumann boundary condition, $[\partial\Psi_{ns} / \partial r]_C = 0$, we have

$$\Psi_{ns} = J_n(y_{ns}(r / a))\begin{Bmatrix}\sin(n\varphi)\\ \cos(n\varphi)\end{Bmatrix},\ J_n'(y_{ns}) = 0 \tag{4b}$$

where y_{ns} is the s-th root of the derivative J_n' of the the ordinary Bessel function of first kind of order n. The eigenvalues of the corresponding wave numbers are $k_{mp} = x_{mp} / a$ and $k_{ns} = y_{ns} / a$, respectively, where a is radius of the cylinder. By taking into account the orthogonality property of the cross-sectional eigenfunctions, we can derive from the inhomogeneous Maxwell equations two sets of coupled first order differential equations for the expansion coefficients. The set $\{a_{mp}, \alpha_{mp}, c_{mp}\}$ is responsible for the dynamics of the TM and longitudinal components of the electromagnetic field. On the other hand, the dynamics of the TE components is governed by the set $\{b_{ns}, \beta_{ns}, \gamma_{ns}\}$. It can be shown that $\alpha_{mp}(z=0,t) = 0$ and $c_{mp}(z=0,t) = 0$, moreover, in the case we shall discuss below a_{mp} vanishes to a good approximation at any vertical position. This means that in the plane of the electron gyration only the TE modes are excited, and that is why henceforth we shall be dealing only with the behaviour of the TE modes.

The coupled system of equations for b_{ns}, β_{ns} and γ_{ns} reads

$$\frac{\partial\beta_{ns}}{\partial z} - \frac{1}{c}\frac{\partial b_{ns}}{\partial t} - \gamma_{ns} = \frac{4\pi}{c}\frac{1}{N_{ns}^2}\int d^2 s \vec{j} \cdot (\vec{e}_z \times \vec{\nabla}_\perp \Psi_{ns}) \tag{5a}$$

$$\frac{\partial b_{ns}}{\partial z} - \frac{1}{c}\frac{\partial\beta_{ns}}{\partial t} = 0 \tag{5b}$$

$$b_{ns} - \frac{1}{ck_{ns}^2}\frac{\partial\gamma_{ns}}{\partial t} = 0 \tag{5c}$$

where $N_{ns}^2 = (\pi / \varepsilon_n)J_n^2(y_{ns})(y_{ns}^2 - n^2)$, and $\varepsilon_1 = 1$ and $\varepsilon_n = 2$ for $n = 2,3,\ldots$.The integration on the rhs of Eq. (5a) is to be evaluated over the cross-section of the cylinder. Having eliminated the functions $\boldsymbol{\beta_{ns}}$ and $\boldsymbol{\gamma_{ns}}$ from Eqs. (5a,b,c) we arrive at an inhomogeneous Klein-Gordon equation for $b_{ns}(z,t)$,

$$\left(\frac{\partial^2}{\partial z^2}-\frac{1}{c^2}\frac{\partial^2}{\partial t^2}-k_{ns}^2\right)b_{ns}=B_{ns}\delta(z)f_n'(t)/c \tag{6a}$$

where

$$B_{ns}\equiv 4e\beta\frac{(y_{ns}/a)J_n'(y_{ns}r_0/a)\varepsilon_n}{J_n^2(y_{ns})(y_{ns}^2-n^2)},\quad f_n(t)\equiv\begin{Bmatrix}\sin[n(\omega_0 t+\varphi_0)]\\ \cos[n(\omega_0 t+\varphi_0)]\end{Bmatrix}u(t) \tag{6b}$$

The Green's function of Eq. (6a) can be derived by using standard methods

$$g(t,z)=-\frac{c}{2}J_0\left(ck_{ns}\sqrt{t^2-z^2/c^2}\right)u\left(t-\frac{|z|}{c}\right),\ t\equiv t_1-t_2\,,\ z\equiv z_1-z_2 \tag{7}$$

With the help of the Green's function in Eq. (7), the solution of Eq. (6a) can be determined by straighforward integrations. The general explicit form of b_{ns} is a complicated expression, however for large values of t a relatively simple expression can be derived (for short, we present only the upper component stemming from the sine oscillations):

$$\begin{aligned}-2b_{ns}(z,t)/B_{ns}&=J_0\left(\omega_{ns}\sqrt{t^2-z^2/c^2}\right)(\sin\varphi_0)u\left(t-\frac{1}{c}|z|\right)\\&+\frac{\nu_{ns}u(\nu_{ns}-1)}{\sqrt{\nu_{ns}^2-1}}\sin n\left[\omega_0\left(t-\frac{1}{c}\sqrt{\nu_{ns}^2-1}\,|z|\right)+\varphi_0\right]\\&+\frac{\nu_{ns}u(1-\nu_{ns})}{\sqrt{1-\nu_{ns}^2}}(\cos n(\omega_0 t+\varphi_0))\exp\left[-\frac{n\omega_0}{c}\sqrt{1-\nu_{ns}^2}\,|z|\right]\end{aligned}\quad, \tag{8}$$

where $\nu_{ns}\equiv n\omega_0/\omega_{ns}$ with $\omega_{ns}\equiv ck_{ns}$ being the TE eigenfrequencies. The first term on the rhs of Eq. (8) represents a (transient) *precursor* with front velocity c, which vanishes as $1/\sqrt{t}$. The second and the third terms correspond to *above-cutoff* and *below-cutoff waves*, respectively, the latter being bound to the $z=0$ plane. At the exact resonances $\nu_{ns}=1$, Eq. (8) loses its validity, and b_{ns} has a qualitatively different form. For short, we present only the upper component of b_{ns} taken at the $z=0$ plane (of the electron's trajectory):

$$-2b_{ns}/B_{ns}=J_0(n\omega_0 t)(\sin n\varphi_0)+(n\omega_0 t)[J_0(n\omega_0 t)(\cos n\varphi_0)-J_1(n\omega_0 t)(\sin n\varphi_0)] \tag{9}$$

On the basis of the asymptotic behaviour of the Bessel functions, it can be shown that for large t times b_{ns} diverge as $\sqrt{t}\times$(oscillatory function). Of course, since some sort of damping is always present in physical systems, such a divergence is not realistic. If we introduce the phenomenological damping term $(-k_{ns}/cQ_{ns})(\partial b_{ns}/\partial t)$ into the Klein-Gordon equation on the lhs of Eq. (6), then the solution has a similar structure as that shown in Eq. (8), but with the essential difference that the potential unphysical divergences are replaced by finite resonance terms. The oscillatory parts will contain resonance denominators of the form $[(\nu_{ns}^2-1)^2+\nu_{ns}^2/Q_{ns}^2]^{1/4}$, where Q_{ns} is spectral Q-factor related to the assumed finite conductivity of the cylinder wall. Thus, close to resonance the amplitudes are increased by a factor of $\sqrt{Q_{ns}}$. For the high harmonics, we are interested in, Q_{ns} can be well approximated by a / δ_{ns}, where $\delta_{ns} = (2 / \mu\sigma\omega_{ns})^{1/2}$ is the skin depth for a spectral component. For example, for silver $\delta_{\omega} \approx 6 \times 10^{-6} cm$ for $\omega / 2\pi \approx 10^{10} Hz$. In the optical region δ_{ω} can well be of two orders of magnitude smaller, thus Q_{ω} can be very large if a is of order of meters, say.

3. Resonance conditions

In the present section we study the question of under what conditions simultaneous resonance can be reached for the 'most of the higher harmonics' in the synchrotron radiation in the cylindrical mirror. The geometrical arrangement we are interested in is shown in Fig. 1. On *geometrical resonance condition* we mean that the radii of the electon's trajectory and of the cylinder are adjusted in such a way, that a signal emanating tangentially at point A, after reflection, gets back to the electron's trajectory at point B exactly at that time when the electron (possibly after $N \geq 1$ complete revolutions, or even 'almost immediately') gets to the same point B. It is clear that, if once this condition is satisfied for the pair of points A and B, then it will be satisfied for the pair A' and B', which can be obtained by rotating the pair A and B by an arbitrary angle. *In this way the electron 'after a while' will continuously move 'in phase' in its own retarded radiation field which has been emitted earlier at different points on the trajectory.* We think that this arrangement would secure an effective feedback for obtaining stimulated emission. Needless to say, the correctness of the expectation suggested by this intuitive picture (which was originally just the starting point of the present investigation) should be checked on the basis of the accurate analytic treatment. By simple kinematic considerations it can be shown that, if the geometrical resonance condition holds, then the ratio a / r_0 satisfies the following transcendental equation

$$\beta\sqrt{x^2 - 1} - \arccos(1 / x) = N\pi \,, \quad x \equiv a / r_0 \,, \tag{10}$$

where N is the number of complete revolutions of the electron before the first ecounter with its own radiation field after one reflection. For $N = 1$ we obtain approximately $a / r_0 \cong 3\pi / 2$, and for $N = 0$ we have $a / r_0 \cong 1 + 3 / 4\gamma^2$. It is clear that for an ultrarelativistic electron the latter 'zeroth resonant condition' can be satisfied if a is only slightly larger than the radius of the trajectory, i.e. the electron moves very close to the inner surface of the cylinder. If we assume $r_0 \approx 10cm$ in a strong confining magnetic field, then the the first resonance condition can be satisfied in a cylinder of radius $a \approx 47cm$. Henceforth we shall study the case $N = 1$.

The wave resonance condition $1 = \nu_{ns} \equiv n\omega_0 / \omega_{ns} = (n / y_{ns})(\beta a / r_0)$, on the basis of the previous section, can be studied by using the asymptotic form of the zeros y_{ns} of the derivatives J_n' of the Bessel functions J_n. Here we restrict the discussion to the case when not only n but also the s are large. (It can be shown that if s considerably differs from n, then the resonance cannot be reached.) For large n we have

$$\frac{a}{r_0} \cong \beta\frac{a}{r_0} = \frac{y_{ns}}{n} = z(\zeta) + O\left(\frac{1}{n^2}\right), \quad \sqrt{z^2 - 1} - \arccos(1 / x) = \frac{2}{3}(-\zeta)^{3/2} \,, \quad \zeta = n^{-3/2} a_s' \,. \tag{11}$$

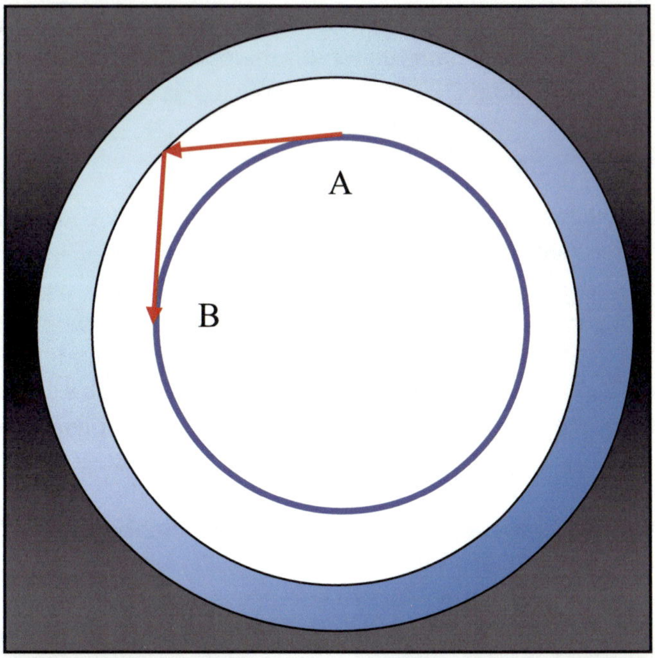

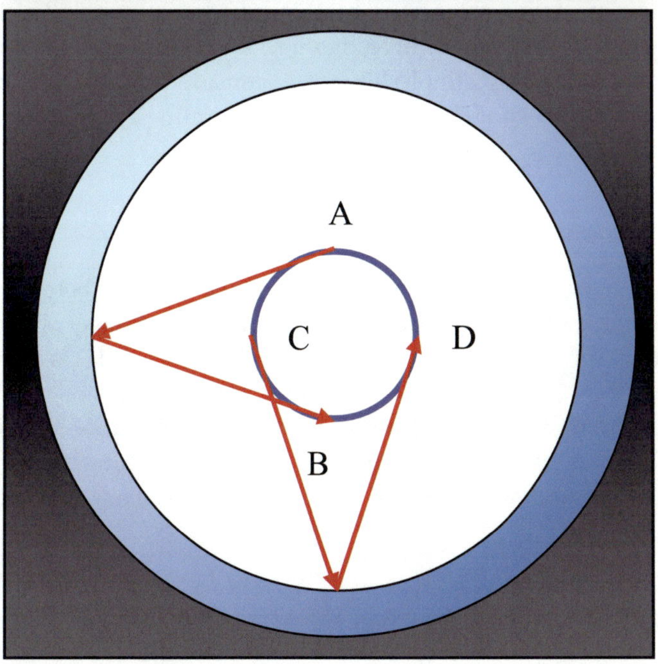

Fig. 1. A ray of radiation emanating tangentially within a narrow cone from the electron at point A, gets reflected on the cylindrical mirror and arrives at point B exacly at the time when the electron arrives there. Left: illustrates the zeroth geometrical resonance condition (N=0). Right: illustrates the first resonance condition (N=1). In the latter case, after one complete revolution the electron encounters with its own radiation field (which was emitted tangentially one turn earlier). For further explanation see the main text.

The function $z(\zeta)$ is the inverse function defined by the second equation of Eq. (11), with a'_s being the s-th negative zeros of the Airy function. Since n is very large , $z = x$, and because $\beta = \upsilon / c$ is practically unity, the left hand sides of Eq. (10) coincides with the left hand side of the second equation of Eq. (11). Now, for large s values the zeros can be approximated by the analytic formula $-\zeta = (3\pi s / 2n)^{2/3}$. By having taken this relation into account, we can easily check that, if the geometrical resonance condition is secured, then the wave resonance condition in Eq. (10) becomes an identity for $s = n$. As a consequence, *the wave resonance condition is independent of the n-values, provided these are large enough.* This means that (if Q_{ns} is a smooth function of n) there exist a broad band of the spectrum which is uniformly 'lifted-up' due to these resonances. *This resonant accumulation process can be interpreted as a result of the constructive interference between the radiation actually emitted and the self- radiation emitted earlier and fed back (by reflection) to the actual position of the electron.*

4. Summary

We have discussed the characteristics of the synchrotron radiation emitted by ultrarelativistic electrons in the interior of a coaxial cylindrical mirror. This is an unconventional geometrical arrangement, where the « superradiance » may be realized in two dimensions. In the first section we have summarized the early history and basic concepts of free-electron lasers, and outlined briefly some still partly open questions concerning the coherence properties of such sources of radiation. In the main part of the present chapter, in sections 2 and 3 we have analysed the characteristics of the synchrotron radiation emitted by a single ultrarelativistic electron in the interior of a coaxial cylindrical mirror. The combined initial-value and boundary-value problem of the Maxwell equations have been solved exactly, but neither the radiation reaction, nor the role of spatio-temporal imperfections have been discussed. It was shown that near the plane of the electron's gyration mostly the TE modes are excited, and the effect of damping at resonance has also been briefly discussed. In section 3 we have introduced the concept of *geometrical resonance*, whose notion is based on a simple ray construction. It was shown that if this condition is satisfied, then the exact *wave resonance condition* does not depend on the excitation indeces of the very high harmonics. At such a 'broad-band resonance' there is an accumulation process taking place, due to which the intensity of the resulting radiation can be increased by orders of magnitudes.

The present analysis supports our original physical picture according to which, if, in the discussed geometrical arrangement, the electron meets with its own radiation field emitted earlier in a tangential narrow cone, and reflected back by the mirror, then there is a constructive interference between this retarded self-field and the actually emitted radiation field.

5. Acknowledgement

This work has been supported by the Hungarian National Scientific Research Foundation OTKA, Grant No. K73728, and by the National Development Agency, Grant No. ELI_09-1-2010-0010, Helios Project.

6. References

[1] G. A. Schott, *Electromagnetic Radiation* (Cambridge University Press, Cambridge, 1912)

[2] J. Schwinger, On the classical radiation of accelerated electrons. *Phys. Rev.* 75, 1912-1925 (1949)

[3] A. A. Sokolov and I. M. Ternov, *Synchrotron Radiation* (Akademie-Verlag GmbH, Berlin, 1968)

[4] H. Winnick and S. Doniach, *Synchrotron Radiation Reseach* (Plenum Press, New York, 1980)

[5] H. Motz, Application of the radiation from fast electron beams. *J. Appl. Phys.* 22, 527-535 (1951). See also H. Motz, Errata. *J. Appl. Phys.* 22, p. 1217 (1951)

[6] H. Motz, W. Thon and R. N. Whitehurst, Experiments on radiation by fast electron beams. *J. Appl. Phys.* 24, 826-833 (1953).

[7] J. Schneider, Stimulated emission of radiation by relativistic electrons in a magnetic field. *Phys. Rev. Lett.* 2, 504-505 (1969)

[8] J. M. J. Madey, Stimulated emission of Bremsstrahlung in a periodic magnetic field. *J. Appl. Phys.* 42, 1906-1913 (1971).

[9] W. Heitler, *The Quantum Theory of Radiation* (Clarendon Press, Oxford, England, 1960), p. 414.

[10] A. Bambini, A. Renieri and S. Stenholm, Classical theory of the free-electron laser in a moving frame. *Phys. Rev. A* 19, 2013-2025 (1979)

[11] F. A. Hopf, T. G. Kuper, G. T. Moore and M. O. Scully, in *Free-electron generation of coherent radiation*. Eds. S. F. Jacobs, M. S. Piloff, M. Surgent III, M. O. Scully and R. Spitzer (Addison-Wesley, London, 1980) pp. 31-57.

[12] W. B. Colson and S. K. Ride, in *Free-electron generation of coherent radiation*. Eds. S. F. Jacobs, M. S. Pieloff, M. Surgent III, M. O. Scully and R. Spitzer (Addison-Wesley, London, 1980) pp. 377-412.

[13] D. A. G. Deacon, L. R. Elias, J. M. J. Madey, G. J. Ramian, H. A. Schwettmann and T. I. Smith, First operation of a free electron laser. *Phys. Rev. Lett.* 38, 892-894 (1977)

[14] M. Billardon, P. Elleaume, J. M. Ortega, C. Bazin, M. Bergher, M. Velghe, Y. Petroff, D. A. G. Deacon, K. A. Robinson and J. M. J. Madey, First operation of a storage-ring free electron laser. *Phys. Rev. Lett.* 51, 1652-1655 (1983)

[15] M. Borenstein and W. E. Lamb, Jr., Classical laser. *Phys. Rev. A* 5, 1298-1311 (1972)

[16] T. Chen and J. M. J. Madey, Study of squeezed state on free electron lasers, In J. Xie and X. Du (Guest Editors), Proceedings of the The Nineteenth International Free Electron Laser Conference and the Fourth FEL Users' Workshop, 18 – 21 August 1997 in Beijing, China. *Nucl. Inst. Meth. Phys. Res. A*–407, 203-209 (1998)

[17] T. Tanabe, Photon statistics of various radiation sources. *Nucl. Inst. Meth. Phys. Res. A*–407, 252-256 (1998)

[18] J. Bergou and S. Varró, Nonlinear scattering processes in the presence of a quantised radiation field: II. Relativistic treatment. *J. Phys. A: Math. Gen.* 14, 1469-1482 (1981)

[19] W. Becker, K. Wódkiewicz and M. S. Zubairy, Squeezing of the cavity vacuum by a charged particle. *Phys. Rev. A* 36, 2167-2170 (1987)

[20] S. Varró, Entangled photon-electron states and the number-phase minimum uncertainty states of the photon field. *New J. Phys.* 10, 053028 (2008)

[21] S. Varró, Entangled states and entropy remnants of a photon-electron system. *Phys. Scr.* T140, 014038 (2010).

[22] S. Varró , Correlations in single-photon experiments. *Fortschritte der Physik – Prog. Phys.* 56, 91-102 (2008)

[23] S. Varró , The role of self-coherence in correlations of bosons and fermions in counting experiments. Notes on the wave-particle duality. *Fortschritte der Physik – Progress of Physics* 59, 296-324 (2011)

[24] M. Yabashi, K. Tamasaku and T. Ishikawa, Intensity interferometry for the study of x-ray coherence. *Phys. Rev. A* 69, 023813 (2004)

[25] E. Ikonen, M. Yabashi and T. Ishikawa, Excess coincidences of reflected and refracted x ray from a synchrotron-radiation beamline. *Phys. Rev. A* 74, 013816 (2006)

[26] Krausz, F. & Ivanov, M. (2009). Attosecond physics. *Rev. Mod. Phys.* 81, 163-234.

[27] S. Varró, Intensity effects and absolute phase effects in nonlinear laser-matter interactions. In F. J. Duarte (Ed.), *Laser Pulse Phenomena and Applications* (InTech, Rijeka, 2010) pp. 243-266.

[28] S. Varró, Attosecond shot noise. 41th Winter Colloquium on the Physics of Quantum Electronics (PQE-41, 2-6 January 2011, Snowbird, Utah, USA) p. 132.

[29] S. Kahn, Free-electron lasers. (a tutorial review) *Journal of Modern Optics* 55, 3469-3512 (2008)

[30] H. Chapman, J. Ullrich and J. M. Rost, Intense x-ray science: the first 5 years of FLASH. (Editorial of the special issue FLASH reseach) *J. Phys. B: At. Mol. Opt. Phys.* 43 (2010) 190201 (1pp)

[31] J. R. Schneider, FLASH – from accelerator test facility to the first single-pass soft x-ray free-electron laser. *J. Phys. B: At. Mol. Opt. Phys.* 43 (2010) 194001 (9pp)

[32] J. L. Hirschfield and J. M. Wachel, Electron cyclotron maser. *Phys. Rev. Lett.* 12, 533-536 (1964)

[33] L. S. Brown, K. Helmerson and J. Tan, Cyclotron motion in a spherical microwave cavity. *Phys. Rev. A* 34, 2638-2645 (1986)

[34] S. J. Han, Stability of a rotating relativistic electron beam in a waveguide near the cyclotron resonance. *Phys. Rev. A* 35, 3952-3955 (1987)

[35] Z. Gy. Horváth and S. Varró, Modes in wall-reflection planar halo lasers. *Optica Acta* 32, 1125-1144 (1985)

[36] S. Varró, Synchrotron radiation in the presence of a perfect cylindrical mirror. *Proceedings of the Third European Particle Accelerator Conference* (Berlin, 24-28 March, 1992) (Editors: H. Henke, H. Homeyer and Ch. Petit-Jean-Genaz, Edition Frontieres, 1992) Volume 1, pp 209-211.

[37] S. Varró, Synchrotron radiation in the presence of a perfect cylindrical mirror. Note on the interference of the direct radiation and the high-order whispering-gallery modes. *arXiv:1101.5811* (physics.acc-ph); (math-ph) (2011)

[38] V. S. Ilchenko, A. A. Savchenko, A. B. Matsko and L. Maleki, Nonlinear optics and crystalline whispering gallery mode cavities. *Phys. Rev. Lett.* 92, 043903 (2004)

[39] J. U. Fürst, D. V. Strekalov, D. Elser, M. Lassen, U. L. Andersen, C. Marquardt and G. Leuchs, Naturally phase-matched second-harmonic generation in a whispering-gallery-mode resonator. *Phys. Rev. Lett.* 104, 153901 (2010)

[40] J. A. Stratton, *Electromagnetic theory* (John Wiley & Sons, Inc., New Jersey, 2007) Ch. 9.
[41] J. Van Bladel, *Electromagnetic fields* (McGraw-Hill Book Company, New York, 1964) Ch. 13.

8

Exotic States of High Density Matter Driven by Intense XUV/X-Ray Free Electron Lasers

Frank B. Rosmej
Sorbonne Universités, Pierre et Marie Curie, Paris,
Ecole Polytechnique, Laboratoire pour l'Utilisation des Lasers Intenses LULI,
Physique Atomique dans les Plasmas Denses PAPD, Palaiseau
France

1. Introduction

XUV and X-ray Free Electron Lasers (XFEL's) have provided the high energy density physics community with outstanding tools to investigate and to create matter under extreme conditions never achieved in laboratories so far. The key parameters of existing and planed XFEL installations [LCLS 2011, XFEL 2011, SACLA XFEL 2011] are micro focusing (to achieve intensities in access to 10^{16} W/cm^2), short pulse lengths (10-100 fs), tunable photon energy (1-20 keV), small bandwidth and high repetition frequency (some 10 Hz, allowing to accumulate thousands of shots to improve signal to noise ratios).

This makes XFEL installations distinct different from well known synchrotron radiation facilities. The brilliance of XFEL's is more than 10-orders of magnitude higher than modern synchrotrons and this allows for the first time to photo ionize inner-shells of almost every atom in a solid crystal in a single pulse. As the pulse duration is of the order of the Auger time scale an exotic state of matter, a "Hollow Crystal" can be created. The decay of crystalline order can be initiated by a burst of Auger electrons with energies in the X-ray range that heat up the hollow crystal [Galtier et al. 2011]. This is distinct different to synchrotrons: Auger electron production is rare compared to the total number of atoms and Auger electrons do not allow to change the physical properties of the crystal.

Next, the tunable photon energy (with small bandwidth) will permit for the first time to pump selected atomic transitions in the X-ray range. Compared to the well known pumping of low energy transitions by optical lasers, X-ray pumping will allow outstanding steps forward: investigations of dense matter via pumped X-ray transitions that can escape without essential absorption. As it has been the case for LIF (Laser-induced fluorescence) with standard optical lasers, a revolutionary impact is expected via the photo pumping of X-ray transitions. In this respect we discuss novel quantum mechanical interference effects that are predicted to be observable via the characteristic X-ray spontaneous emission of hole states in dense matter.

As synchrotrons might neither allow selective nor efficient pumping (drastic change of atomic populations) XFEL facilities will open a new world for scientific activity.

2. Atomic kinetics driven by intense short pulse radiation fields

Radiation field quantum mechanics in second quantization is the most general approach to study the interaction of radiation fields with atoms. On a unique footing it allows describing atomic population and coherences and provides all necessary matrix elements to take into account elementary atomic processes (cross sections) that influence on the atomic populations and interference effects.

Under the assumption of broadband illumination and/or large collisional broadening the non-diagonal density matrix elements are negligible compared to the diagonal ones (atomic populations) and the so-called atomic population kinetic approach becomes valid [Loudon 2000]. In its most general form, the atomic population kinetics theory describes the transient evolution of any atomic population (e.g., ground states, excited states, multiple excited states, hollow ion states,..) under the influence of any collisional-radiative process. This theory will be outlined below, paying particular attention to external intense sort pulse radiation fields to describe the XFEL interaction with matter.

2.1 Non-equilibrium atomic population kinetics in collisional-radiative regimes

In dense non-equilibrium plasmas, collisions, radiative processes and time dependent evolution are equally important and have therefore to be treated on the same general footing. It is also necessary to include all ionization stages, ground states and excited states (single, multiple, hollow ion states) via the elementary collisional-radiative processes combined with the time dependent evolution operator. We note that time scales of typical free electron laser radiation are of the order of some 10 fs and those of hollow ion transitions scale down to 1-fs. Therefore, simulations of the radiative properties have to include photon relaxation effects together with collisional-radiative population kinetics and radiation field physics and any approximations for the time dependent evolution operator are highly questionable. We therefore consider the exact time evolution of the atomic populations which is given by the following set of differential equations:

$$\frac{dn_{j_Z}}{dt}=-n_{j_Z}\sum_{Z'=0}^{Z_n}\sum_{i_{Z'}=1}^{N_{Z'}}W_{j_Z i_{Z'}}+\sum_{Z'=0}^{Z_n}\sum_{k_{Z'}=1}^{N_{Z'}}n_{k_{Z'}}W_{k_{Z'}j_Z} \tag{2.1.1}$$

n_{j_Z} is the atomic population of level j in charge state Z, Z_n is the nuclear charge, N_Z is the maximum number of atomic levels in charge state Z and $W_{j_Z i_{Z'}}$ is the population matrix that contains the rates of all elementary processes from level j of charge state Z to level i of charge state Z′.

In general, eq. (2.1.1) is a system on non-linear differential equations because the population matrix might contain the populations by itself. Only for special cases the populations matrix W does not depend on the atomic populations and equations (2.1.1) become linear. Equations (2.1.1) provide N differential equations where N is given by:

$$N=\sum_{Z=0}^{Z_n}N_Z \tag{2.1.2}$$

Looking more carefully to the symmetry relations of eq. (2.1.1) one finds that the system contains only (N-1) independent equations for the N atomic populations. We are therefore seeking for a supplementary equation. If we consider atomic populations in the framework of probabilities (like in quantum mechanics) the probability to find the atom in any state is equal to 1:

$$\sum_{Z=0}^{Z_n} \sum_{j_Z=1}^{N_Z} n_{j_Z} = 1 \tag{2.1.3}$$

Eq. (2.1.3) is the desired N^th equation and is called the "boundary condition". The distribution of atomic populations over the various charge stages is readily obtained from the solution of eqs. (2.1.1):

$$n_Z = \sum_{j_Z=1}^{N_Z} n_{j_Z} \tag{2.1.4}$$

n_Z is the population of the charge stage Z. The population matrix is given by

$$W_{ij} = W_{ij}^{col} + W_{ij}^{rad} + W_{ij}^{FEL} \tag{2.1.5}$$

The collisional processes are described by

$$W_{ij}^{col} = n_e C_{ij} + n_e I_{ij} + n_e^2 T_{ij} + n_e R_{ij} + n_e D_{ij} + Cx_{ij} + n_{HP} C_{ij}^{HP} + n_{HP} I_{ij}^{HP} \ldots.. \tag{2.1.6}$$

and the matrix describing the radiative and autoionizing processes is given by

$$W_{ij}^{rad} = A_{ij} + \Gamma_{ij} + P_{ij}^{abs} + P_{ij}^{em} + P_{ij}^{rr} + P_{ij}^{iz} \tag{2.1.7}$$

A_{ij}: Spontaneous radiative decay rate, Γ_{ij}: Autoionization rate, P_{ij}^{abs}: Stimulated photo absorption, P_{ij}^{em} : Stimulated photoemission, P_{ij}^{rr} : Stimulated radiative emission, P_{ij}^{iz} : Photo ionization, C_{ij} : Electron collisional excitation/de-excitation, I_{ij}: Electron collisional ionization, T_{ij}: 3-body recombination (electrons), R_{ij}: Radiative recombination, D_{ij}: Dielectronic capture, Cx_{ij}: Charge exchange, C_{ij}^{HP} : Excitation/de-excitation by heavy particle collisions, I_{ij}^{HP} : Ionization by heavy particle collisions. The radiation field matrix elements W^{FEL} for the external laser radiation (e.g., the XFEL radiation) are given by

$$W_{ij}^{FEL} = W_{ij}^{FEL,PI} + W_{ij}^{FEL,SR} + W_{ij}^{FEL,SA} + W_{ij}^{FEL,SE} \tag{2.1.8}$$

$$W_{ij}^{FEL,PI} = \int_{\hbar\omega_{ij}}^{\infty} d(\hbar\omega)\, \sigma_{ij}^{PI}(\hbar\omega)\, c\tilde{N}(\hbar\omega) \tag{2.1.9}$$

$$W_{ji}^{FEL,SR} = n_e \frac{\pi^2 c \hbar^3}{\sqrt{2}\, m_e^{3/2}} \frac{g_i}{g_j} \int_0^{\infty} dE \frac{F(E)}{\sqrt{E}} \sigma_{ij}^{PI}\left(\hbar\omega_{ij} + E\right) \tilde{N}\left(\hbar\omega_{ij} + E\right) \tag{2.1.10}$$

$$W_{ij}^{FEL,SA} = \pi^2 c^3 \hbar^3 A_{ji} \frac{g_j}{g_i} \int_0^{\infty} d(\hbar\omega) \frac{\phi_{ij}(\omega)}{\hbar} \frac{\tilde{N}(\hbar\omega)}{(\hbar\omega)^2} \tag{2.1.11}$$

$$W_{ji}^{FEL,SE} = \pi^2 c^3 \hbar^3 A_{ji} \int_0^{\infty} d(\hbar\omega) \frac{\phi_{ji}(\omega)}{\hbar} \frac{\tilde{N}(\hbar\omega)}{(\hbar\omega)^2} \tag{2.1.12}$$

$W^{FEL,PI}$ describes photo ionization, $W^{FEL,SR}$ stimulated radiative recombination, $W^{FEL,SA}$ stimulated photo-absorption, $W^{FEL,SE}$ stimulated photo-emission. σ^{PI} is the photo ionization cross section, F(E) the energy distribution function of the continuum electrons, A the Einstein coefficient of spontaneous emission, φ the line profile, c the velocity of light, $\hbar$ is the Planck constant, g the statistical weight of a bound state, ω the angular frequency of the external radiation field, ω_{ij} the atomic transition frequency and $\tilde{N}$ is the number of external photons (those of the Free Electron Laser) per unit volume and energy.

The population matrix elements are not independent from each other. They are connected by first principles of quantum mechanics: the CPT-invariance of the Hamiltonian. This invariance results in the principle of micro-reversibility: to each elementary process there must exist an inverse process. In thermodynamics, this principle is known as "detailed balance". It states that each elementary process is balanced by its inverse. The difference to the principle of micro-reversibility is that the general effect of a process is considered rather than the detailed cross sections by itself. A general set of atomic population equations for non-Maxwellian plasmas need to be based on the principle of micro-reversibility; the interesting reader is refereed to the article of [Rosmej and Lisitsa 2011].

In optically thin plasmas, the spectral intensity distribution of an atomic transition j→ i with frequency ω_{ji} is given by:

$$I_{j_Z i_Z}(\omega) = \frac{\hbar\omega_{j_Z i_Z}}{4\pi} n_{j_Z} A_{j_Z i_Z} \varphi_{j_Z i_Z}\left(\omega, \omega_{j_Z i_Z}\right) \tag{2.1.13}$$

n_{j_Z} is the population of the upper level j, $A_{j_Z i_Z}$ is the spontaneous transition probability for the transition j→ i and $\varphi_{j_Z i_Z}\left(\omega, \omega_{j_Z i_Z}\right)$ is the associated local emission line profile. Eq. (2.1.13) indicates a strong interplay between the atomic structure (means transition probabilities $A_{j_Z i_Z}$) and atomic populations kinetics (population densities n_{j_Z}). The total spectral distribution is given by

$$I(\omega) = \sum_{Z=0}^{Z_n} \sum_{j_Z=1}^{N_Z} \sum_{i_Z=1}^{N_Z} I_{j_Z i_Z}(\omega) \tag{2.1.14}$$

Eq. (2.1.14) is of outstanding importance: it is the spectral distribution that is accessible via measurements (spectroscopy).

In plasmas where opacity in line transitions is important the spectral distribution according eq. (2.1.14) can be modified employing the escape probability Λ and a generalized optically thick line profile [Rosmej 2012]. If also continuum radiation is important, the radiation transport equation has to be solved. The interesting reader is refereed to [Mihalas 1978, Rosmej 2012] for further reading on these subjects.

2.2 Shocking atomic systems by XFEL radiation fields

The high peak brilliance of current/planed XFEL installations allows changing atomic populations of even highly charged ions. The coupling of the XFEL radiation to the atomic system is essentially via photo ionization and photo excitation.

2.2.1 XFEL radiation

Le us assume that time and energy dependence of the XFEL radiation are independent ($f_{FEL}(t)$ and $\tilde{N}_{FEL}(E)$, respectively) from each other. The number of photons per volume/time/energy is then given by:

$$\tilde{N}_{FEL}(E,t) = \tilde{N}_{FEL}(E) f_{FEL}(t) \tag{2.2.1}$$

$$\int_{-\infty}^{+\infty} f_{FEL}(t)dt=1 \tag{2.2.2}$$

We assume a Gaussian energy dependence to simulate the narrow bandwidth of the XFEL:

$$\tilde{N}_{FEL}(E) = \tilde{N}_0 \frac{1}{\sqrt{\pi}\Gamma_{FEL}} \exp\left(-\frac{(E-E_{FEL})^2}{\Gamma_{FEL}^2}\right) \tag{2.2.3}$$

$$\Gamma_{FEL} = \delta E / 2\sqrt{\ln 2} \tag{2.2.4}$$

E_{FEL} is the central energy of the radiation field, $\tilde{N}_{FEL}(E)$ is the number of photons / volume / energy, $\tilde{N}_0$ is the peak number of photons / volume, δE is the bandwidth. Assuming a Gaussian time dependence the number of photons $N_{tot,\tau}$ per pulse length τ is given by

$$N_{tot,\tau} = \int_0^{\infty} dE \int_{volume} dV \int_{-\tau/2}^{+\tau/2} dt \tilde{N}_{FEL}(E,t) \approx 2Ac\tau\tilde{N}_0\, erf\left(\sqrt{\ln 2}\right) \approx 0.76Ac\tau\tilde{N}_0 \tag{2.2.5}$$

A is the focal spot area, τ is the XFEL pulse width (FWHM). The laser intensity $\tilde{I}_{FEL}(E,t)$ per bandwidth energy and time interval is related to the photon density $\tilde{N}_{FEL}(E,t)$ via

$$\tilde{I}_{FEL}(E,t)dEdAdt = \tilde{N}_{FEL}(E,t)E \cdot dEdVdt \tag{2.2.6}$$

Integrating the XFEL beam over a full width at half maximum with respect to energy and time, $\overline{I}_{FEL,\delta E,\tau}$ (energy/time/surface) is given by (assuming a Gaussian time dependence):

$$\overline{I}_{FEL,\delta E,\tau} = \int_{-\delta E/2}^{\delta E/2} dE \int_{-\tau/2}^{\tau/2} cdt E \cdot \tilde{N}(E,t) \approx 4E_{FEL}c\tilde{N}_0\, erf^2\left(\sqrt{\ln 2}\right) \approx 0.58 \cdot c \cdot E_{FEL} \cdot \tilde{N}_0 \tag{2.2.7}$$

or, in convenient units

$$\overline{I}_{FEL,\delta E,\tau} \approx 2.8x10^{-9}\left(\frac{\tilde{N}_0}{cm^3}\right)\left(\frac{E_{FEL}}{eV}\right)\left[\frac{W}{cm^2}\right] \tag{2.2.8}$$

The number of photons $N_{tot,\tau}$ is related to the intensity $\overline{I}_{FEL,\tau}$ via (d is the focal spot diameter)

$$\overline{I}_{FEL,\tau} = 2\cdot erf\left(\sqrt{\ln 2}\right)\cdot\frac{N_{tot,\tau}\cdot E_{FEL}}{\pi\tau\cdot d^2/4} \approx \frac{N_{tot,\tau}\cdot E_{FEL}}{\tau\cdot d^2} \tag{2.2.9}$$

2.2.2 Photo ionization

In order to change atomic populations, photo ionization rates need to be larger than corresponding electron ionization rates and, in case of photo pumping, photo excitation rates need to be larger than corresponding spontaneous radiative decay rates. In order to obtain analytical formulas, we consider a hydrogen-like atom with effective charge Z and an atomic level with principal quantum number n and energy

$$E_n = \frac{Z^2 Ry}{n^2} \tag{2.2.10}$$

where Ry = 13.6 eV. For the case of photo ionization this leads to the following relation:

$$\int_{-\tau/2}^{+\tau/2} dt \int_{E_n}^{\infty} dE\,\sigma_n^{iz}(E)\,c\,\tilde{N}_{FEL}(E,t) > n_e I_n \tag{2.2.11}$$

$\sigma_n^{iz}(E_{FEL})$ is the photo ionization cross section from level n, n_e is the electron density, I_n is the electron collisional ionization rate, c the velocity of light. Employing the Kramers classical cross section for the photo ionization

$$E > E_n : \sigma_n^{iz}(E) = 2.9\cdot 10^{-17}\frac{E_n^{5/2}}{Z\cdot E^3}\left[cm^2\right] \tag{2.2.12}$$

and the Lotz-formula for the electron collisional ionization

$$I_n = 6\cdot 10^{-8}\left(\frac{Ry}{E_n}\right)^{3/2}\sqrt{\beta_n}e^{-\beta_n}\ln\left[1+\frac{0.562+1.4\beta_n}{\beta_n(1+1.4\beta_n)}\right]\left[cm^3 s^{-1}\right] \tag{2.2.13}$$

$$\beta_n = \frac{E_n}{kT_e} \tag{2.2.14}$$

eqs. (2.2.11-2.2.14) provide the following estimate (peak intensity $I_{FEL} = cE_{FEL}\tilde{N}_0$):

$$I_{FEL} > 3\cdot 10^{-8} n_e\left(cm^{-3}\right)Z\sqrt{\beta_n}e^{-\beta_n}\ln\left[1+\frac{0.562+1.4\beta_n}{\beta_n(1+1.4\beta_n)}\right]\left[\frac{W}{cm^2}\right] \tag{2.2.15}$$

assuming $F_{EEL}=E_n+3\delta E$ ($\delta E<<E_{FEL}$) for effective photo ionization. For $n_e = 10^{21}$ cm^{-3}, Z=13, $\beta_n = 2$ eq. (2.2.15) delivers $I_{FEL} > 3x10^{13}$ W/cm^2. Let us now consider the relations for photo pumping of X-ray transitions.

2.2.3 Photo excitation

In order to influence via photo excitation on the atomic populations, photo excitation rates need to be about larger than corresponding spontaneous radiative decay rates:

$$\int_{-\tau/2}^{+\tau/2} dt \int_{\Delta E_{nm}-\delta\tilde{E}/2}^{\Delta E_{nm}+\delta\tilde{E}/2} dE\, \sigma_{nm}^{abs}(E)\, c\, \tilde{N}_{FEL}(E,t) \geq A_{mn} \tag{2.2.16}$$

$\sigma_{nm}^{abs}(E)$ is the photo absorption cross section for the transition from level n to level m and A_{mn} is the spontaneous radiative decay rate from level m to level n, $\delta\tilde{E}$ will be defined below (eq. (2.2.21)). The photo absorption cross section is given by

$$\sigma_{nm}^{abs}(E) = \frac{E}{4\pi} B_{nm} \varphi_{nm}(E) \tag{2.2.17}$$

B_{nm} is the Einstein coefficient of stimulated absorption that is related to the Einstein coefficient of spontaneous radiative decay according

$$B_{nm} = \frac{4\pi^3 \hbar^3 c^2}{E^3} \frac{g_m}{g_n} A_{mn} \tag{2.2.18}$$

$\varphi_{nm}(E)$ is the normalized local absorption line profile :

$$\int_{-\infty}^{+\infty} \varphi_{nm}(E)\, dE = 1 \tag{2.2.19}$$

We assume a Gaussian line profile ($FWHM = 2\sqrt{\ln 2}\,\Gamma_G$, $\Gamma_G = \Delta E_{nm}\sqrt{2kT_i / Mc^2}$ if a Doppler profile is considered) to obtain analytical estimates:

$$\varphi_{nm}(E) = \frac{1}{\sqrt{\pi}\,\Gamma_G} \exp\left[-\left(\frac{E-\Delta E_{nm}}{\Gamma_G}\right)^2\right] \tag{2.2.20}$$

$$\frac{4\ln 2}{\delta\tilde{E}^2} = \frac{1}{\Gamma_G^2} + \frac{1}{\Gamma_{FEL}^2} \tag{2.2.21}$$

If the XFEL photon energy is exactly tuned to the transition energy, e.g., $E_{FEL} = \Delta E_{nm}$, eqs. (2.2.3-4, 2.2.16-21) provide the following estimate:

$$I_{FEL} > 2 \cdot 10^5 \Delta E_{nm}^3 \frac{g_n}{g_m} \sqrt{FWHM_{FEL}^2 + FWHM_G^2} \left[\frac{W}{cm^2}\right] \tag{2.2.22}$$

with ΔE_{nm} and FWHM in [eV]. For H-like Al Ly_α, ΔE_{nm} = 1728 eV, g_n=2, g_m=6, we obtain I_{FEL} > 4x10[15] W/cm[2] (assuming $\sqrt{FWHM_{FEL}^2 + FWHM_G^2} = 10eV$). The relation (2.2.22) indicates an important scaling law:

$$I_{FEL} \propto Z^6 \tag{2.2.23}$$

Therefore, extremely high brilliance of XFEL's are needed to pump X-ray transitions. Assuming a spot diameter of d = 2 μm, pulse length τ = 100 fs, photon energy E_{FEL}=1.7 keV and a laser intensity of $\overline{I}_{FEL,\tau} = 10^{16} W / cm^2$, a minimum of about $N_{tot,\tau} \approx 2x10^{11}$ photons in the XFEL pulse is requested according eq. (2.2.9) to effectively move atomic populations in the X-ray energy range. Currently operating/planed Free Electron Laser facilities fulfill these requirements. As relation (2.2.22) does not depend on the electron density, the estimate for the requested XFEL intensity holds equally for low and high-density plasmas. We note that even in case of photo pumping, considerable effects on the ionic fractions take place, as collisional ionization from pumped excited states is important in dense plasmas.

2.2.4 Simulations of the interaction of XFEL with dense plasmas

Fig. 2.1 shows a principal experimental scheme for a typical pump probe experiment. A ps-ns optical laser is irradiating a solid target to create a dense plasma plume and the XFEL is used to pump X-ray transitions of ions in the plume. Corresponding simulations of the XFEL interaction with the dense plasma are carried out with the MARIA-code [Rosmej 1997, 2001, 2006] that includes all the radiation field physics described above. A detailed LSJ-split atomic atomic/ionic level system is employed to calculate the populations of different ion charge stages, ground, single and multiple excited states as well as hollow ion states.

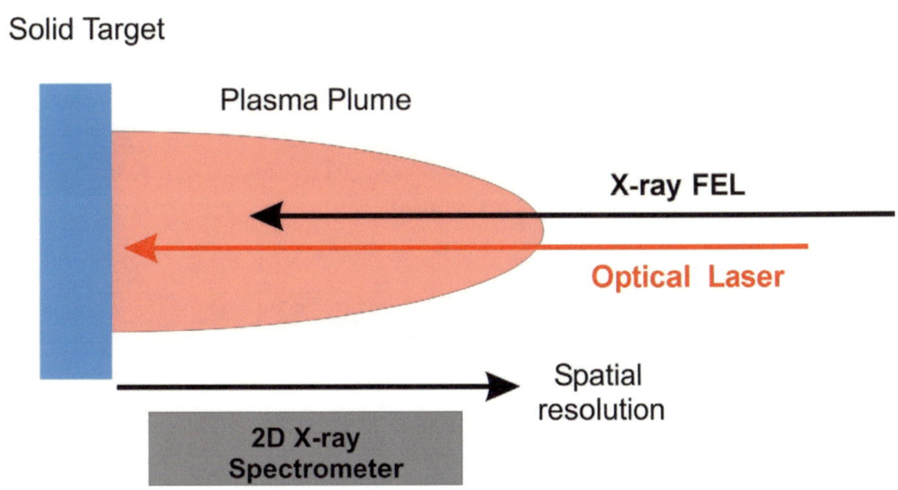

Fig. 2.1. Schematic scheme of a pump probe/photo ionization experiment. The optical laser irradiates a solid target and creates a plasma plume. The XFEL is used to pump selected X-ray transitions in the plume. A high resolution (high spectral and spatial) X-ray spectrometer is employed to record the spectral distribution of the pumped x-ray transitions and to investigate the spatial variations.

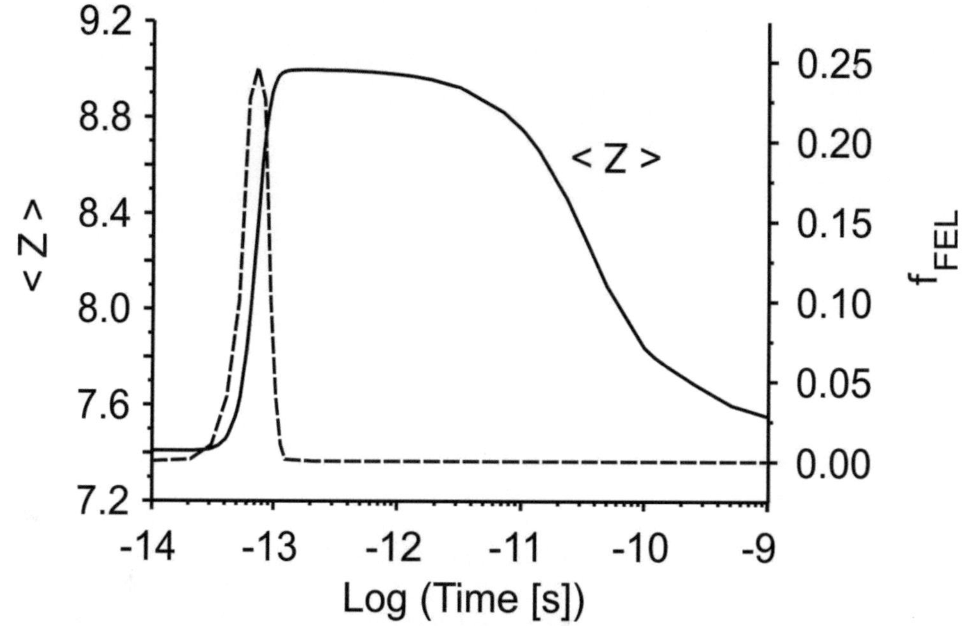

Fig. 2.2. MARIA simulations of the temporal evolution of the XFEL pulse and the average charge state of a dense Mg plasma, E_{FEL} = 1850 eV, τ=100 fs, I_{max}= 2.2x10^{17} W/cm^2, n_e = 10^{21} cm^{-3}, kT_e = 40 eV, L_{eff} = 10 μm.

Fig. 2.2 shows the evolution of the average charge (solid curve) when an intense pulsed radiation field (dashed curve) is interacting with dense magnesium plasma:

$$\langle Z\rangle = \sum_{Z=0}^{Z_n} n_Z Z \qquad (2.2.24)$$

where n_z is the ionic population of charge Z (see eq. (2.1.4)). The plasma density is n_e = 10^{21} cm^{-3}, the temperature kT_e=40 eV. Opacity effects of the internal atomic/ionic radiation are included via an effective photon path length of L_{eff} = 10 μm [Rosmej 2012]. The XFEL pulse duration is τ = 100 fs, photon energy E_{FEL} = 1850 eV and the photon density is $\tilde{N}_0 = 10^{23}\, cm^{-3}$. The maximum laser intensity is related to these quantities according

$$I_{\max} = c\,E\, f_{FEL,\max}\,\tilde{N}_0 = 4.8x10^{-9}\, f_{FEL,\max}\left(\frac{\tilde{N}_0}{cm^3}\right)\left(\frac{E}{eV}\right)\,\left[\frac{W}{cm^2}\right] \qquad (2.2.25)$$

where $f_{FEL,max}$ is the maximum value ($f_{FEL,max}$ = 0.246 in Fig. 2.2) of the normalized time dependent function of the laser intensity (see eq. (2.2.1, 2.2.2)), I_{max} = 2.2x10^{17} W/cm^2.

Before the XFEL pulse interacts with the Mg plasma plume, the average charge state is about $\langle Z\rangle \approx 7.4$ that rises dramatically during the interaction with the XFEL pulse. The system shows shock characteristics: after laser pulse maximum, the average charge state is still increasing (at about t=10^{-13} s), then stays almost constant for a few ps, then decreases on a 100 ps time scale followed by a very slow final equilibration phase (10-100 ns).

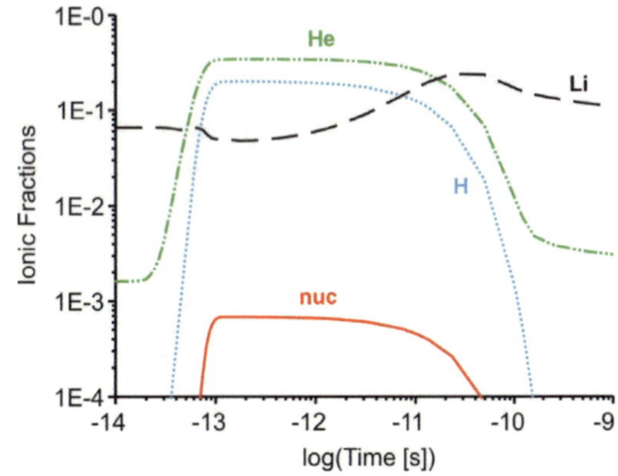

Fig. 2.3. MARIA simulations of the temporal evolution of the ionic fractions after interaction of the XFEL pulse with a dense Mg plasma plume, E_{FEL} = 1850 eV, τ=100 fs, $\tilde{N}_0 = 10^{23} cm^{-3}$, I_{max}= 2.2x10^{17} W/cm^2, n_e = 10^{21} cm^{-3}, kT_e = 40 eV, L_{eff} = 30 μm.

Let us follow the shock characteristics in more detail. Fig. 2.3 shows the charge state evolution of the bare nucleus (nuc), H-like ions (H), He-like ions (He) and Li-like ions (Li). Before the XFEL pulse the ionic fractions nuc, H and He are negligibly small due to the low electron temperature of the plasma plume. With the onset of the XFEL pulse, He-like and H-like ionic fractions rise rapidly because the photon energy E_{FEL} = 1.85 keV is larger than the ionization potential of the He-like Mg ground state ($E_i(1s^2\ ^1S_0)$ = 1762 eV).

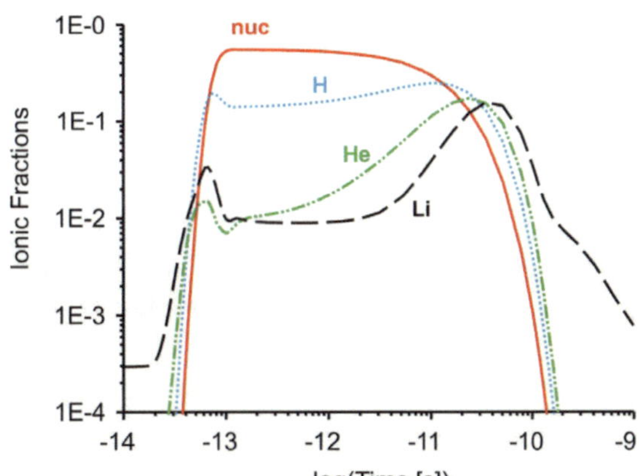

Fig. 2.4. MARIA simulations of the temporal evolution of the ionic fractions after interaction of the XFEL pulse with a dense Mg plasma plume, E_{FEL} = 3100 eV, τ=100 fs, $\tilde{N}_0 = 10^{23} cm^{-3}$, I_{max}= 3.7x10^{17} W/cm^2, n_e = 10^{21} cm^{-3}, kT_e = 30 eV, L_{eff} = 30 μm.

Fig. 2.4 shows a simulation when the photon energy is larger than the ionization potential of the H-like ground state. As in Fig. 2.3, before the XFEL pulse the ionic fractions of the bare nucleus, H- and He-like ions are negligibly small due to the low electron temperature of the plasma plume. With the onset of the XFEL pulse, Li-like, He-like and H-like ionic fractions rise rapidly. At about laser pulse maximum, the fraction of H-, He- and Li-like ions drop again because the XFEL photons are photoionizing the H-like ground state 1s $^2S_{1/2}$ because the photon energy of E_{FEL} = 3.1 keV is larger than the ionization potential of H-like Mg ground state ($E_i(1s\ ^2S_{1/2})$ = 1963 eV). The depletion of almost all electrons from the atomic system makes the plume transparent to the XFEL radiation as no more absorption is possible: the absorption is saturated (see also paragraph 5). When the pulse is off, H-like, He-like and Li-like ionic fractions increase as recombination starts from the bare nucleus. At even later times (about $t = 10^{-10}$ s), all ionic fractions (nuc, H, He, Li) decrease due to the overall cooling of the plume (rise of ionic fractions of low Z -ions not shown in the figures).

Figs. 2.3 and 2.4 indicate, that in the photo ionization regime, the tuning of the XFEL beam allows selection of charge states and investigation of specific shock regimes.

3. Beating the Auger-clock: Hollow crystal and hollow ion formation

3.1 Photo ionization versus autoionization

Photo ionization of inner atomic shells creates multiple excited states that can decay via non-radiative transitions. Let us consider the photo ionization from the K-shell:

$$K^2L^XM^YN^Z + h\nu_{XFEL} \rightarrow K^1L^XM^YN^Z + e_{photo} \tag{3.1.1}$$

(for example titan is described by the configuration $K^2L^8M^{10}N^2$). The photo ionized state is multiple excited and can decay via radiative and non-radiative (autoionization, known as Auger effect in solid state physics) transitions. Let us consider a simple example (Y=0, Z=0):

$$K^1L^X \rightarrow \left\{ \begin{array}{ll} radiative\,decay: & K^2L^{X-1} + h\nu_{K_\alpha} \\ non-radiative\,decay: & K^2L^{X-2} + e_{Auger} \end{array} \right\} \tag{3.1.2}$$

Radiative and non-radiative decay processes in the x-ray energy range have extensively been studied in the very past [Flügge 1957]. Particularly synchrotrons have been employed for advanced studies of X-ray interaction with solid matter. Synchrotron radiation, however, is not very intense, allowing occurrence of photo ionization of inner-shells only as a rare process (means a negligible fraction of the atoms in the crystal are photo ionized thereby leaving the solid system almost unperturbed).

This situation is quite different for XFEL's: their brilliance is more than 10 orders of magnitude higher than those of most advanced synchrotrons. Photo ionization of inner-shells may therefore concern almost every atom in the crystal structure leading to essential perturbations and corresponding dramatic changes in the physical properties (see below).

In terms of elementary processes XFEL driven photo ionization rates allow to compete even with the Auger rates (autoionizing rates Γ are very large, order of 10^{12} - 10^{16} s^{-1}). The necessary XFEL intensities to „compete" with the Auger effect can be estimated according (see also eqs. (2.2.10-2.2.12))

$$\int_{-\tau/2}^{+\tau/2} dt \int_{E_n}^{\infty} dE\, \sigma_n^{iz}(E)\, c\, \tilde{N}_{FEL}(E,t) > \Gamma \tag{3.1.3}$$

Assuming a photon energy E_{FEL} of the XFEL which is sufficient to proceed towards effective photo ionization, namely $E_{FEL} = E_n+3dE$, ($dE<<E_n$, E_n is the ionization energy of the inner shell with principal quantum number "n") we obtain the following estimate:

$$I_{FEL} > 4 \cdot 10^{-1} \cdot \Gamma \cdot \frac{Z^4}{n^3} \left[\frac{W}{cm^2}\right] \tag{3.1.4}$$

As autoionizing rates scale approximately like $\Gamma \propto Z^0$ (means almost independent of Z in the hydrogenic approximation) the Z-scaling of eq. (3.1.4) is approximately given by

$$I_{FEL} \propto Z^4 \tag{3.1.5}$$

Let us consider the photo ionization of the K-shell of Al I as an example: $Z\approx10.8$, $n=1$, $\Gamma \approx 10^{14}\ s^{-1}$, $I_{FEL} > 5x10^{17}\ W/cm^2$.

As micro-focusing is now a standard setup at the XFEL installations, intensities in excess of $10^{17}\ W/cm^2$ can be achieved and photo ionization of inner-shells can compete with the Auger rate. We note that this competition means that the change in atomic populations due to photo ionization is essential compared to the Auger rate that destroys the inner-shell hole.

3.2 Auger clock and hollow ion formation

Apart the threshold intensity (eq. (3.1.4) the characteristic Auger time scale is another important issue. Before XFEL's became available for dense plasma physics experiments [Rosmej & Lee 2006, 2007] proposed on the basis of simulations carried out with the MARIA code [Rosmej 1997, 2001, 2006] that "beating the Auger clock" will allow massive creation of hollow ions and permit their observation via the characteristic X-ray emission.

Let us consider the relevant physics via an example: creation of hollow ion K^0L^X-configurations and corresponding characteristic inner-shell X-ray emission. We start from the K^2L^X configurations. Photo ionization of the K-shell creates the state

$$K^2L^X + h\nu_{XFEL} \rightarrow K^1L^X + e_{photo} \tag{3.2.1}$$

In order to proceed with interesting processes from the XFEL produced single hole state K^1L^X, the duration of the XFEL pulse (being responsible for the first photo ionization) must be of the order of the characteristic Auger time scale. As planed/operating VUV/X-ray FEL facilities propose the requested pulse durations (order of 10-100 fs) photo ionization may further proceed from the single K-hole state to produce a second K-hole (hollow ion):

$$K^1L^X + h\nu_{XFEL} \rightarrow K^0L^X + e_{photo} \tag{3.2.2}$$

The existence of the double K-hole configuration K^0L^X can easily be identified via the characteristic hollow ion X-ray transitions that are located approximately between Ly_α and He_α of highly charged ions [Faenov et al. 1999]:

$$K^0L^X \rightarrow K^1L^{X-1} + h\nu_{Hollow\ ion} \tag{3.2.3}$$

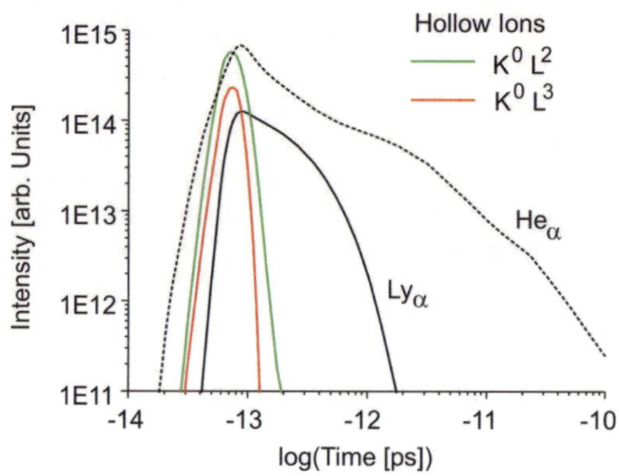

Fig. 3.1. MARIA simulations of the temporal evolution of the various line intensities after interaction of the XFEL pulse with a dense Mg plasma plume, E_{FEL} = 3100 eV, τ=100 fs, $\tilde{N}_0 = 10^{23} cm^{-3}$, I_{max} = 3.7x10^{17} W/cm^2, n_e = 10^{21} cm^{-3}, kT_e = 30 eV.

Ab initio calculations with the MARIA-code that include radiation field physics outlined in paragraph 2 demonstrate that hollow ion production is effective and observable levels of characteristic X-ray emission are achieved. These simulations have lead to a proposal for hollow ion research in dense plasmas at planed XFEL installations [Rosmej and Lee 2006].

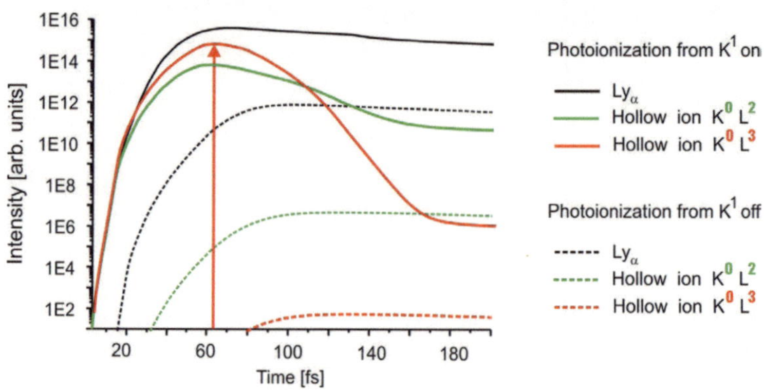

Fig. 3.2. MARIA simulations of the temporal evolution of the various line intensities after interaction of the XFEL pulse with a dense Mg plasma plume, E_{FEL} = 3100 eV, τ=100 fs, $\tilde{N}_0 = 10^{23} cm^{-3}$, I_{max} = 3.7x10^{17} W/cm^2, n_e = 10^{21} cm^{-3}, kT_e = 30 eV.

Fig. 3.1 shows the time evolution of the characteristic X-ray emission of Ly_α (2p-1s), He_α (1s2p $^1P_1$-1s^2 $^1S_0$) as well as the X-ray emission originating from hollow ions: K^0L^2-K^1L^1 and K^0L^3-K^1L^2. The MARIA simulations have been carried out for an intense XFEL beam that is interacting with a dense Mg plasma (see Fig. 2.1) with electron density n_e = 10^{21} cm^{-3} and electron temperature kT_e = 30 eV. The photon energy is E_{FEL} = 3100 eV, pulse duration τ=100 fs and a photon density $\tilde{N}_0 = 10^{23} cm^{-3}$ (corresponding to an intensity of I_{max} = 3.7x10^{17} W/cm^2).

As can be seen from Fig. 3.1 the intensity of the hollow ion X-ray emission is of the order of the resonance line emissions (Ly_α and He_α) that are known to be observable. Let us clearly identify the real importance of the successive photo ionization for the hollow ion X-ray emission (eqs. (3.1.1-3.1.3)). Fig. 3.2 shows the temporal evolution when all photo ionization channels are included in the simulations (solid curves) and when photo ionization from and to the states that involve a K^1-electron are artificially switched off (dashed curves in Fig. 3.2). It can clearly be seen that the hollow ion X-ray emission is practically absent when photo ionization from K^1 is off: the remaining intensities are due to collisional effects. This means that in a proof of principal simulation with the MARIA code hollow ion production and corresponding X-ray emission have been identified as driven by successive photo ionization from K^2 and K^1-electron states (see flash in Fig. 3.2). This is equivalent to say that the XFEL allows beating the Auger clock to proceed towards successive K-shell ionization before the autoionization/Auger effect disintegrates the state. We note that above-predicted double K-hole states have recently been observed [Cryan et al. 2010].

4. X-ray bursts from hollow ions and fast x-ray switches

The X-ray emission of hollow ions discussed in the forgoing paragraph provides outstanding possibilities to investigate exotic states of matter that are just produced during the XFEL pulse. Fig. 4.1 shows the temporal evolution of the hollow ion X-ray emission $K^0L^3 \rightarrow K^1L^2 + h\nu_{Hollow\ ion}$ on a linear intensity scale.

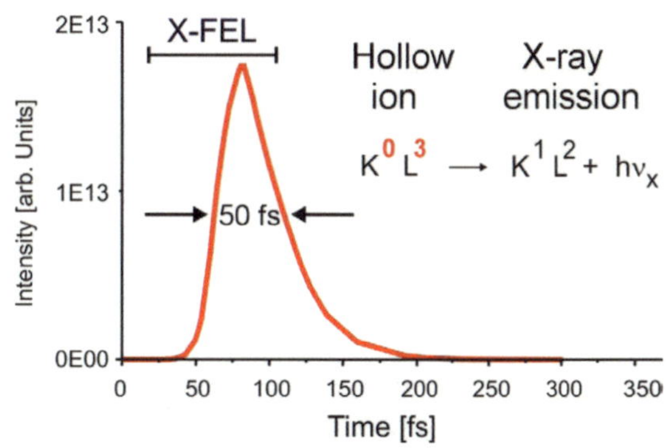

Fig. 4.1. MARIA simulations of the temporal evolution of the hollow ion X-ray emission induced by the interaction of a XFEL pulse with a dense Mg plasma plume, E_{FEL} = 3100 eV, τ=100 fs, $\tilde{N}_0 = 10^{23} cm^{-3}$, I_{max} = 3.7x10^{17} W/cm^2, n_e = 10^{21} cm^{-3}, kT_e = 30 eV.

The simulations demonstrate that the FWHM of the X-ray emission is only 50 fs and temporally located very close to the XFEL pulse. Therefore, dense matter properties that are just produced during the XFEL interaction can be studied via this X-ray emission produced by the matter itself. Moreover, we meet outstanding properties of the characteristic hollow ion X-ray emission for the K^0L^X-configurations [Rosmej et al. 2007]:

- Opacity is very small as the absorbing lower states K^1L^X are autoionizing with corresponding small populations even in dense plasmas.
- Radiative recombination effects are negligible. Therefore emission from the long lasting recombination regime at low density does not mask the high density physics during the XFEL interaction with matter. In this respect we note that particularly resonance line emission is perturbed by radiative recombination.
- Even dielectronic recombination is small as effective dielectronic caputre proceeds from ground states and not from the K^1L^X-states.

- The short time scale (some 10 fs) of the characteristic hollow ion x-ray emission acts as an effective X-ray switch that allows to study high density physics and exotic matter just after its creation by short pulse XFEL radiation. We note, that X-ray streak cameras may help to suppress emission from the recombination regime; however, they will hardly be able to streak down to 50 fs (current limits are about 0.5 ps).

5. Transparent materials and saturated absorption

A material is transparent to photons at certain energies, if neither photo absorption nor photo pumping is effective at these photon wavelengths. This is related to the density of the atomic populations: in the case of photo ionization this is the population density of the state that is photo ionized, in the case of photo pumping it is the lower state of the atomic transition that is pumped.

As has been shown in the forgoing paragraphs (eqs. (2.2.11-15), (2.2.16-22), (3.1.3-4)) XFEL radiation allows to effectively change atomic populations in the X-ray energy range. This permits to selectively deplete atomic populations. If these populations are related to photo ionization/photo pumping transparency to the XFEL radiation itself is induced and a so called "saturated absorption regime" is achieved.

Observation of saturated absorption has been claimed recently [Nagler et al. 2009] irradiating solid Al foils with a 92 eV FEL beam in the photo ionization regime:

$$1s^2 2s^2 2p^6 3s^2 3p^1 + h\nu_{XFEL} \rightarrow 1s^2 2s^2 2p^5 3s^2 3p^1 + e_{photo} \tag{5.1}$$

As photo ionization of a 2p-electron from the $2p^6$-configuration is the most effective (see eq. (2.2.12)) and a second photo ionization (means the creation of a $2p^4$-configuration) seems energetically not probable the ionization of almost all $2p^6$-configurations will induce transparence to the 92 eV XUV-laser radiation. Solid aluminum has therefore turned transparent for 92 eV photons. We note that effects of transparency are limited by the principle of detailed balance: stimulated photoemission (eq. 2.1.12) and stimulated radiative recombination (eq. 2.1.10) sets a definite limit to that what can actually be observed. Also 3-body recombination in dense matter will destroy the hole states thereby driving the saturation regime to higher intensities.

Saturated absorption implies enhanced homogeneity of the irradiated material, as no more geometrical energy deposition peaks exist. This effect is well known from the stopping of relativistic heavy ion beams in matter: if the Bragg peak is placed outside the target, almost homogenous parameter conditions are meet [Kozyreva et al. 2003, Tauschwitz et al. 2007].

The term "transparent aluminum" is also known in the non-scientific society from the science fiction series "Stark Trek" [Wiki 2011]: the chief engineer M. Scott has invented transparent aluminum to fabricate windows that have the strength and density of solid aluminum (in particular for its use to transport whales in an aquarium). This has moved XFEL research to the frontiers of science fiction [Larousserie 2009].

6. Auger electron heating

The possibility to fulfill the relations (2.2.11) and (3.1.3) on the Auger time scale allows a sudden almost maximum depletion of internal atomic shells due to photo ionization. As almost every atom is transformed to an autoionizing state, a massive burst of Auger electrons is following.

In the x-ray energy range, the Auger electrons carry a high kinetic energy. For example, the energies of the KL-decay of the configurations $K^1L^XM^YN^Z$ are of the order of $Z^2Ry/2$, means in the keV range. As almost every atom in the crystal structure is concerned, a huge kinetic energy is released on a 10-fs time scale. This results in a rapid heating of the hollow crystal and subsequent disintegration of crystalline order followed by the creation of Warm Dense Matter and dense strongly coupled plasmas. Fig. 6.1 illustrates schematically the relevant steps in the evolution of matter after irradiation with intense XFEL radiation. We note that if the photo ionization energy is just tuned to the edge, kinetic energy of photo electrons is negligible and material heating starts from the kinetic energy of the Auger electrons.

In a proof of principle experiment Auger electron heating has been identified via high-resolution spectroscopy and introduced to the XFEL community as an important heating mechanism [Galtier et al. 2011]. We note that for optical lasers, Auger heating is irrelevant, as low photon energies are not producing holes in inner atomic shells.

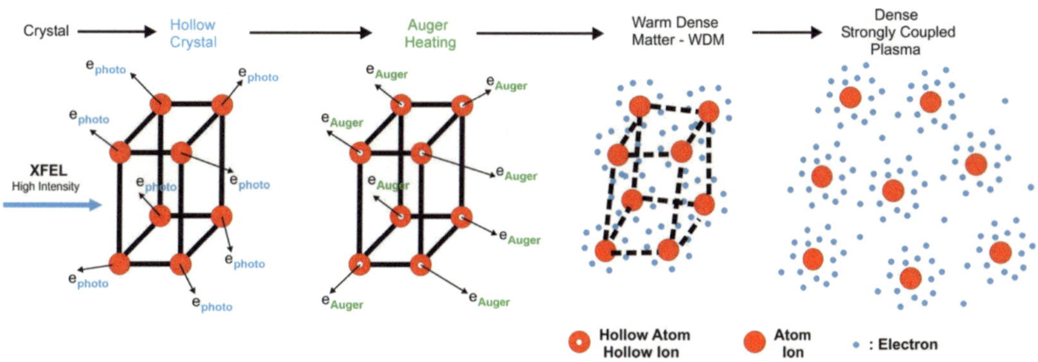

Fig. 6.1. Schematic mechanism to create hollow crystals followed by Auger electron bursts and heating, formation of Warm Dense Matter and dense strongly coupled plasmas.

In principle synchrotron radiation may produce Auger electrons via photo ionization of inner shells, however, the low intensity makes Auger emission a rare process compared to the huge number of atoms that are not concerned (note, that this is not a contradiction to the fact that Auger electron spectra can be well measured). Therefore no heating of the crystal is induced. Moreover, synchrotron radiation does not allow photo ionization on the Auger time scale and is therefore in principle not able to create exotic states of matter such as "hollow crystals", "transparent solids" etc.

7. Hole states in dense plasmas and Excited Sates Coupling (ESC) effects

7.1 Re-creation of hole states in dense plasmas after hollow crystal formation

First principles predict a re-population of hole states after photo ionization at times when the hollow crystal turns to a dense strongly coupled plasma, see Fig. 6.1. The principle of micro-reversibility that is based on the CPT-invariance of the Hamiltonian [e.g. Dawydow 1981] requests that autoionization is followed by its inverse process (inverse Auger or dielectronic capture):

$$D_{jk} = \frac{\pi^2}{\sqrt{2}} \frac{\hbar^3}{m_e^{3/2}} \frac{g_j}{g_k} \frac{\Gamma_{jk}}{\sqrt{E_{jk}}} F\left(E_{jk}, E\right) \quad (7.1.1)$$

D_{jk} is the dielectronic capture from state k to the autoionizing state j, Γ_{jk} is the corresponding autoionization rate, E_{jk} is the dielectronic capture energy, g_j and g_k are the statistical weights of states j and k, $F(E_{jk}, E)$ is the electron energy distribution function. In order to proceed with analytical estimates, let us assume that collisions between the Auger electrons are so frequent that a Maxwellian electron energy distribution is quickly established (fs-time scale):

$$D_{jk} = \frac{(2\pi)^{3/2} \hbar^3}{2m_e^{3/2}} \frac{g_j}{g_k} \Gamma_{jk} \frac{\exp\left(-E_{jk} / kT_e\right)}{\left(kT_e\right)^{3/2}} = 1.656 \cdot 10^{-22} \frac{g_j \Gamma_{jk}}{g_k} \frac{\exp\left(-E_{jk}(eV) / T_e(eV)\right)}{\left(T_e(eV)\right)^{3/2}} \frac{cm^3}{s} \quad (7.1.2)$$

As discussed in paragraph 6, capture energies are of the order of $Z^2Ry/2$ implying that in a regime of saturated absorption where almost every atom is accompanied by an Auger electron, the electron temperature is expected to be of similar order (note, that after autoionization Auger electron kinetic energy is redistributed amongst the remaining electrons; for metals the electrons in the conduction band). Therefore the exponential factor in eq. (7.1.2) is not small and the inverse Auger effect that is effectively re-creating hole states in a dense plasma (see also Fig. 6.1) can be detected via the characteristic X-ray emission.

7.2 Excited States Coupling (ESC) of inverse Auger effect

In atomic physics, the characteristic X-ray emission of hole states is known as dielectronic satellite emission that is of considerable importance to study dense plasmas physics via independent (from plasma simulations) diagnostics of temperature, density, charge states, supra-thermal electrons, non-equilibrium effects etc. The interesting reader is refereed to the reviews of [Boiko et al. 1985, Rosmej 2012] for further reading on this subject.

Fig. 7.1. Schematic energy level diagram of Al III and Al IV including hole states. Possible autoionization channels to ground and excited states are indicated.

In a dense plasma (Fig. 6.1) the inverse Auger effect is not only related to the original state "k" (see eq. (7.1.1)) but to excited states too. This can entirely change the picture of the inverse Auger effect. We illuminate excited state effects with an example: L-shell photo ionization of singly ionized aluminum ($K^2L^8M^2$-configuration). Photo ionization creates the autoionizing state j = $K^2L^7M^2$ that decays towards k = K^2L^8. The principle of micro-reversibility predicts therefore dielectronic capture according $K^2L^8 + e \rightarrow K^2L^7M^2$.

In dense plasmas, however, excited states are strongly populated via electron collisional excitation: $K^2L^8 + e \rightarrow K^2L^7M^1 + e$. This opens up the possibility to proceed towards dielectronic capture from excited states if the energy level structure energetically does permit this channel [Rosmej et al. 1998]. The so called "Excited States Coupling - ESC" changes almost all properties of the radiation emission and even overall satellite intensity by itself is not anymore related to the fundamental exponential factor of eq. (7.1.2).

	$1s^22s^22p^53l3l'$	$1s^22s^12p^63l3l'$
$1s^22s^22p^6$	$2.2x10^{12}$	$1.1x10^{12}$
$1s^22s^22p^53l$	$2.1x10^{13}$	$5.4x10^{14}$
$1s^22s^12p^63l$	-	$2.1x10^{13}$

Table 7.1. Averaged autoionzing rates Γ in [s^{-1}] of Al III related to ground and excited states.

In order to illuminate the situation, let us consider the relevant energy level diagram of Al III and Al IV in more details, Fig. 7.1. It can be seen that the levels $1s^22s^12p^63l3l'$ are not only coupled to the ground state $1s^22s^22p^6$ (Γ_{gr}) but to excited states $1s^22s^22p^53l$ too (Γ_{ex}) and even partially to the states $1s^22s^12p^63l$ ($\Gamma_{ex'}$). Similar relations hold true for the $1s^22s^22p^53l3l'$-levels: ground state coupling and partially excited states $1s^22s^22p^53l$ ($\Gamma_{ex'}$) coupling.

	$1s^22s^22p^53l3l'$	$1s^22s^12p^63l3l'$
$1s^22s^22p^63l$	$3.8x10^9$	$2.4x10^9$
$1s^22s^22p^53l3l'$	-	$4.1x10^{10}$

Table 7.2. Averaged spontaneous radiative decay rates A in [s^{-1}] of Al III.

Table 7.1 illustrates that autoionizing rates from excited states are even more important than those from ground states: about 1-2 orders of magnitude. Table 7.2 shows the relevant spontaneous radiative decay rates that are 2-5 orders of magnitude smaller than autoionizing rates. The statistically averaged data in tables 7.1-2 have been calculated with the FAC code [Gu 2008] employing a multi-configuration relativistic atomic structure, fine structure (LSJ-split levels), intermediate coupling and configuration interaction.

The data depicted in tables 7.1 and 7.2 imply that characteristic line emission from hole states that are produced by dielectronic capture from the ground state are barely visible due to a non-favorable branching factor for spontaneous radiative emission:

$$B_{ji} = \frac{A_{ji}}{\sum_i A_{ji} + \sum_k \Gamma_{jk}} \tag{7.2.1}$$

In dense plasmas, however, the excited states are strongly populated and dielectronic capture may proceed from excited states (also the branching factor is modified by collisions). In this case, quite different relations are encountered for characteristic emission of hole states. In a single level approximation, the line intensity is given by

$$I_{ji}\left(K^2L^7M^2\right) = n_e n\left(K^2L^8\right) B_{ji} D_{j,K^2L^8} + \sum_{K^2L^7M^1} n_e n\left(K^2L^7M^1\right) B_{ji} D_{j,K^2L^7M^1} \tag{7.2.2}$$

As the excited states autoionizing rates are much larger than the radiative decay rates and the ground state autoionizing states, the first term in eq. (7.2.2) almost vanishes. The second term is almost independent of the autoionzing rates as the branching factor multiplied with the excited state autoionizing rate is of the order of 1:

$$I_{ji}\left(K^2L^7M^2\right) \approx \sum_{K^2L^7M^1} n_e n\left(K^2L^7M^1\right) A_{ji} \frac{2^{3/2}\pi^{3/2}\hbar^3}{2m_e^{3/2}} \frac{g_j}{g_{K^2L^7M^1}} \frac{\exp\left(-E_{j,K^2L^7M^1} / kT_e\right)}{\left(kT_e\right)^{3/2}} \quad (7.2.3)$$

The dependence on the excited state density $n(K^2L^7M^1)$ implies that the intensity (eq. (7.2.3)) depends strongly on the electron density and not only on temperature as originally proposed by [Gabriel 1972].

This is demonstrated with simulations of the characteristic line emission from $K^2L^7M^2$ hole states together with the corresponding resonance line emission $K^2L^7M^1$-K^2L^8, Fig. 7.2. Simulations have been carried out employing all LSJ-split levels of the K^2L^8-configuration (1 level), $K^2L^7M^1$-configuration (36 levels), $K^2L^8M^1$-configuration (5 levels) and the $K^2L^7M^2$-configuration (237 levels) and including intermediate coupling and configuration interaction. Corresponding atomic population kinetics include electron collisional excitation/de-excitation, ionization/three body recombination, spontaneous radiative decay, autoionization and dielectronic capture [Rosmej et al. 2011].

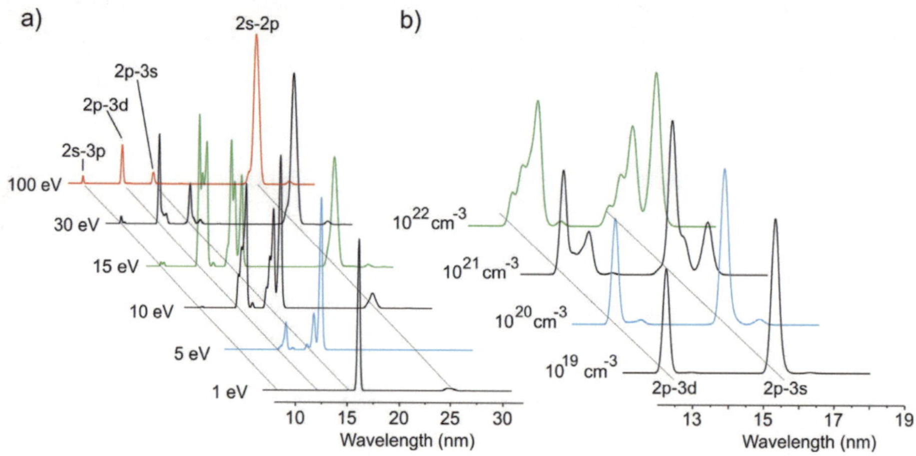

Fig. 7.2. a) Simulations of the spectral emission of Al IV and Al III emission (normalized to peak), a) temperature dependence, $n_e = 10^{22}$ cm^{-3}, b) density dependence, $kT_e = 10$ eV.

Fig. 7.2a shows the electron temperature dependence of the spectral distribution (see eq. (2.1.14)). The dashed lines indicate the positions of the resonance transitions in Al IV (2s-3p, 2p-3d and 2p-3s) as well as the intra-shell transitions (2s-2p). Other spectral features are due to the characteristic line emission of hole states originating from the $K^2L^7M^2$-configuration. Two observations can be made: first, with increasing electron temperature, the intensity of the intra-shell transition rises considerably, second, strong satellite emission is observed for temperatures around 15 eV. For very low electron temperatures (e.g., 1 eV in Fig. 7.2a), the line emission consists only from satellite transitions (note, however, that absolute line intensity is very low making experimental observation rather difficult). Fig. 7.2b shows the spectral range near the resonance transitions 2p-3d and 2p-3s (indicated by dashed lines) for different electron densities, other emission features are due to characteristic line emission from the hollow ion configuration $K^2L^7M^2$. At low electron densities, $n_e = 10^{19}$ cm^{-3}, satellite emission is barely visible. With increasing electron density, the satellite emission rises considerably due to excited state coupling effects (see discussion of eq. (7.2.3)).

We note, that also collisional redistribution between the autoionizing levels leads to intensity changes of the spectral distribution of satellite transitions (included in the present simulations), however, this concerns essentially deformations of the spectral distribution [Rosmej 2012] and not an overall drastic intensity increase as observed in Fig. 7.2b.

With respect to the overall temporal evolution of matter irradiated by XFEL radiation (Fig. 6.1), Fig. 7.2 demonstrates (see curve for $n_e = 10^{22}$ cm^{-3}) that the characteristic line emission from hole states might be even more important than usual resonance line emission. Satellite emission plays therefore an exceptional role to explore radiative properties of high density matter under extreme conditions.

8. Quantum mechanical interference effects

8.1 Pumping characteristic X-ray transitions in autoionizing hole states

Let us now consider novel effects in the spectral line broadening of characteristic X-ray line emission from hole states. The newly emerging XFEL installations will permit outstanding observations of quantum mechanical interference effects as the XFEL can be employed to directly pump the characteristic x-ray emission in dense matter.

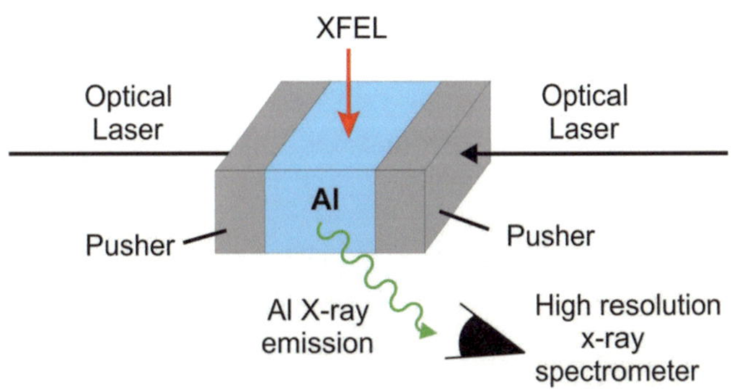

Fig. 8.1. Schematic experimental pump-probe scheme to investigate quantum mechanical interference effects in near and above (compressed) solid density matter with X-ray Free Electron Lasers and high resolution X-ray spectroscopy.

A principle experimental scheme is depicted in Fig. 8.1. The radiating test element (Al in Fig. 8.1) is compressed with optical laser beams. When the matter is effectively compressed a 100 fs XFEL pulse will further heat the compressed matter (e.g., Auger electron heating and photo electron heating, see paragraph 6) and pump X-ray transitions by effective wavelength tuning. Let us consider hole states in Li-like ions. The XFEL frequency is adjusted in such a way to pump X-ray transitions from the Li-like states $1s^22l$ to excite the multiple excited states $1s2l2l'$: $1s^22l + h\nu_{XFEL} - 1s2l2l'$. After excitation, the soft x-ray emission $1s2l2l' - 1s^22l' + h\nu_{satellites}$ is observed with a high-resolution x-ray spectrometer.

The short time scale of the X-ray pump provides practically a snap shot of the parameter situation thereby avoiding too many complications due to time integration. The pumping of the satellite transitions has namely the great advantage of extremely short time scale for the radiation emission itself because the relevant characteristic emission time $\tau_{eff,j}$ of a certain autoionizing atomic level "j" is not given by the spontaneous transition probabilities but rather by the sum of autoionizing and radiative decay rates [Rosmej 2012]:

$$\tau_{eff,j}^{-1} \approx \sum_i A_{ji} + \sum_k \Gamma_{jk} \tag{8.1.1}$$

For the 1s2l2l'-satellites of Al, the strongest radiative decay and autoionizing rates are of the order of $1\text{x}10^{13}$ s^{-1} and $1\text{x}10^{14}$ s^{-1}, respectively, implying effective response times of the order of $\tau_{eff}(1s2l2l') \approx 10 - 100$ fs. Even for 1s2l3l'-satellites we encounter very fast response times: the strongest radiative decay and autoionizing rates are of the order of $2\text{x}10^{13}$ s^{-1} and $3\text{x}10^{13}$ s^{-1}, respectively, implying $\tau_{eff}(1s2l3l') \approx 30 - 500$ fs. Therefore, time integration effects due to characteristic photon emission times are very small as τ_{eff} is smaller than hydrodynamic time scales.

8.2 Line broadening and interference effects of hollow ion X-ray emission

Spectral line broadening that is due to the interaction of a radiating atom with surrounding particles is closely connected with the theory of atomic collisions and extensive reviews have been published on this issue [e.g., Griem 1974, 1997, Sobelman et al. 1995].

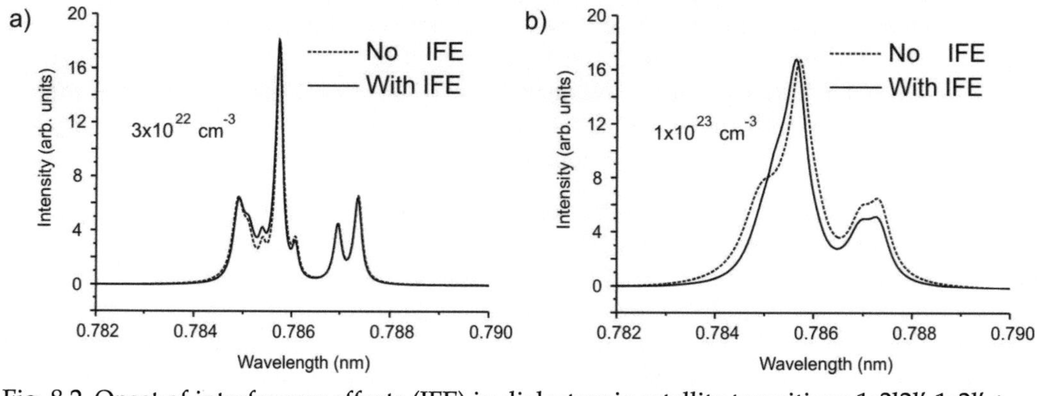

Fig. 8.2. Onset of interference effects (IFE) in dielectronic satellite transitions 1s2l2l'-1s2l' + hν of Li-like aluminum, a) $n_e = 3\text{x}10^{22}$ cm^{-3}, $kT_e = 100$ eV, b) $n_e = 1\text{x}10^{23}$ cm^{-3}, $kT_e = 100$ eV.

The general theory of impact broadening is based on the density matrix and quantum kinetic approach and considers the scattering amplitudes and phases thereby allowing to consider quantum mechanical interference effects. In the line broadening theory, interference effects arise due to transition frequencies that coincide or are so closely spaced that the corresponding spectral lines overlap. In some cases, the interference effects are so important that they alter the entire picture of the line broadening and it has been noted long time ago that interference effects may lead to a considerable line narrowing [Aleseyev and Sobelman 1969, Sobelman et al. 1995].

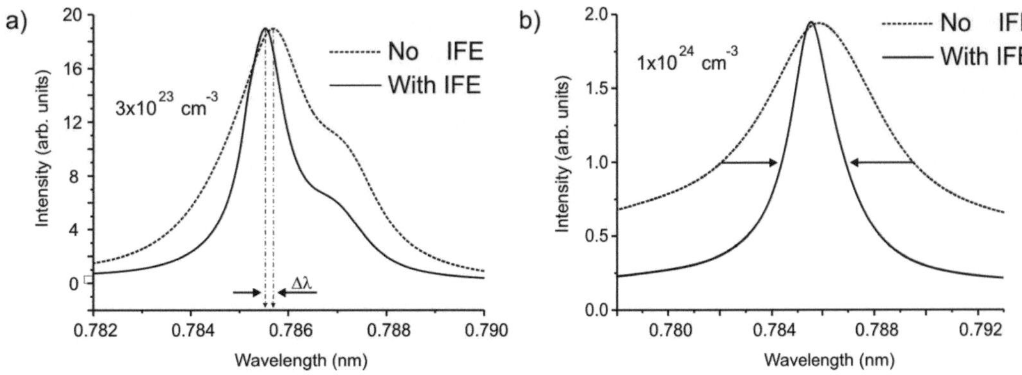

Fig. 8.3. Strong interference effects (IFE) in dielectronic satellite transitions 1s2l2l'-1s2l' + hν of Li-like aluminum, a) $n_e = 3x10^{23}$ cm^{-3}, $kT_e = 100$ eV, b) $n_e = 1x10^{24}$ cm^{-3}, $kT_e = 100$ eV. Interference effects lead to a considerable line narrowing (see arrows in (b)) and also to wavelengths shift Δλ (see (a)) of the emission group.

In order to apply Stark broadening to real experimental conditions, opacity broadening and sensitivity to a low density recombination regimes has to be avoided: this is difficult employing resonance lines of H- and He-like ions. Moreover, as the respective ground states are the states 1s $^2S_{1/2}$ and 1s^2 $^1S_0$, interference effects do not arise. All these problems are circumvented employing the dielectronic statellite transitions to Ly$_\alpha$ of He-like ions 2lnl' - 1snl' + hν$_{sat}$ and to He$_\alpha$ of Li-like ions 1s2lnl' - 1s^{2}nl' + hν$_{sat}$ (one the most frequently used transitions to diagnose hot dense plasmas [Rosmej 2012]). As lower states are numerous interference effects can arise. Moreover, their short emission times scales (see discussion of eq. (8.1.1)) confines the emission near the XFEL interaction times where density is highest. We note that optical laser produced plasmas suffer from limited plasma density (order of the critical densities) and the experimental scheme depicted in Fig. 8.1 will be extremely challenging to probe near and above (compressed) solid density matter by XFEL pumping.

Stark broadening calculations of the characteristic line emission that originate from hollow ions involves very complex configurations with the corresponding need to calculate millions of Stark transitions. At present, one of the most general and powerful methods to calculate line profiles for such complex transitions has been developed with the PPP-code that is based on the frequency fluctuation model [Talin et al. 1995, Calisti et al. 2006, 2010]. PPP allows rapid Stark broadening calculations of millions of Stark transitions and includes the possibility to calculate ion dynamics and interference effects.

Fig. 8.2 shows the Stark broadening simulations carried out with the PPP-code for the dielectronic satellite transitions of Li-like Al: 1s2l2l' - 1s^{2}2l' + hν$_{sat}$. We note that for the present calculations the requested dipole matrix elements do include configuration interaction, intermediate coupling and LSJ-split level structure. In order not to mask interference effects with population kinetic effects, a statistical population between the levels has been employed. The curves in Fig. 8.2a are calculated for an electron density of $n_e = 3x10^{22}$ cm^{-3} without (dashed curve) and with (solid curve) interference effects (IFE). It can be seen, that interference effects are barely visible at these electron densities. At $n_e = 1x10^{23}$ cm^{-3} interference effects start to show up.

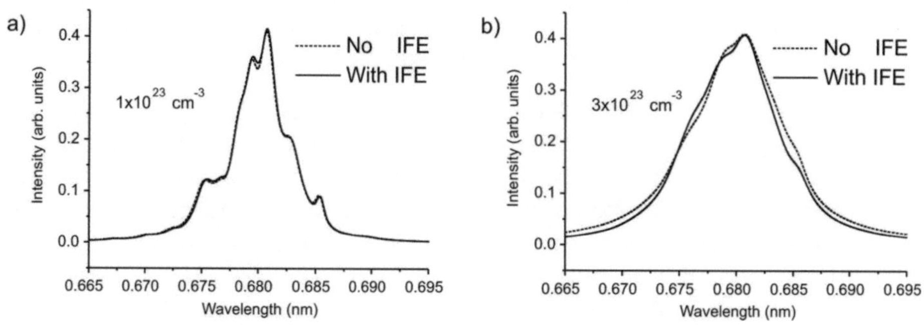

Fig. 8.4. Investigation of interference effects (IFE) in dielectronic satellite transitions 1s2l3l′-$1s^2$2l + hν of Li-like aluminum near He_β, a) $n_e = 1x10^{23}\ cm^{-3}$, kT_e = 100 eV, b) $n_e = 3x10^{23}\ cm^{-3}$, kT_e = 100 eV. Interference effects lead only to small changes in the overall spectral distribution.

Figure 8.3 shows parameters, where strong interference effects are expected. At $n_e = 3x10^{23}$ cm^{-3} IFE result in a serious narrowing of the emission group and also to a qualitative distortion of the spectral distribution. Also a strong wavelength shift of the emission group (indicated by the peak center shift Δλ in Fig. 8.3a) is observed. At even higher densities, n_e = $3x10^{23}\ cm^{-3}$ IFE have reduced the overall width of the emission group by a factor of 2-3 (see arrows in Fig. 8.3b). This indicates that Stark broadening simulations that do not include IFE considerably underestimate the electron density when applied to experimental data.

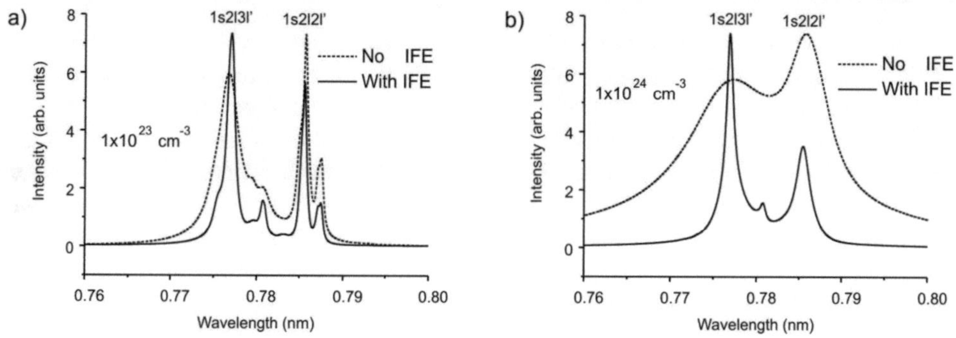

Fig. 8.5. Investigation of interference effects (IFE) in dielectronic satellite transitions 1s2l3l′-$1s^2$3l′ + hν of Li-like aluminum near He_α, a) $n_e = 1x10^{23}\ cm^{-3}$, kT_e = 100 eV, b) $n_e = 1x10^{24}\ cm^{-3}$, kT_e = 100 eV. Interference effects lead to large changes in the overall spectral distribution.

As the XFEL radiation can equally be tuned to the β-transitions it is quite challenging to look for the Li-like satellite transitions 1s2l3l′ - $1s^2$2l + $h\nu_{He\beta\text{-sat.}}$ near the He_β resonance line (He_β = 1s3p $^1P_1$ - $1s^2$ $^1S_0$): broadening effects are expected to be visible for lower densities (as compared to the 1s2l2l′-satellites) and opacity effects are even more reduced. Fig. 8.4 shows the corresponding simulations. As can be seen from Fig. 8.4a/b, interference effects lead only to rather small changes in the overall spectral distribution. The reason for these rather small effects is connected with the relatively limited number of lower states ($1s^2$2l′-configuration). The situation is dramatically different when considering interference effects of the 1s2l3l′-satellites near He_α (1s2p $^1P_1$ - $1s^2$ $^1S_0$): 1s2l3l′ - 1s2nl′ + $h\nu_{He\alpha\text{-sat.}}$ (lower states $1s^2$3l′). Fig. 8.5 shows the corresponding simulations. Dramatic changes in the overall spectral distribution are observed, in particular, interference effects lead to a considerable shape narrowing of the group emission. We note that the simulations of Fig. 8.5 are rather complex involving the calculation of some 10 million Stark transitions.

The dramatic difference of the interference effects of He_{α}- and He_{β}-satellites (see Figs. 8.4 and 8.5) originating from 1s2l3l'-configurations allows direct experimental verification: the negligible interference effects in the He_{β}-satellites serve as an experimental reference broadening allowing to detect the group narrowing due to interferences in the 1s2l3l' He_{α}-satellites. In this respect we note, that high resolution X-ray spectroscopy has proven to provide sufficient resolution to separate the 1s2l3l' He_{α}-satellites from the He_{α}-resonance line itself even in dense laser produced plasmas [Rosmej et al. 1998]. Also observable line intensities are expected as successful pumping of Li-like satellite transitions in a dense plasma plume has already been demonstrated in recent experiments at the LCLS [LCLS 2011] XFEL facility [Seely et al. 2011].

The XFEL pumping of characteristic X-ray transitions in hollow ions provides therefore outstanding experimental conditions to study novel high-density matter physics. : first, the pump allows selectively increasing the satellite transitions to obtain good signal to noise ratio, second, the short time scale (of XFEL pump and satellite transitions itself) avoids ambiguities due to time integration effects (integration over different plasma parameters during evolution).

9. Conclusion

High intensity short pulse XUV/X-FEL Free Electron Laser radiation provides to the scientific community outstanding tools to investigate matter under extreme conditions never obtained in laboratories so far. We have presented novel effects in the solid-to-plasma transition considering irradiation of solid matter with high intensities and short XUV/XFEL pulses. Exotic states of matter such as transparent metals, hollow crystals and X-ray bursts from hollow ions have been investigated. Novel effects in atomic physics have been studied: Auger electron bursts from hollow crystals, 10-fs atomic X-ray switches, excites states coupling effects induced by dense matter and quantum mechanical interference effects in the characteristic X-ray line emission from hole states.

A new heating mechanism was discussed: "Auger electron heating" followed by the decay of crystalline order, formation of Warm Dense Matter and strongly coupled plasmas. Finally we have explored the exceptional role of characteristic X-ray emission (satellites) from hole states/hollow ions to study radiative properties of dense matter under extreme conditions.

10. Acknowledgment

Support from the project "Èmergence-2010: *Métaux transparents créés sous irradiations intenses émises par un laser XUV/X à électrons libres*" of the University Pierre and Marie Curie and the "Extreme Matter Institut - EMMI" are greatly appreciated.

11. References

Aleseyev, V.A. & Sobelman, I.I. (1969). *Influence of Collisions on Stimulated Random Scattering in Gases,* JETP 28, 991.

Boiko, V. A.; Vinogradov, A. V.; Pikuz, S. A.; Skobelev, I. Yu. & Faenov, A. Ya. (1985). *X-ray spectroscopy of laser produced plasmas*, J. Sov. Laser Research 6, 82.

Calisti, A.; Ferri, S.; Mossé-Sabonnadière, C. & Talin, B. (2006). *Pim Pam Pum*, Report Université de Provence, Marseille, France.

Calisti, A.; Galtier, E.; Rosmej, F.B.; Ferri, S.; Mossé, C.; Talin, B. & Lisitsa, V.S. (2010). *Detailed Stark calculations of aluminum emission from hole states induced by XUV-Free Electron Laser irradiations*, 14th International Workshop on Radiative Properties of Hot Dense Matter (4-8/10/2010), Marbella, Spain.

Cryan, J.P.; Glownia, J. M.; Andreasson, J.; Belkacem, A.; Berrah, N.; Blaga, C.I.; Bostedt, C. Bozek, J.; Buth, C.; DiMauro, L.F.; Fang, L.; Gessner, O.; Guehr, M.; Hajdu, J.; Hertlein, M.P.; Hoener, M.; Kornilov, O.; Marangos, J.P.; March, A.M.; McFarland, B. K.; Merdji, H.; Petrovic´, V.S.; Raman, C.; Ray, D.; Reis, D.; Tarantelli, F.; Trigo, M.; White, J. L.; White, W.; Young, L.; Bucksbaum, P.H. & Coffee, R.N. (2010). *Auger Electron Angular Distribution of Double Core-Hole States in the Molecular Reference Frame*, Phys. Rev. Lett. 105, 083004.

Dawydow, A.S. 1981. *Quantenmechanik*, VEB Deutscher Verlag der Wissenschaften, Berlin.

Faenov, A.Ya.; Magunov, A.I.; Pikuz, T.A.; Skobelev, I.Yu.; Pikuz, S.A. ; Urnov, A.M.; Abdallah, J.; Clark, R.E.H.; Cohen, J.; Johnson, R.P. ; Kyrala, G.A.; Wilke, M.D.; Maksimchuk, A.; Umstadter, D.; Nantel, R. Doron, N.; Behar, E.; Mandelbaum, P.; Schwob, J.J.; Dubau, J. ; Rosmej, F.B. & Osterheld, A. 1999. *High-Resolved X-Ray Spectra of Hollow Atoms in a Femtosecond Laser-Produced Solid Plasma*", Physica Scripta T80, 536 (1999).

Flügge, E. (1957). *Encyclopedia of Physics*, vol. XXX (X-rays), Springer.

Gabriel, A.H. 1972. *Dielectronic satellite spectra for highly charged He-like ion lines*. Mon. Not. R. astro. Soc. 160, 99.

Galtier, E.; Rosmej, F.B.; Riley, D.; Dzelzainis, T.; Khattak, F.Y.; Heimann, P.; Lee, R.W.; Vinko, S.M.; Whitcher, T.; Nagler, B.; Nelson, A.; Wark, J.S.; Tschentscher, T.; Toleikis, S.; Fäustlin, R.; Sobierajski, R.; Jurek, M.; Juha, L.; Chalupsky, J.; Hajkova, V.; Kozlova, M. & Krzywinski, J. (2011). *Decay of cristaline order and equilibration during solid-to-plasma transition induced by 20-fs microfocused 92 eV Free Electron Laser Pulses*, Phys. Rev. Lett. 106, 164801.

Griem, H.R. (1974). *Specral Line Broadening by Plasma*, Academic Press, New York.

Griem, H.R. (1997). *Principles of Plasma Spectroscopy*, Cambridge University Press, New York, ISBN 0-521-45504-9.

Gu, M.F. 2008. The flexible atomic code FAC, Canadian Journal of Physics 86 (5), 675.

Kozyreva, A.; Basko, M.; Rosmej, F.B.; Schlegel, T.; Tauschwitz, A. & Hoffmann, D.H.H. (2003). *Dynamic confinement of targets heated quasi-isochorically with heavy ion beam*, Phys. Rev. E 68, 056406.

Larousserie, D. 2009. *L'aluminium devient transparent aux rayons X*. Science et Avenir, vol. 09/2009, p. 20.

LCLS 2011. http://www-ssrl.slac.stanford.edu/lcls/

Loudon, R. (2000). *The quantum theory of light*, Oxford University Press, New York, ISBN 0-19-8501767-3.

Mihalas, D. 1978. *Stellar Atmospheres*, W.H. Freeman, San Francisco, 2nd edition.

Nagler, B.; Zastrau, U; Fäustlin, R; Vinko, S.M.; Whitcher, T.; Nelson, A. J.; Sobierajski, R.; Krzywinski, J.; Chalupsky, J.; Abreu, E.; Bajt, S.; Bornath, T.; Burian, T.; Chapman, H.; Cihelka, J.; Döppner, T.; Düsterer, S.; Dzelzainis, T.; Fajardo, M.; Förster, E.;

Fortmann, C.; Galtier, E.; Glenzer, S. H.; Göde, S.; Gregori, G.; Hajkova, V.; Heimann, P.; Juha, L.; Jurek, M.; Khattak, F.Y.; Khorsand, A.R.; Klinger, D.; Kozlova, M.; Laarmann, T.; Lee, H.J.; Lee, R.W.; Meiwes-Broer, K.-H.; Mercere, P.; Murphy, W.J.; Przystawik, A.; Redmer, R.; Reinholz, H.; Riley, D.; Röpke, G.; Rosmej, F.B.; Saksl, K.; Schott, R.; Thiele, R.; Tiggesbäumker, J.; Toleikis, S.; Tschentscher, T.; Uschmann, I.; Vollmer, H. J. & Wark, J. (2009). *Transparency induced in solid density aluminum by ultra-intense XUV Radiation*, Nature Physics 5, 693.

Rosmej, F.B. (1997). *Hot Electron X-ray Diagnostics*, J. Phys. B. Lett.: At. Mol. Opt. Phys. 30, L819.

Rosmej, F.B.; Faenov, A.Ya.; Pikuz, T.A.; Flora, F.; Lazzaro, P. Di; Bollanti, S.; Lizi, N.; Letardi, T.; Reale, A.; Palladino, L.; Batani, O.; Bossi, S.; Bornardinello, A.; Scafati, A. & Reale, L. (1998). *Line Formation of High Intensity He_b-Rydberg Dielectronic Satellites 1s3lnl' in Laser Produced Plasmas*, J. Phys. B Lett.: At. Mol. Opt. Phys. 31, L921.

Rosmej, F.B. (2001). "*A new type of analytical model for complex radiation emission of hollow ions in fusion and laser produced plasmas*", Europhysics Letters 55, 472.

Rosmej, F.B. (2006). *An alternative method to determine atomic radiation*", Europhysics Letters 76, 1081.

Rosmej, F.B. & Lee, R.W. (2006). *Hollow ion emission*, in "XFEL Technical Design Report", Chapter 6, p. 251-253, DESY 2006, http://xfel.desy.de/tdr/tdr

Rosmej, F.B. & Lee, R.W. (2007). *Hollow ion emission driven by pulsed x-ray radiation fields*, Europhysics Letters 77, 24001.

Rosmej, F.B. ; Lee, R.W. & Schneider, D.H.G (2007). *Fast x-ray emission switches driven by intense x-ray free electron laser radiation*, High Energy Density Physics 3, 218.

Rosmej, F.B.; Petitdemange, F. & Galtier, E. (2011). *Identification of Auger electron heating and inverse Auger effect in experiments irradiating solids with XUV Free Electron Laser Radiation at intensities larger than 10^{16} W/cm^2*, SPIE International Symposium on Optical Engineering and Applications "X-Ray Lasers and Coherent X-Ray Sources: Development and Applications (OP321)" (21–25/08/2011) San Diego, USA.

Rosmej, F.B & Lisitsa, V.S (2011): *"Non-equilibrium radiative properties in fluctuating plasmas"*, Plasma Physics Reports 37, 521.

Rosmej, F.B. (2012). *X-ray emission spectroscopy and diagnostics of non-equilibrium fusion and laser produced plasmas"*, in "Highly Charged Ions, editors Y. Zou and R. Hutton, Taylor and Francis 2012, ISBN 9781420079043.

SACLA 2011. http://xfel.riken.jp/eng

Seely, J.; Rosmej, F.B.; Shepherd, R.; Riley, D. & Lee, R.W. (2011). *Proposal to Perform the 1st High Energy Density Plasma Spectroscopic Pump/Probe Experiment*, accepted beam time proposal n° L332 at LCLS.

Sobelman, I.I.; Vainshtein, V.S.; Yukov & E.A. (1995). *Excitation of Atoms and Broadening of Spectral Lines*, Springer, Berlin, ISBN 3-540-58686-5.

Talin, B. ; Calisti, A. ; Godbert, L.; Stamm, R.; Lee, R. W. & Klein, L. (1995). *Frequency-fluctuation model for line-shape calculations in plasma spectroscopy*, Phys. Rev. A 51, 1918.

Tauschwitz, An. ; Maruhn, J.A. ; Riley, D. ; Shabbir Naz, G. ; Rosmej, F.B. ; Borneis, S. & Tauschwitz, A. 2007. *Quasi-isochoric ion beam heating using dynamic confinement in spherical geometry for X-ray scattering experiments in WDM regime"*, High Energy Density Physics 3, 371.

WIKI 2011. *http://en.wikipedia.org/wiki/List_of_Star_Trek_materials#Transparent_aluminum*

XFEL 2011. http://xfel.desy.de/

9

Digital RF Control System for the Pulsed Superconducting Linear Accelerator

Valeri Ayvazyan
Deutsches Elektronen-Synchrotron DESY, Hamburg*
Germany

1. Introduction

The FLASH (Free Electron Laser in Hamburg) (Ackermann et al., 2007) is a user facility providing high brilliant laser light for experiments. It is also a unique facility for testing the superconducting accelerator technology for the European XFEL (Weise, 2004) and the International Linear Collider (Ross et al., 2011). The LLRF (low level RF) system is used to maintain the beam stabilities by stabilizing the RF field in the superconducting cavities with feedback and feed-forward algorithms. In the RF system for FLASH linear accelerator each klystron supplies RF power to up to 16 cavities. The superconducting cavities are operating in pulsed mode and high accelerating gradients close to the performance limit. The RF control of the cavity fields to the level of 10^{-4} for amplitude and 0.1 degree for phase however presents a significant technical challenge due to the narrow bandwidth of the cavities which results in high sensitivity to perturbations of the resonance frequency by mechanical vibrations (microphonics) and Lorentz force detuning. FLASH accelerator LLRF control system employs a completely digital feedback system to provide flexibility in the control algorithms, precise calibration of the accelerating field, and extensive diagnostics and exception handling capabilities. The LLRF control algorithms are implemented in FPGAs (field programmable gate array) firmware and DOOCS (Distributed Object Oriented Control System) (Rehlich, 2007) control system servers.

2. Brief description of FLASH facility

The FLASH consists of a 256 m long electron linear accelerator followed by a photon transport line equipped with photon diagnostics and 5 experimental beam lines for users (Vogt et al., 2011). A schematic layout of the facility is shown in fig. 1. The injector consists of a laser-driven photocathode in a 1.5-cell RF cavity operating at 1.3 GHz with a peak accelerating field of 40 MV/m on the cathode. The electron injector section is followed by a total of seven 12.2 m long accelerating modules each containing eight 9-cell L-band superconducting niobium cavities. The maximum accelerating gradient in the best cavities is 35 MV/m. The average gradient is about 22 MV/m. With seven accelerating modules the maximum energy of the FLASH linac is about 1250 MeV.

**for the LLRF team*

In order to remove the RF induced curvature, a module with four superconducting cavities operating at 3.9 GHz (third harmonic of 1.3 GHz) is installed downstream of the first accelerating module (Harms et al., 2009). Downstream of the last accelerating module FLASH has a transverse collimation section and a dispersive "dog–leg" for energy collimation upstream of the undulators. At intermediate energies of 150 MeV and 500 MeV the electron bunches are longitudinally compressed, thereby increasing the peak current. The FEL radiation is produced by six 4.5 m long undulators. The undulators consist of permanent NdFeB magnets with a fixed gap of 12 mm, a period length of 27.3 mm and peak magnetic field of 0.47 T. The wavelength of the FEL radiation depends on the energy of the accelerated electrons. It can be tuned between 4.1 nm and 44 nm. In addition, an experiment for seeded FEL radiation, sFLASH, is integrated to the FLASH linac. One bunch per RF pulse can be seeded in 4 tunable (variable gap) undulators.

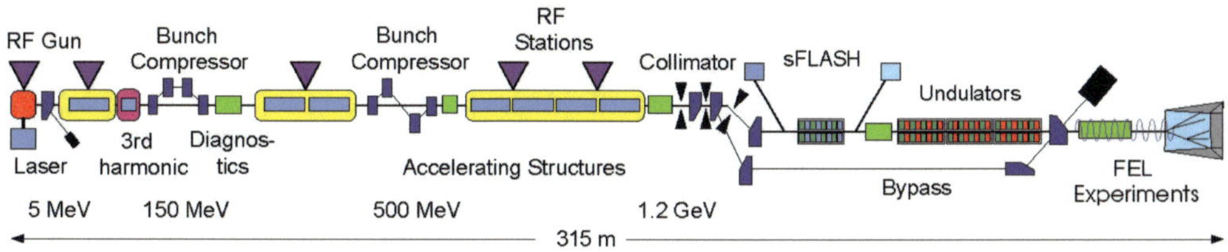

Fig. 1. Layout of the FLASH facility

3. FLASH RF system overview and sources for field errors

The accelerating modules are powered by four RF stations consisting a klystron (three 5 MW klystrons and one 10 MW multi-beam klystron), a high voltage pulse transformer and a pulsed power supply (modulator). Third harmonic module is driven by a custom–made klystron and modulator. In addition, the RF gun has its own RF station with a 5 MW klystron. The cavities in FLASH are operating in a pulsed mode. The pulse repetition rate is 10 Hz. The RF pulse length is 1300 μs from which 500 μs are required to build up the RF field in the cavities. The beam is accelerated for 800 μs with a constant accelerating field.

The low level RF system provides field control for the RF gun and the superconducting cavities with the vector-sum group of cavities. Fig. 2 shows main components of the FLASH RF control scheme. At FLASH, with the vector-sum concept, one klystron powers a set of up to 16 cavities (for European XFEL project, this number will be 32). This is cost efficient compared to a solution where each cavity has its own RF station consisting of one controller and one amplifier. RF power is distributed from a high power klystron to the cavities via a forked set of waveguides. The synchronism of the incident waves to the cavities is ensured by waveguide tuners in front of each cavity which allow an adjustment of phases for individual cavities. A frequency tuning mechanism is used to operate the cavities on resonance. Fig. 2 also shows the reference master oscillator and the timing system which plays a central role in the LLRF system. The goal of LLRF control can be stated as locking the field inside the resonators to the reference.

Each accelerating cavity has one input coupler for RF power and a small pickup antenna to measure the cavity field. Probes measure the field directly inside the cavities by coupling out a small fraction of the power. The measured fields are digitized and further processed by FPGAs in order to calculate the vector-sum and from that, by comparison with the set-point, the drive that is applied to the klystron. The stability of the accelerating field depends critically on the precision of the calibration of the vector-sum which is based on transient beam loading.

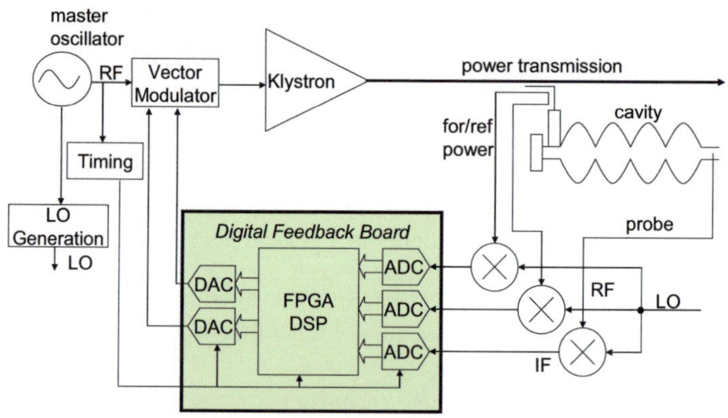

Fig. 2. Schematic of one RF station of FLASH including the LLRF system

Energy fluctuations of an accelerated beam along the linac are caused by various phenomena. Three different types of energy fluctuations have to be distinguished: macro-pulse to macro-pulse, bunch-to-bunch and intra-bunch energy fluctuations. The main sources for field perturbations are fluctuations of the cavity resonance frequency and beam current. Changes in resonance frequency result from deformations of cavity walls induced by mechanical vibrations (microphonics) or the gradient dependent Lorentz force.

3.1 Microphonics

Mechanical vibrations caused by the accelerator environment such as vacuum pumps at the cryogenics facility are always present and are transferred to the accelerating structures trough the beam pipe and transfer lines as well as trough ground motion. The amplitude of the excitation and associated frequency spectrum depend strongly on the coupling of the cavities and the mechanical resonance frequencies and their mechanical quality factor. Measurements in FLASH accelerating module show typical excitation amplitudes of the order of 5-10 Hz rms (Schilcher, 1998; Ayvazyan et al., 2007) and frequencies ranging from 0.1 Hz up to a few hundred Hz reflecting the convolution of mechanical resonances with the spectral components of the sources of excitation.

3.2 Lorentz force detuning

The static detuning of resonator due to Lorentz force is proportional to the square of the accelerating field. In the case of pulsed RF fields the mechanical resonances of the cavities will be excited resulting in a time varying detuning even during the flat-top portion of the RF pulse where the gradient is constant (Ayvazyan & Simrock, 2004). The lower the mechanical quality factor and the higher the resonance frequency (only longitudinal modes should be excited), the less likely is the enhancement of peak cavity detuning caused by resonant excitation by the Lorentz force. Stiffening rings at the iris are used to reduce the Lorentz force detuning constant and increase the mechanical resonance frequencies.

3.3 Beam loading

The loaded quality factor is usually chosen for matched conditions so that all of the generator power (with the exception of a small amount dissipated in the cavities) is transferred to the beam under design operating conditions. In the case of on-crest operation the magnitude of the beam induced voltage is half of the generator induced voltage.

4. RF control requirements

The RF control requirements for amplitude and phase stability are usually derived from the desired beam parameters. It is also important to include operational issues such as the turn-on of the RF system, calibration of gradient and phase, and control of the waveguide and frequency tuners.

4.1 Amplitude and phase stability

The requirements for amplitude and phase stability of the vector-sum group of cavities are driven by the maximum tolerable energy spread for the FLASH linac. The goal is an rms energy spread of $\sigma_E/E=10^{-4}$. The requirements for gradient and phase stability are therefore of the order of 10^{-4} and 0.1 degree respectively.

The amplitude and phase errors to be controlled are of the order of 5% for the amplitude and 20 degrees for the phase as a result of Lorentz force detuning and microphonics. These errors must be suppressed by a factor of more than 100 which implies that the loop gain must be adequate to meet this goal. Fortunately, the dominant source of errors is repetitive (Lorentz force and beam loading) and can be reduced by use of feed-forward significantly.

The requirements for the phase stability become also more severe for off-crest operation. In the case of the control of the vector-sum of several cavities driven by one klystron, the requirement for the phase calibration of the vector-sum components may become critical depending on the magnitude of microphonics.

4.2 Operational requirements

Besides field stabilization the RF control system must provide diagnostics for the calibration of gradient and beam phase, measurement of the loop phase, cavity detuning, and control of the cavity frequency tuners. Exception handling capability must be implemented to avoid unnecessary beam loss. Features such as automated fault recovery will help to maximize accelerator up-time. The RF control must be fully functional over a wide range of operating parameters such as gradients and beam current. For efficiency reasons the RF system should provide sufficient control close to klystron saturation.

5. Design choices for the RF control

Control of the RF power normally takes place on the low power level before the amplification of the signals. The main components are detectors for amplitude and phase of the individual cavity fields, the controller for the feedback itself, and actuators to regulate the incident wave to the klystron and thus to the cavities. There are different ways to drive an RF cavity (Schilcher, 1998): the generator driven resonator, the self-excited loop and the direct RF feedback. In the first two cases one can choose to control amplitude and phase or the real and imaginary parts of the incident RF vector.

Apart from the direct RF feedback all other introduced RF control systems include elements to detect amplitude and phase of the RF field. In general an RF field is described by a vector in the complex plane. This RF vector can be represented either by its amplitude and phase with respects to a reference oscillator or by its real and imaginary parts. These are sometimes called the I (in-phase, 0 degree) and the Q (quadrature, 90 degree) components of a vector. The appropriate choice of representation depends on the requirements. An amplitude detector is usually a Schottky diode and a widely used phase detector is a double balanced RF mixer. The determination of real and imaginary parts is completely based on RF mixers. Phase noise from the master oscillator directly influences the measurement of real and imaginary parts while in the traditional amplitude and phase control only the phase measurement is affected. If the disturbance of the RF field is mainly on the amplitude, gradient and phase control might be preferable. On the other hand, vector control by real and imaginary parts is preferred in systems with large errors in amplitude and phase where sufficiently high feedback gains have to be used. It should be noted that a phase controller can cause a phase correction in a wrong quadrant if the gain is too high. Besides that the phase is not well defined if the input signal of the phase detector is too small. As a result it is not possible to control the zero set point with amplitude and phase controllers.

6. LLRF control implementation at FLASH

In the linac of FLASH the cavities are operated with a driven feedback system. Lorentz force detuning and microphonics lead to considerable amplitude and phase disturbances in the superconducting cavities with their high quality factors. The expected errors are in the order of a few percent for amplitude and above 20 degrees for the phase during the 800 μs long beam pulse. Therefore a vector control has been chosen. RF mixers decompose the cavity RF vector into the real and imaginary parts. The incident RF wave is regulated by a vector modulator. In principle the feedback controller can be realized as an analog or a completely digital system. In the case of a digital system ADCs digitize the decomposed RF signal from the cavity. A following digital signal processing stage performs the feedback algorithm before the calculated actuator signals are reconverted to analog signals. For the RF control in the FLASH a fully digital system has been implemented and it has several stages of major developments (Simrock et al., 1996; Schilcher, 1998; Ayvazyan et al., 2005, 2007, 2010).

6.1 Principles for LLRF control

The RF system signal flow is shown in fig. 3. The cavity probe signal is down converted from the cavity frequency of 1.3 GHz (3.9 GHz) to an intermediate frequency (IF) of 250 kHz for superconducting modules and 54 MHz for the 3rd harmonic module. This lower IF holds the original amplitude and phase information of the field inside the cavity.

The control algorithm employs tables for feed-forward, set-point and feedback gain settings to allow time varying of those parameters. The down converted signal is digitized with ADCs (sampling rates of 1 MHz or 81 MHz are used). The digitized signal is going to the digital field detector which extracts the I and Q components out of the input stream. We use two different methods: IQ-sampling and so-called non-IQ-sampling or IF-sampling. The resulting field vector of each cavity is multiplied by a rotation matrix to calibrate amplitude and phase. Finally the field vectors of 8 cavities are summed up for the vector-sum of a whole cryogenic module, and those of 2 cryogenic modules are summed up to the vector-sum of the RF station which is driven by single klystron. The vector-sum of the 16 cavity fields represents the total voltage and phase seen by the beam. This signal is regulated by a feedback control algorithm which calculates corrections to the driving signal of the klystron: the measured vector-sum is subtracted from the set-point table and the resulting error signal is amplified and filtered to provide a feedback signal to the vector modulator controlling the incident wave. A feed-forward signal is added to correct the averaged repetitive error components. Beam current information (measured by toroids) is used to scale the feed-forward table to provide fast feed-forward corrections if the beam current varies. In addition beam diagnostics signals are in use for fast intra-pulse feedback (Koprek et al., 2010). The cavity detuning is determined from forward power reflected power, and probe signal and is used to control the fast piezo tuners to reduce cavity detuning errors to less than a tenth of the cavity bandwidth.

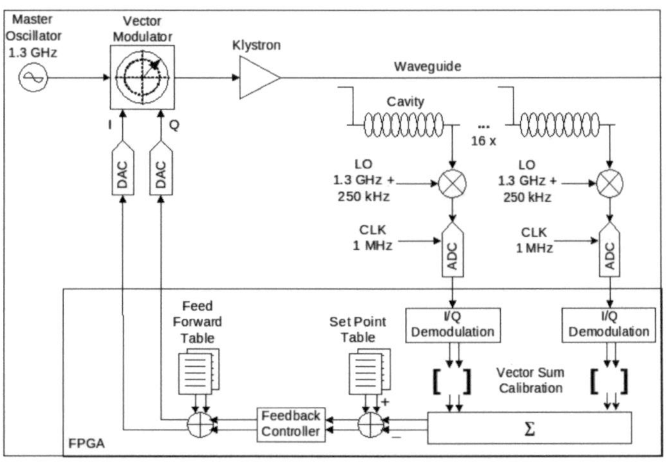

Fig. 3. Control algorithm block diagram

6.2 Digital feedback hardware

As a test facility, the accelerator undergoes a constant modification and expansion. After last upgrade in 2009/2010 the RF gun and all accelerating modules are controlled by similar modern FPGA based controller boards with unified firmware and software. The digital feedback hardware consists of Simcon-DSP (simulator & controller) board (fig. 4) which has a VME interface, 10 ADCs to digitize the intermediate frequency signal from the field probe signals, FPGAs and DSPs to execute the control algorithms and 8 DACs, 2 of them drives the vector modulator for field control. Other components include a timing and synchronization module. The field detection hardware consists of a down converter which converts the cavity field frequency of 1.3 GHz (3.9 GHz) to an intermediate frequency. Additional features included variable input attenuators for level adjustment, an input for a calibration signal and a local oscillator distribution system. The challenging requirements of the down converters are low noise, good linearity over large dynamic range, and small crosstalk. A function generator with VME interface drives a vector-modulator which produces the local oscillator signal generated from the 1.3 GHz. It switches the local oscillator phase by 90 degrees every microsecond.

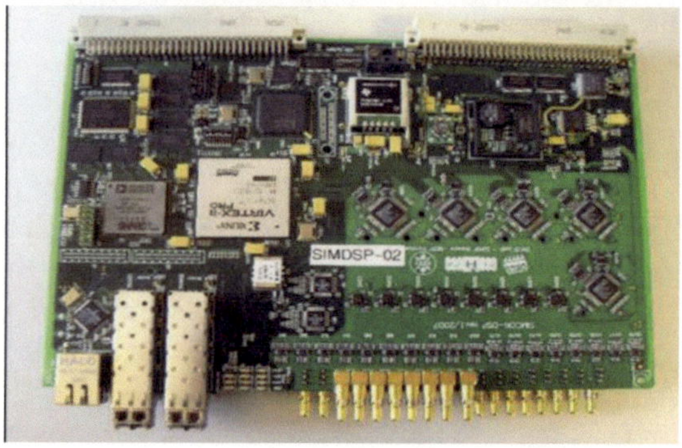

Fig. 4. Simcon-DSP board

6.3 Digital feedback software

The cavity field controller algorithm consists of the field detection scheme (fig. 5), calculation of the calibrated vector-sum, the field error measurement, the controller filter, a feed-forward signal, and the drive signal generation. Beam loading compensation through feed-forward and real time beam measurements are supported. The LLRF control system is integrated with FLASH control system DOOCS (Hensler et al., 2010) by a development of device and middle layer servers. Furthermore the DOOCS standard server is used for automation, like simple FSMs (finite state machine), and the FLASH data acquisition system for bunch-to-bunch monitoring tasks, e.g. quench detection. The control system for the cavities which are driven by a single klystron is considered as a functionally complete unit of the RF system. The feedback algorithm is implemented in the FPGA system. The digital signal processing in turn gets its parameters from the controller server.

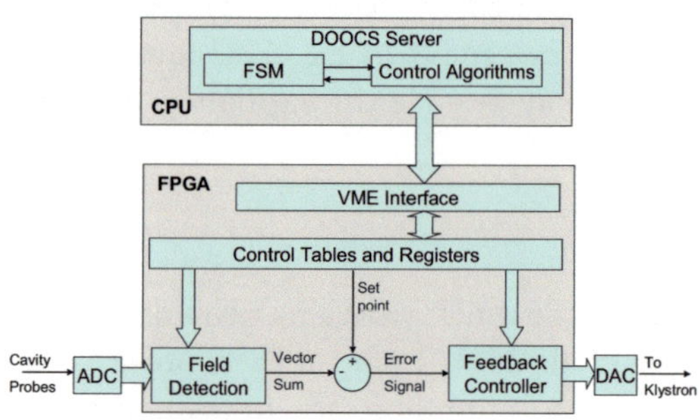

Fig. 5. Controller firmware and software architecture

The controller server software handles: generation of set-point, feed-forward and feedback gain tables from basic settings, rotation matrices for I and Q of each cavity, loop phase constant, start-up configuration files, feedback parameters and exception handler control parameters. The functionality of the server gives the user the opportunity to down/upload data into the FPGA and download and start the controller firmware. The server calculates and adjusts the set of the feedback algorithm parameters in accordance with the required field gradient and phase value. The interrupt service routines are used to start the data reading from the controller board. The parameters of the feedback algorithm are modified by the FPGA programs in the time slot between beam macro-pulses.

6.4 Piezo control

The cavities operating with high gradient are deformed due to Lorentz force that causes detuning of the order of the cavity bandwidth from resonance frequency. Detuned cavity reflects the supplied RF power that requires excessive RF driving. For the compensation of Lorentz force detuning (LFD) the piezo actuator is used to excite the cavity mechanically. Each cavity in new accelerating modules (1st, 6th and 7th) is equipped with double piezos that allow compensating of LFD and measurement of cavity vibrations simultaneously (Przygoda et al., 2010).

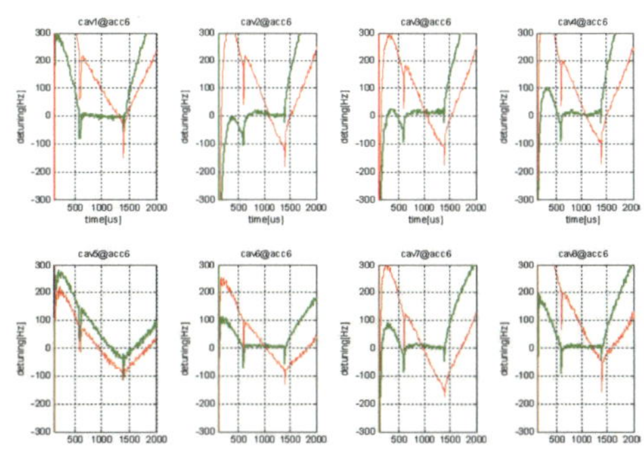

Fig. 6. Lorentz force detuning compensation in 6th accelerating module (green - detuning with piezo compensation, red - without)

The piezo control system is able to compute detuning in each cavity basing on RF signals and calculates the parameters of compensating piezo excitation pulse. The signal from programmable generator is amplified by high power piezo driver. The results of LFD compensation in 6th module is presented in fig. 6. Using piezos the dynamic and static detuning was compensated to only few Hz during flattop in all cavities except the 5th one where piezo is not fixed properly.

6.5 Application software

A set of generic and especially devoted programs provides the tools for the operators to control the RF system (Geng et al., 2011). Some of them are created based on the MATLAB, others, for example, vector-sum calibration are implemented as DOOCS middle layer servers. The adaptive feed-forward is implemented on a front end server, to allow pulse to pulse adaptation.

The application software includes automated operation of the frequency tuners, calibration, phasing of cavities, and adjustment of various control system parameters such as feedback gains, feed-forward tables, and set-point correction during cavity filling. Extensive diagnostics inform the operator about cavity quenches, cavities requiring tuning, and an excessive increase in control power.

6.5.1 Adaptive control

The RF field regulation is subject to various, random and deterministic disturbance sources. Both disturbance contributions are reduced in closed loop operation by applying a feedback compensator. However repetitive disturbances are particularly suppressed by adaptation of the system input drive, using the known system response from previous pulses. The reference for the RF field is in general not changed very frequently, so the control task can be seen as a repetitive process for the pulsed operation mode of this accelerator. The basic update algorithm (Schmidt, 2010) is given by

$$u_{k+1}(t) = u_k(t) + L(t)\, e_k(t)$$

where u_k *and* e_k are defined as the system input and the deviation of the measured RF output to the given set-point for the pulse number k, respectively. *L(t)* is a linear, non-causal, time varying filter based on the identified system model. The current implementation of the system allows changes of all controller tables inside the FPGA between two consecutive pulses. With the minimum computation time necessary for this algorithm, as well as fast data transfer is fast enough, the adaptation can be performed synchronized to the repetition rate of FLASH. Therefore three steps have to be performed between two pulses: read from previous pulse the error and feed-forward signals e and u, compute the feed-forward signal of next pulse, and write the feed-forward signals to FPGA tables.

6.6 Beam based calibration

In this section, the algorithm for system calibration based on the moderate beam (such as 30 bunches with 3nC bunch charge) is presented, including: vector-sum calibration, cavity gradient and phase calibration, forward and reflected power calibration.

For this calibration scheme, the beam pulse should be short compared to the time constant of the cavity, so that the beam induced voltage can be calculated with approximation, for the beam transient measurement (Brandt, 2007). Another more practical measurement method of the beam transient follows the steps below:

- Measure the cavity probe signals without beam as reference.
- Switch on the beam, and then measure the cavity probe signals with beam.
- Compare the two measurements, and the beam transient can be estimated by subtracting the reference from the cavity probe signals with beam.

Note the microphonics may introduce some errors, because the reference is changing by it. Averaging can be done to reduce the errors.

The output of the beam transient measurement will include:

- Beam induced voltage in digit unit:

$$\vec{V}_{ind\_digit} \tag{1}$$

- Cavity voltage in digit unit:

$$\vec{V}_{c\_digit} \tag{2}$$

6.6.1 Cavity incident phase adjustment

The beam phase with respect to RF in each cavity can be calculated from (1) and (2) with the equation of

$$\varphi_b = 180^\circ - \left(\angle\vec{V}_{ind\_digit} - \angle\vec{V}_{c\_digit}\right) \tag{3}$$

It is shown in the fig. 7 (Schilcher, 1998).

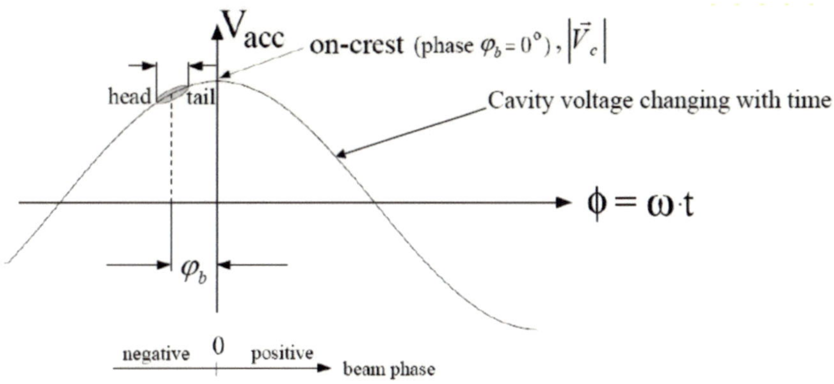

Fig. 7. The beam phase with respect to RF

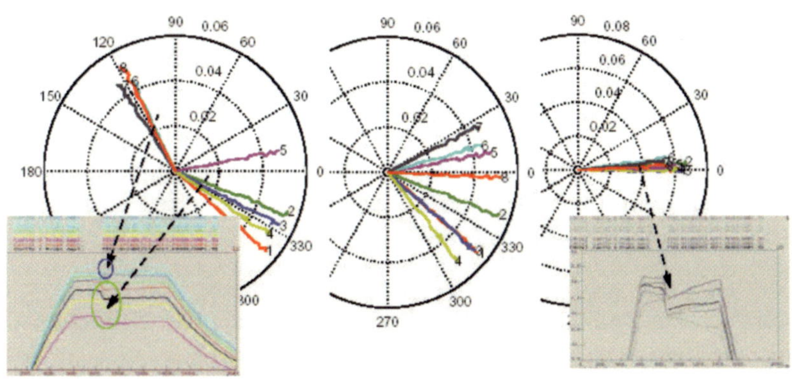

Fig. 8. Transient detection applied to the adjustment of the incident phase accelerating module 1. The transient portion is shown in the polar plots which show the result of the progressive tuning of the incident phase from left to right

In principle, the beam phase with respect to RF in all cavities of one RF station should be adjusted to be the same. Based on equation (3), waveguide tuners can be used to adjust the incident phase of each cavity. An example of the transient based measurements is shown in fig. 8 (Simrock, 2007).

6.6.2 Vector-sum calibration

This is to decide the rotation matrices for each cavity for vector-sum measurement. First we assume that the r/Q of each cavity should be the same, or has been known in advance, and the beam should be lossless through all the cavities. In this case, the absolute values of the beam induced voltage and its phase should be the same for all the cavities, so if the vector of the first cavity acts as reference, the rotation gain and rotation angle of the nth cavity are

$$g_{rot,n} = \left| \frac{\vec{V}_{ind\_digit,1}}{\vec{V}_{ind\_digit,n}} \right| \qquad (4)$$

$$\phi_{rot,n} = \angle \vec{V}_{ind\_digit,1} - \angle \vec{V}_{ind\_digit,n}$$

6.6.3 Cavity voltage calibration to physical unit

The cavity voltage can be calibrated to the physical unit (MV) if the beam current of $\left|\vec{I}_{b0}\right|$ in the physical unit (Ampere) is known.

The cavity equation for the superconducting cavity is

$$\frac{d\vec{V}_c}{dt} + \left(\omega_{1/2} - j\Delta\omega\right)\vec{V}_c = 2\omega_{1/2}\vec{V}_{for} + 2\omega_{1/2}R_L\vec{I}_{b0} \qquad (5)$$

If we only study the beam induced voltage, we can remove the driving term from the klystron (because the system is linear, we can study the effect of different driving terms separately)

$$\frac{d\vec{V}_c}{dt} + \left(\omega_{1/2} - j\Delta\omega\right)\vec{V}_c = 2\omega_{1/2}R_L\vec{I}_{b0} \qquad (6)$$

If the beam pulse is very short (such as 30 microseconds), the beam induced cavity voltage will be small, while the beam induced cavity voltage is changing very fast and approximately changing linear with time, so we can drop the second item of the left side, the equation (6) changes will be

$$\frac{d\vec{V}_c}{dt} = 2\omega_{1/2}R_L\vec{I}_{b0} \qquad (7)$$

And because

$$\omega_{1/2} = \frac{\omega_0}{2Q_L} = \frac{\pi f_0}{Q_L}$$

$$R_L = \frac{1}{2}\left(\frac{r}{Q}\right)Q_L \qquad (8)$$

So the Q_L will be cancelled by each other and the cavity equation changes will be

$$\frac{d\vec{V}_c}{dt} = 2\omega_{1/2}R_L\vec{I}_{b0} = \left(\frac{r}{Q}\right)\pi f_0\vec{I}_{b0} \qquad (9)$$

After integration of the equation, the beam induced voltage in physical unit (MV) can be calculated as

$$\vec{V}_{ind\_physical} = \left(\frac{r}{Q}\right)\pi f_0 \vec{I}_{b0}\Delta t \tag{10}$$

If the beam current, beam pulse length, r/Q of the cavity are all known, the beam induced voltage can be calculated in the physical unit of MV.

With equation (10), (1) and (2), the cavity voltage in physical unit of MV can be calculated

$$\vec{V}_{c\_physical} = \vec{V}_{ind\_physical} \cdot \frac{\vec{V}_{c\_digit}}{\vec{V}_{ind\_digit}} \tag{11}$$

The coefficient between the digit unit and physical unit can be calculated as

$$C_{digit2physical} = \frac{\vec{V}_{ind\_physical}}{\vec{V}_{ind\_digit}} \tag{12}$$

6.6.4 Forward and reflected power calibration

Now we have got the waveform of the cavity voltage in the physical unit of MV, and based on the forward and reflected signal directivity calibration, the forward power and reflected power can be calculated.

With the directivity calibration, the forward and reflected signal can be measured as

$$\begin{aligned} \vec{V}_{for\_digit} &= a\vec{V}_{for\_m} + b\vec{V}_{ref\_m} \\ \vec{V}_{ref\_digit} &= c\vec{V}_{for\_m} + d\vec{V}_{ref\_m} \end{aligned} \tag{13}$$

Note that the unit of the forward and reflected signal is digit now. And from equation (12), the forward and reflected voltage in physical unit of MV can be calculated.

$$\begin{aligned} \vec{V}_{for\_physical} &= C_{digit2physical} \cdot \vec{V}_{for\_digit} \\ \vec{V}_{ref\_physical} &= C_{digit2physical} \cdot \vec{V}_{ref\_digit} \end{aligned} \tag{14}$$

The forward power and reflected power can be calculated as

$$\begin{aligned} P_{for} &= \frac{\left|\vec{V}_{for\_physical}\right|^2}{2R_L} \\ P_{ref} &= \frac{\left|\vec{V}_{ref\_physical}\right|^2}{2R_L} \end{aligned} \tag{15}$$

The loaded resistance R_L can be calculated from equation (8).

7. RF system modeling

The RF control relevant electrical and mechanical characteristics of the cavity are described in form of time varying state space models. Included are perturbations from Lorentz force detuning and microphonics with the appropriate parameters for several mechanical resonances. The power source is calibrated in terms of actual power and includes saturation characteristics and noise. An arbitrary time structure can be imposed on the beam current to reflect a macro-pulse structure and bunch charge fluctuations. For RF feedback several schemes can be selected: generator driven system or self-excited loop, traditional amplitude and phase control or IQ control. The choices for the feedback controller include analog or digital approaches and various choices of filters for the gain stages. Feed-forward can be added to further suppress repetitive errors. An analysis tool has been developed for RF control for superconducting cavities (Vardanyan et al., 2002).

The concept of the analysis tool for RF control systems is based on library of sub-system blocks used in SIMULINK. The RF system is usually modeled as a combination of such sub-system blocks as shown in fig. 9. The key properties of the RF system analysis tool are:

- Main blocks for major RF sub-subsystems.
- Well defined interface for inputs and outputs for the sub-systems with possibility to add ports and new parameters.
- Initialization files for sub-systems parameters and simulation parameters.
- Variable interconnection structure between sub-systems.
- Possibility of adding various new sub-subsystems.

Once the library is established the user can simply select the desired sub-systems from the RF component library to develop a new model. The user has the possibility to use also other SIMULINK library blocks.

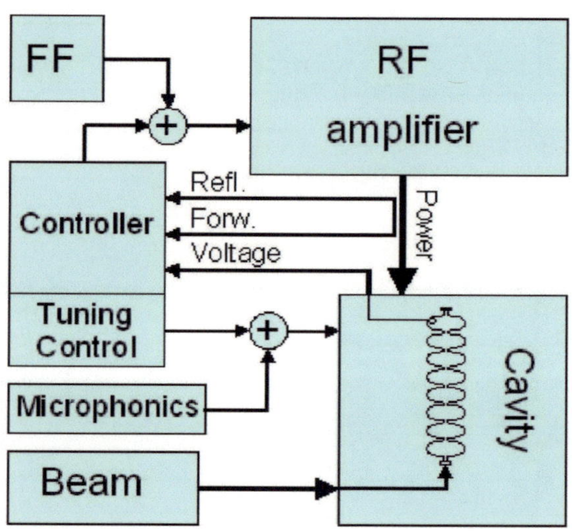

Fig. 9. RF sub-systems and interconnections

The RF sub-systems contain user defined blocks with accompanying initialization files. These RF system specific blocks are added to a library. The beam can be modeled with arbitrary time varying amplitude and phase. It is also possible to use pulse generators for a repetitive pulse structure and add a noise source to simulate bunch charge fluctuations. For the amplifier it is possible to use the measured non-linear characteristics of a klystron to create an input-output mapping table. The present klystron library contains a linear amplifier with a table of real klystron saturation curve which can be normalized for maximum output and drive power.

The use of superconducting cavities implies that Lorentz force detuning and microphonics are included in the model. The state space representation of a cavity including Lorentz force detuning. The controller (fig. 10) is generic for different type of controls (self-exited loop, generator driven resonator amplitude and phase detectors and controllers, and IQ detectors and controller) composed of the sub-system's feedback, feed-forward, and tuner control. It is possible to mix different types of controls during simulation and/or simultaneously.

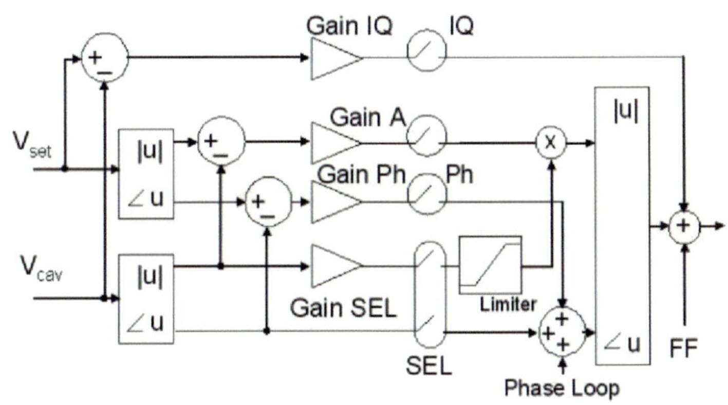

Fig. 10. Controller scheme

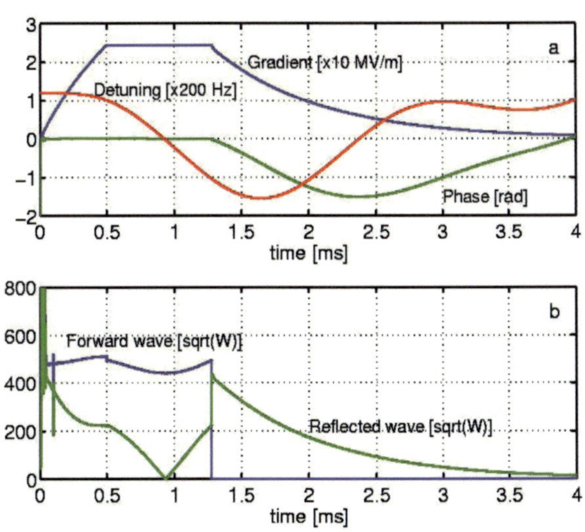

Fig. 11. FLASH cavity control: a) gradient, phase, detuning; b) forward and reflected waves

Fig. 11 shows the pulsed operation of a FLASH cavity with an IQ controller. During the cavity filling time of 500 μs, the gradient is regulated on a time varying filling curve computed from the time constant of the cavity. The additional klystron power is required during the filling and flat-top as a result of the dynamic Lorentz force detuning.

8. Conclusion

A digital RF control system has been developed to control the vector-sum of the accelerating field of group of superconducting cavities powered by a single klystron. The RF control system is realized as a driven feedback system and has proven that the phase and amplitude stability requirements can be meeting even in the case of control of the vector-sum of multiple cavities. The goal to provide a constant accelerating field in order to minimize the energy spread has been successfully reached. The major advantages of the system are the built-in diagnostics, configurationally flexibility which is essential for the extension of up to 32 cavities driven by one klystron. The digital boards have been upgraded to latest generation controller hardware. All modules are controlled by similar modern FPGA based controller boards with unified firmware and software. They are generic and flexible enough to be usable for a variety of control and data processing applications. Algorithms are improved: beam loading compensation, feed-forward waveform generation, etc. Beam diagnostics signals are in use for fast intra-pulse feedback. For cavity frequency fast tuning piezo control has been implemented. FLASH achieved beam energy above 1.25 GeV and lasing below 4.12 nm with a remarkably improved LLRF control performance. Basic automation schemes have been implemented for loop phase, loop gain, cavity detuning and loaded quality factor measurement as well as fault recovery. These features allow a high availability which has been demonstrated during long term user operation.

9. Acknowledgment

I gratefully acknowledge the contributions from LLRF team members for their dedicated support. I also want to express my thanks to the FLASH operation team for their valuable comments and helpful discussions.

10. References

Ackermann, W., Asova, G., Ayvazyan, V. et al. (2007). Operation of Free-Electron Laser from the Extreme Ultraviolet to the Water Window, *Nature Photonics* 1, June 2007, pp. (336-342)

Ayvazyan, V. & Simrock, S. N. (2004). *Dynamic Lorentz Force Detuning Studies in TESLA Cavities,* Proceedings of EPAC 2004, Lucerne, Switzerland, July 2004

Ayvazyan, V., Petrosyan, G., Rehlich, K., Simrock, S. N. & Vetrov, P. (2005). Hardware and Software Design for the DSP Based LLRF Control, Proceedings of PCaPAC 2005, Hayama, Japan, March 2005

Ayvazyan, V., Brandt, A., Choroba, S., Petrosyan, G., Rehlich, K., Simrock, S. N. & Vetrov, P. (2007). *Digital RF Control System for the DESY FLASH Linear Accelerator,* Proceedings of EUROCON 2007, Warsaw, Poland, September 2007

Ayvazyan, V., Czuba, K., Geng, Z. et al. (2010). *LLRF Control System Upgrade at FLASH,* Proceedings of PCaPAC 2010, Saskatoon, Canada, October 2010

Brandt, A. (2007). *Development of a Finite State Machine for the Automated Operation of the LLRF Control at FLASH,* PhD thesis, ISSN 1435-8085, DESY, Hamburg, Germany, July 2007

Geng, Z., Ayvazyan, V. & Simrock, S. (2011). *Architecture Design of the Application Software for the Low-Level RF Control System of the Free-Electron Laser at Hamburg,* Proceedings of ICALEPCS 2011, Grenoble, France, October 2011

Harms, E., Edwards, H., Arkan, T. et al. (2009). *Third Harmonic System at FERMILAB/FLASH,* Proceedings of SRF 2009, Berlin, Germany, September 2009

Hensler, O., Koprek, W., Schlarb, H., Ayvazyan, V., Schmidt, C. & Geng, Z. (2010). *Consolidating the FLASH LLRF System Using Standard Server and the FLASH DAQ,* Proceedings of PCaPAC 2010, Saskatoon, Canada, October 2010

Koprek, W., Behrens, C., Bock, M. K. et al. (2010). *Intra-train Longitudinal Feedback for Beam Stabilization at FLASH,* Proceedings of FEL 2010, Malmö, Sweden, August 2010

Przygoda, K., Pozniak, T., Napieralski, A. & Grecki, M. (2010). *Piezo Control for Lorentz Force Detuned SC Cavities of DESY FLASH,* Proceedings of IPAC 2010, Kyoto, Japan, May 2010

Rehlich, K. (2007). *Status of the FLASH Free Electron Laser Control System,* Proceedings of ICALEPCS 2007, Knoxville, Tennessee, USA, October 2007

Ross, M., Walker, N. & Yamamoto, A. (2011). *Present Status of the ILC Project and Developments,* Proceedings of IPAC 2011, San Sebastian, Spain, September 2011

Schilcher, T. (1998). *Vector Sum Control of Pulsed Accelerating Fields in Lorentz Force Detuned Superconducting Cavities,* PhD thesis, TESLA 98-20, DESY, Hamburg, Germany, August 1998

Schmidt, C. (2011). *RF System Modeling and Controller Design for the European XFEL,* DESY-THESIS-2011-019, ISSN 1435-8085, DESY, Hamburg, Germany, June 2011

Simrock, S. N., Altmann, I., Rehlich, K. & Schilcher, T. (1996). *Design of the Digital RF Control System for the TESLA Test Facility,* Proceedings of EPAC 1996, Barcelona, Spain, June 1996

Simrock, S. N. (2007). *Measurements for Low Level RF Control Systems,* Meas. Sci. Technol. 18, July 2007, pp. (2320-2327)

Vardanyan, A., Ayvazyan, V. & Simrock, S. N. (2002). *An Analysis Tool for RF Control for Superconducting Cavities,* Proceedings of EPAC 2002, Paris, France, June 2002

Vogt, M., Faatz, B., Feldhaus, J., Honkavaara, K., Schreiber, S. & Treusch, R. (2011). *Status of the Free-Electron Laser FLASH at DESY,* Proceedings of IPAC 2011, San Sebastian, Spain, September 2011

Weise, H. (2004). *The TESLA X-FEL Project,* Proceedings of EPAC 2004, Lucerne, Switzerland, July 2004

10

Radiation Safety

Yoshihiro Asano
Safety Design Group, SPring-8/RIKEN, Japan

1. Introduction

Radiation safety is one of the most important issues for free electron laser [FEL] facilities and the users. Many FEL facilities are under operation worldwide, and their number is growing. The wave length of FEL ranges from a few millimeters to about 0.5 nm, and the electron energy is from a few MeV to about 20 GeV. Furthermore, various kinds of electron accelerator types are constructed for FEL facilities, such as Linac, and storage rings. In this chapter, radiation means ionizing radiation, excluding non-ionizing radiation. Radiation sources in relation to FEL facilities are γ (X) ray (high energy photons), electron, neutron due to thephoto-nuclear reaction, and muons. Activation is another important problem for high-energy accelerators.

An radiation safety system depends strongly on the machine and the energy. Based on ALARA* (As Low As Reasonably Achievable) principle, therefore, the safety system must be constructed to be better suited for the machine as much as possible. Radiation safety for FEL mainly consists of, (1) the beam containment system, including shield, (2) the access control system, and (3) the radiation monitoring system. These systems are linked organically by on interlock system to prevent the hazardous conditions. There are fundamentally no differences between FEL facilities and high-energy electron accelerators, especially synchrotron radiation facilities.

2. Safety system

2.1 Outline of basic quantities for radiation protection

Basic quantities for radiation safety are the absorbed dose, *D*, and effective dose, *E*. The absorbed dose, *D*, is a physical quantity, and the quotient of $d\bar{\varepsilon}$ by dm is as the follows:

$$D = d\bar{\varepsilon} / dm\,, \tag{1}$$

where $d\bar{\varepsilon}$ is the mean energy imparted to a tissue organ of mass *dm* for radiation safety. The unit is Joule·kg^{-1} and the special name for the unit is Gray (Gy) (Another unit is erg·g^{-1} and the unit of rad (1rad=10^{-2}Gy)). The effective dose *E* is a quantity for radiation protection control, and the special name for the unit is Sievert (Sv). (Another unit is rem (1 rem =10^{-2}

*ALARP: "As Low As Reasonably Practicable", principle is used in UK

Sv)). Since the biological effectiveness of radiation exposure is different for various types of radiation, irradiation conditions, and irradiated tissues or organs, the effective dose, E, is the unified quantity to discuss and utilize for radiation protection against the different biological effectiveness. The effective dose, E, is calculated in three steps, as follows (1): (1) The mean absorbed doses, D_T, in a specified tissue or organ T of the human bodies calculate in the first step as follows:

$$D_T = \frac{1}{m_T}\int_{m_T} D \cdot dm\ , \tag{2}$$

where m_T is the mass of the tissue or organ, T.

(2) The equivalent dose, H_T, which is corrected dose for different biological effectiveness due to different type of radiation, R, by using radiation weighting factor, w_R, calculates in the second step as follows,

$$H_T = \sum_R w_R \cdot D_{T,R}\ , \tag{3}$$

where w_R is given by the ICRP recommendations (2), which based on the function of linear energy transfer, LET. For neutrons, w_R is given as a function of energy, as shown in Fig.2-1. The w_R of the γ (X) ray, electron, and muon are given as 1.

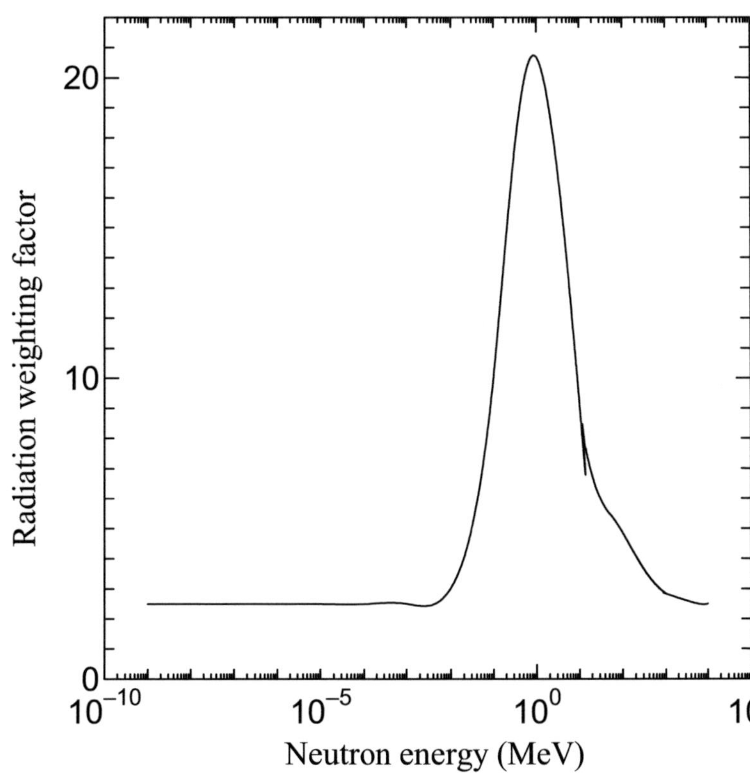

Fig. 2.1. Radiation weighting factor for neutrons.

(3) Effective dose, E, is the sum of the weighted equivalent doses in all tissues and organs of the body by using weighting factor of the tissue or organ, w_T and an anthropomorphic phantom, as follows:

$$E = \sum_T w_T \cdot H_T \,, \tag{4}$$

where w_T is given by the ICRP recommendations (2) for each of 14 organs and others as given in Table 2-1.

As mentioned above, the effective dose, E, is not easily calculated,so that the tables and figures of the effective doses, E, are presented for the unit fluence of the typical irradiation conditions as a function of the radiation energy (3,4). In order to measure and simplify the doses, operational doses, such as the ambient dose equivalent, directional dose equivalent, and personal dose equivalent are defined (5) for practical use, and these doses are fundamentally conservative in comparison with the effective dose, E.

Tissue or organ	Tissue weighting factor w_T
Bone marrow (red)	0.12
Colon	0.12
Lung	0.12
Stomach	0.12
Breast	0.12
Gonads	0.08
Bladder	0.04
Liver	0.04
Oesophagus	0.04
Thyroid	0.04
Skin	0.01
Bone surface	0.01
Brain	0.01
Salivary glands	0.01
Remainders*	0.12

*Remainders: adipose tissue, adrenals, connective tissue, extrathoracic, airways, gall bladder, heart wall, kidney, lymphatic nodes, muscle, pancreas, prostate, small intestine wall, spleen, thymus, and uterus/cervix

Table 2.1. Tissue weighting factors (2)

2.2 Outline of dose limitation system

Based on the fundamental rules of justification and optimization, dose limitation systems are performed under the achievement of consensus. The main dose limitations that International Commission on Radiation Protection (ICRP) recommends are listed in Table 2-2.

Radiation worker	100mSv/5y, 50mSv in any year, 20mSv/y averaged 5y
	H_T;150mSv/y (lens eye), 500mSv/y(skin, legs, hands),
Radiation worker(pregnant)	H_T ; 1mSv/ embryo/fetus,
Public	E;1mSv/y
	H_T ; 15mSv/y (lens eye), 50mSv/y(skin),

Table 2.2. Main ICRP recommendations for dose limitation. (2)

Based on the ICRP recommendations, each country or facility employs the design criteria of dose limitation. For example, the design criteria of the dose limitation at SPring-8 are 8 μSv/h (40hours for 1 week radiological worker occupancy), 2.5 μSv/h, and 100 μSv/y, for a radiation-controlled area, the boundary of the controlled area, and the site boundary of SPring-8, respectively (5). At Stanford national linear accelerator center, SLAC, the design criteria are 10mSv/y (5mSv for 2000h radiological worker occupancy), 1mSv/y (0.5mSv for 2000h non-radiological worker occupancy) for normal operation, 4mSv/h for miss steering conditions, 30mSv/event for system failure (6).

2.3 Radiation sources

Radiation corresponding to the safety at FEL facilities is (1) X-rays produced by synchrotron radiation, (2) X-rays due to high density of lasers, (3) γ-rays due to high energy electron-beam loss, including gas bremsstrahlung, (4) neutrons due to photo-nuclear reactions, and (5) muons. The importance of each type of radiation for safety strongly depends on the energy of the electron machine and the power of the laser, as described below:

1. A free-electron laser is produced by undulators with oscillation and the coherency of the synchrotron radiation so that the spontaneous emission light of the undulator is also produced. Generally, the spontaneous emission light is more important for safety because of strong penetration of high-energy photons (7).
2. The interaction of a high-intensity laser with matter generates plasma and associated produced high-energy electrons and ions. These high-energy electrons interact with ions in the target and induce bremsstrahlung X-rays (8). X-rays will be produced when the laser is focused on a target at peak intensities of over 10^{12} W/cm^2 (9), and neutrons and protons may be produced at the peak intensities over 10^{19} W/cm^2 (10)
3. Interactions of high-energy electrons with matter, such as electron beam transport pipe or shield materials produce an electro-magnetic shower and high-energy γ rays (11). In this case, the energy of the photons distribute up to the maximum accelerated electron energy. Interactions of high-energy electrons with residual gas molecules within the electron beam transport pipe of undulators produce high-energy γ rays, so called "gas bremsstrahlung" with high directivity in special cases (12).
4. The interactions of γ rays with the energy of over about 10 MeV with thick materials produce neutrons due to the nuclear reactions of the giant resonance, quasi-deutron and photo-pion reaction process.
5. Interactions of high-energy γ-rays with thick materials also produce muons. In this process, because the threshold energy of the muon production is 211 MeV, the muons must care at high-energy machines.

2.4 Radiation monitoring

Radiation monitoring is classified into two categories: one for personnel monitoring and the other for area monitoring. Because the personnel monitoring is to measure and confirm individually the personal dose, badge types are employed to equip on radiation workers. Now, OSLs (Optical Stimulate Luminescence, TLDs (Thermo Luminescence Detectors), film badges (13), and Glass dosimeters (14), etc are used to measure the γ (X)ray dose. For a dose due to high energy neutrons, CR-39 neutron track detectors are employed, mainly.

The area monitoring is to clarify the circumstances of the radiation fields at the facility so that the monitors will be assigned at the position where the dose will be increased in comparison with the dose limitation. Ionization chambers are usually used for γ (X)ray area monitors. Plastic scintillation counters are employed at some facilities (15) to measure pulsed radiations. For neutron monitoring, Helium-3 or BF_3 proportional counters with moderators are used. Some facilities employ superheated drop detectors (bubble detectors) (16).

2.5 Safety interlock system

A radiation safety interlock system is one of the key issues for accelerator facilities, including FELs. The safety interlock system must be constructed with high reliability and fail-safe system, however, the system depends strongly on each facility design and the safety philosophy. The conceptual main frame of the safety interlock system is shown in Fig.2-2. The equipment of the interlock system is linked fundamentally to a PLC (Programmable Logic Controller) using hard wires to stop the machine operation infallibly. Multiplex systems are often employed such as a door keep system, emergency button and a beam shutter system. It depends on the level of the importance or hazardous and the credibility of the system to construct to the multiplex system.

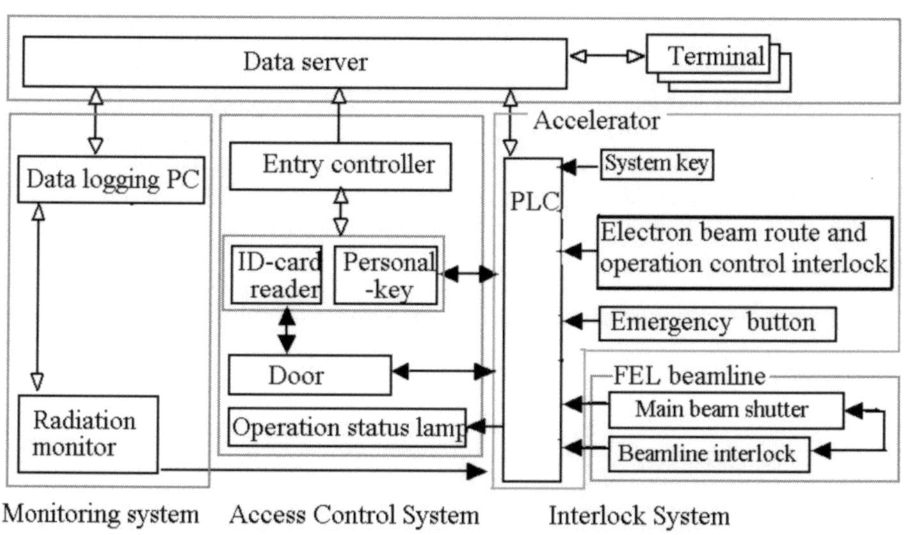

Fig. 2.2. Conceptual main framework of the safety interlock system (PLC: Programmable Logic Controller, Black full arrows mean to connect interlock system)

3. Shielding design

There are mainly two methods for shielding design. One is to use Monte Carlo simulation codes; the other is to use analytical and empirical method. Normally, close checks between Monte Carlo simulations and the results of empirical methods are required to confirm the design.

3.1 Monte Carlo code

The Monte Carlo codes can simulate radiation transport with complicated geometry. Typical Monte Carlo codes to perform the shielding design of FEL facilities are summarized in Table 3-1. The codes have characteristic differences from each other, such as the available nuclear reaction and the maximum electron energy etc so that one must pay attention to choose the code.

Code	Type	References
EGS4 or EGS5	Electro-magnetic shower (γ,e), exclude (γ,n) reaction	17
Penelope	Electro-magnetic shower (γ,e) , exclude (γ,n) reaction	18
MCNPX	Multipurpose transport	19
MARS15	Multipurpose transport and heavy ion transport	20
FLUKA	Multipurpose transport including induced activity	21
GEANT4	Multipurpose transport including induced activity	22
PHITS	Multipurpose particle and heavy ion transport	23

Table 3.1. Monte Carlo simulation code for using the shielding design of the FEL facility

3.2 Analytical and empirical methods

3.2.1 Synchrotron radiation

The leakage dose due to synchrotron radiation spontaneous photon emission can be calculated by using the STAC8 code (24). This code can calculate doses outside the shield material from an undulator source to the leakage dose due to scattering photons, sequentially. Doses due to FEL can be calculated by using STAC8 with the calculated FEL source spectrum by using something such as SIMPLX (25). Generally, FEL can be neglected because of low energy-photons in comparison with the synchrotron-radiation spontaneous emission photons.

3.2.2 Plasma X-ray

Intense laser over 10^{12} W/cm^2 can produce plasma and associated X-rays. The dose can be calculated as follows (26):

$$H_x \approx 6.0 \times (P_{ef} / R^2) \times T \qquad (T \geq 3MeV)\,, \tag{10}$$

$$H_x \approx 2.0 \times (P_{ef} / R^2) \times T^2 \qquad (T \leq 3MeV)\,, \tag{11}$$

$$T \approx 6.0 \times 10^{-5} \left[I\lambda^2 \right]^{1/3} \qquad (10^{12} W / cm^2 \mu m^2 < I\lambda^2 < 10^{17} W / cm^2 \mu m^2)\,, \tag{12}$$

$$T = M_e \times (-1.0 + \sqrt{1.0 + I\lambda^2 / 1.37 \times 10^{18}} \quad (I\lambda^2 > 10^{18} W / cm^2 \mu m^2), \tag{13}$$

Here, H_x is the photon dose in Sv in the forward direction, P_{ef} is the laser energy to the electron energy conversion efficiency, R is the distance from the target to the measurement point in cm. T is the hot electron temperature in MeV. I is the laser intensity in W/cm², λ is the laser wave length in μm.

3.2.3 High energy γ ray and neutron due to electron beam loss

Radiation due to electron beam loss is an anisotropic distribution with strongly high in forward direction so that the shielding design is normally considered manner for two directions, forward and lateral directions, from the electron beam.

a. Forward direction (11)

γ-ray leakage doses can be calculated as follows:

$$H_\gamma = (P / r^2) \times 20 \times E_e^2 \times \exp(-d / \lambda_\gamma) \quad (1 < E_e < 20MeV) \text{ (Sv/h)}, \tag{14}$$

$$H_\gamma = (P / r^2) \times 300 \times E_e \times \exp(-d / \lambda_\gamma) \quad (E_e \geq 20MeV) \text{ (Sv/h)}, \tag{15}$$

$$H_\gamma = (P / r^2) \times 10^6 \times E_e \times \exp(-d / \lambda_\gamma) \quad (E_e \geq 1.0GeV) \text{ (Sv/h)}. \tag{16}$$

Also the neutron leakage dose in the forward direction can calculates as follows:

$$H_{nh} = (P / r^2) \times 4.0 \times \exp(-d / \lambda_{nh}) \quad (E_e \geq 50MeV) \text{ Sv/h)}, \tag{17}$$

$$H_{nt} = (P / r^2) \times 22.7 \times \exp(-d / \lambda_{nt}) \text{ (Sv/h)}, \tag{18}$$

Here H_γ is the γ-ray dose in Sv/h, P is the electron beam loss power in kW, r is the distance from the loss point to the measurement point in m, d is the thickness of the shielding material in cm; λ_γ, λ_{nh}, and λ_{nt} are the attenuation length of γ-rays, high-energy neutron and the giant resonance neutron, respectively. These attenuation lengths are summarized in Table 3-2. H_{nh} and H_{nt} in Sv/h mean the dose due to high-energy neutrons and the dose due to giant resonance neutrons, respectively.

Shield material	Attenuation length (cm)		
	λ_γ	λ_{nh}	λ_{nt}
Lead	2.1	66.7	23.8
Iron	4.5	58.8	18.5
Ordinary Concrete*	19.0	41.8	17.7

*Density: 2.2g/cm³

Table 3.2. Attenuation length the shield materials for the forward direction

b. Lateral direction

The neutron and γ-ray doses for the lateral direction can be calculated by using an analytical code, SHIELD11, which was developed by Stanford National Linear Accelerator Center (27). This code can calculate the leakage dose due to beam loss, ranging from over 1 GeV to about 20 GeV electron energy.

The γ-ray doses due to electrons with an energy of less than about 20 GeV can be calculated as follows (28):

$$H_{\gamma} = P \cdot 50 \cdot \frac{1}{r^2} \cdot \exp(-d/\lambda_{\gamma}) \qquad 50\ \text{MeV} \le \text{Ee} \le 150\ \text{MeV}\ \ (\text{Sv/h}), \tag{19}$$

$$H_{\gamma s} = 3.6 \times 10^{-14} \cdot J \cdot E_e \cdot \frac{1}{r^2} \cdot \left(\frac{133 \cdot \exp\{(-d\cos ec\phi)/\lambda_{\lambda}\}}{(1-0.98\cos\theta)^{1.2}} + \frac{f1 \times 0.267 \cdot \exp\{(-d\cos ec\phi)/\lambda_{\gamma}\}}{(1-0.72\cos\theta)^2} \right) \tag{20}$$
$$(30^\circ < \theta < 130^\circ)(\text{Sv}/\text{h})$$

The neutron dose due to electrons with the energy of less than About 20 GeV can be calculated as follows (28):

$$H_n = P \cdot 22.7 \cdot \frac{1}{r^2} \cdot \exp(-d/\lambda_1) \qquad (\approx 10MeV \le E_e \le 50MeV)\ (\text{Sv/h}), \tag{21}$$

$$H_n = 3.6 \times 10^{-14} \cdot J \cdot E_e \cdot \frac{1}{r^2} \cdot \left(\frac{f1 \cdot \exp\{(-d\cos ec\phi)/\lambda_1\}}{(1-0.72\cos\theta)^2} + \frac{f2 \times 10 \cdot \exp\{(-d\cos ec\phi)/\lambda_3\}}{(1-0.75\cos\theta)} + 3.79 \cdot Z^{0.73} \cdot \exp\{-d\cos ec\phi/\lambda_2\} \right) \tag{22}$$
$$(30^\circ < \theta < 130^\circ)(\text{Sv}/\text{h})$$

Here, λ_{γ}, λ_1, λ_2, and λ_3 are the attenuation length of γ-rays, high-energy neutrons, giant resonance neutrons, and intermediate neutrons, as summarized in Table 3-3, respectively. J is the amount of electron beam loss in s^{-1}. f_1 and f_2 are correction factors of source reduction for high-energy and intermediate-energy neutrons (29), respectively. θ and ϕ are inclined degrees from the electron beam axis to a measurement point and the shield material, respectively. Z is the atomic number of a target.

Shield material	Density	Attenuation length (cm)			
	(g/cm³)	λ_{γ}	λ_1	λ_2	λ_3
Lead	11.3	2.1	22.7	10.0	18.3
Iron	7.8	4.3	21.3	6.8	12.4
Ordinary Concrete	2.2	18.9	54.6	13.7	25.0

λ_{γ}, photon; λ_1:, high energy neutron; λ_2, giant resonance neutron; λ_3, intermediate energy neutron

Table 3.3. Attenuation length for the lateral direction

3.2.4 Gas bremsstrahlung and associated neutrons

The combination of a long straight section for undulators and high-energy electrons generates gas bremsstrahlung by interactions of the accelerated electrons with residual gas molecules in a vacuum chamber. The dose due to gas bremsstrahlung is in proportion to the current of the electrons, the pressure of the residual gas molecules, and the length of the straight section, as follows (30):

$$D = 2.5 \times 10^{-27} \cdot \left[\frac{E_e}{m_0 c^2} \right] \cdot \frac{L}{d(L+d)} \cdot I \cdot \frac{P}{P_0}, \tag{23}$$

where D is the maximum dose rate due to gas bremsstrahlung in Gy/h, m_0c^2 is electron rest mass (0.511MeV), E_e is the electron energy in MeV, L is the length of the straight section in m, and d is the distance from end of the straight section in m. I is the current of electrons in $e^- \cdot s^{-1}$, P is the pressure of the straight section in Pa, and P_0 is 1.33×10^{-7} Pa. Photo-neutrons are produced when gas bremsstrahlung photons have a sufficiently high energy. In this case, the dose can be estimated by using Liu's data (31). The doses due to gas bremsstrahlung and associated neutrons are negligibly small at almost the FEL facilities.

3.2.5 Muon

Muons are produced by high-energy photons in the Colomb field of the target directly and the decay products of photo-produced π and K mesons. The dose due to muons can be used to calculate as follows (11):

$$H_\mu = \frac{25}{25 + X/X_0} \times \frac{X(E_e)\text{-}X}{X(E_e)} \times H_0 \quad (X(Ee) > X) \ (Sv/h), \tag{24}$$

$$H_u = 0.0 \qquad (X(E_e) \le X), \tag{25}$$

$$H_0 = 8.0 \times 10^{-15} \cdot J \cdot E_e / r^2 \ (Sv/h) \tag{26}$$

where X is the thickness of the shield material, and X_0 is the radiation length of the shield material and summarized in Table 3-4. $X(E_e)$ is the maximum possible muon range of the shield material, as summarized in Table 3-5 for various energies. The muon dose is negligibly small at facilities with an electron energy of less than about 10 GeV.

Shield material	Radiation length		Density
	(g/cm²)	(cm)	(g/cm³)
Lead	6.4	0.6	11.3
Iron	13.8	1.8	7.8
Ordinary Concrete	25.7	11.7	2.2

Table 3.4. Radiation length of shield materials

Shield materials	Maximum Range (g/cm²)				density
	0.3GeV	1GeV	7GeV	10GeV	(g/cm³)
Lead	233	809	4857	6696	11.3
Iron	176	629	3968	5522	7.8
Ordinary Concrete	144	526	3324	4624	2.2

Table 3.5. Maximum possible muon range

3.2.6 Maze streaming

An maze is usually constructed at the shield tunnel of accelerators for entryways. The leakage dose outside the maze (see next section) can be calculated as follows.

For the γ -ray dose attenuation ratio of the labyrinths, g_i (32) is

$$g_1 = 0.22 \times (d + L_1)^{-3.0} , \tag{27}$$

$$g_i = 0.26 \times (d + L_i)^{-2.6} \qquad (i = 2,3,*,*,*) , \tag{28}$$

and for neutrons (33),

$$n_i = 2 \times \frac{\exp(-L_i / 0.45) + 0.022A^{1.8} \cdot \exp(-L_i / 2.35)}{1 + 0.022A^{1.8}} \qquad (i = 1,2,3,*,*,*) . \tag{29}$$

Here, L_i is the length of the ith labyrinth in m, d is the half width of the maze in m, and A is the cross section of the access in m². For the last labyrinth, the factor 2 in formula (29) is ignored.

3.3 Induced activity

High-energy electrons of about over 10 MeV (depending on the material, for example, over about 1.68 MeV for Beryllium) can produce radioactive materials so that high-energy high-power machine must take care the induced activities within the machine components, air, cooling water, and shield materials including concretes. The induced activity produced by photo-nuclear reactions can be estimated by using the saturation activity and the electron beam loss power of the machine (11). For example, the activity in air derives as follows:

3.3.1 Photo-nuclear reaction

$$S_A = \lambda \cdot A_{AS} \cdot P \cdot X_m \cdot F_1 , \tag{30}$$

$$A(t) = \frac{S_A}{\lambda}\{1 - \exp(-\lambda \cdot T) \ \} \cdot \exp(-\lambda \cdot t) , \tag{31}$$

where S_A is the production rate (GBq·h⁻¹),

A_{AS}; the saturation activity (GBq·kW^{-1}·m^{-1}),
P; the electron beam loss power (kW),
X_m; the photon traversable distance within air (m),
λ; the decay constant (h^{-1}),

A(t); the activity after t hours from the shutdown (GBq),
T; the irradiation time (h),
t; the time after the shutdown (h), and
F_1; the ratio of the contribution of electron beam loss to induced activity.

3.3.2 Neutron absorption (dominant nuclei; Argon-41)

$$S_A = \lambda \cdot 10^{-24} \cdot \sigma \cdot N \cdot Y \cdot e \cdot X_{cm} \cdot F_1 \cdot 10^{-9} \tag{32}$$

Here σ is the neutron absorption cross section (cm^2; Ar=0.64 b),
N; the number of atoms (cm^{-3}: $^{40}Ar=2.3 \times 10^{17} cm^{-3}$),
Y; the neutron yield per electron (neutrons/electrons),
$Y \approx 2.4 \times 10^{-3}$ (Ee-0.2) Ee: electron energy (GeV),
e; the electron loss rate (s^{-1}), and
X_{cm}; the neutron traversable distance within air(cm).

The air activity concentration under the operation of the ventilation system is described as follows:

$$Q\frac{dC_i}{dT} = S_{Ai} - \lambda_i \cdot Q \cdot C_i - q \cdot C_i \,, \tag{33}$$

where Q ; is the volume of air (m^3),
S_{Ai}; the production rate of i nuclei ($GBq \cdot h^{-1}$),
λ_i ; the decay constant of I nuclei (h^{-1}),
C_i ; the activity concentration of i nuclei ($GBq \cdot m^{-3}$), and
q ; the air volume of ventilation ($m^3 \cdot h^{-1}$).

Thus, the activity concentration of nuclei indicates the following:

$$C_i = \frac{S_{Ai}/Q}{\lambda_i + q/Q} \cdot \left[1 - \exp\{-(\lambda_i + q/Q) \cdot T\}\right] \tag{34}$$

The saturation activities are summarized in reference (11). The important nuclei of air and water-induced activity are (^3H, ^7Be, ^{11}C, ^{13}N, ^{15}O, ^{16}N, ^{38}Cl, ^{39}Cl, ^{41}Ar), and (^{14}O, ^{15}O, ^{13}N, ^{11}C, ^{10}C, ^7Be, ^3H), respectively.

4. Radiation safety at SCSS prototype FEL facility

The SPring-8 Compact Self Amplification of Spontaneous Emission Source (SCSS) prototype FEL facility (34) has been employed to analyze practical cases of the radiation condition in comparison with the design. The SCSS prototype FEL facility was constructed in 2005 to demonstrate the feasibility of an X-ray FEL based on three new technologies, one for a low-emittance thermionic gun, one for C-band accelerators of up to 250 MeV and 30nC/s, and the other for in-vacuum type undulators. The shortest wave length of this 60 m SCSS prototype system is 49nm. Continuous lasing has been successfully performed, as shown in Fig. 4-1, and photo 4-1.

Photo 4.1. SCSS prototype FEL facility (view from the gun to the dump)

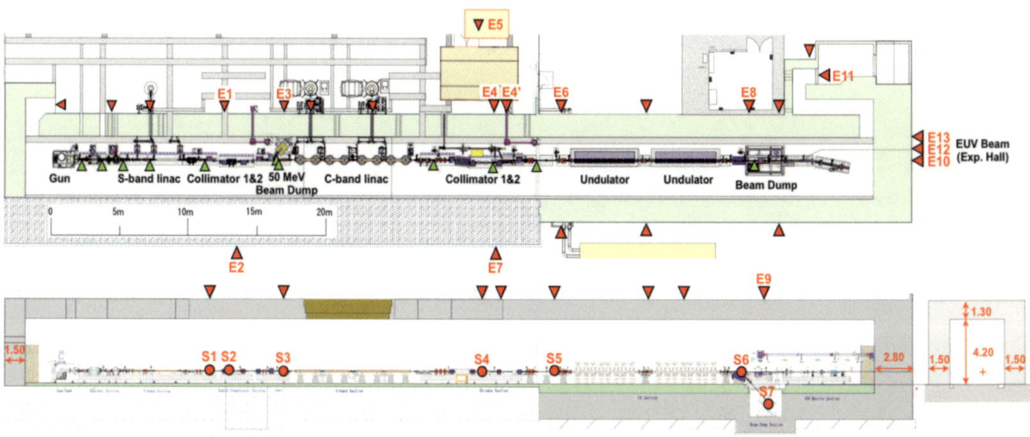

Fig. 4.1. Layout of the SCSS prototype accelerator (up, top view; down, side view; ▲, leakage dose measurement & estimation points; ▲, CT(Current Transformer); ●, beam loss point).

The SCSS prototype has two beam dumps (one for 50 MeV electrons and the other for 250 MeV electrons), two bunch compressors with local shields, and two in-vacuum type undulators, as shown in Fig. 1. The two dumps are double-cylindrical structures with graphite for the inner cylinder and iron for the outer cylinder to reduce the production of photo-neutrons. The 250 MeV dump was constructed with an inclination angle of 45 degrees, and was embedded underground. The beam-loss distributions were measured by using subtractions of the current transformers (CT) output. The CTs are installed upstream of the dumps, the undulators, and the bunch compressors, as indicated in Fig. 4-1. The FEL light beamline including the SCSS prototype should be constructed in the forward direction of the accelerated electron beam and electron beam dump. It is thus important to prevent high-energy radiation due to electron beam loss in designing the light beamline. For the SCSS prototype, the offset between the electron beam axis and the light beamline at the end of the shield tunnel wall and the distance from the scatterer (mirror) were designed at to be

40cm and 4.5 m, respectively. The thicknesses of the shield tunnel of the SCSS prototype are ordinary concrete of 1.5 m in the lateral direction, 2.8 m in the forward direction, 1.3 m for the roof.

Two area monitors are set up outside of the shield tunnel near the downstream collimator and the beam dump. The access control systems with personal keys are installed at the two entrances. These are linked to a safety interlock system.

4.1 Shielding design

The shielding design of the SCSS prototype was achieved by using analytical methods and the FLUKA Monte Carlo code (35). The leakage dose distributions outside the shield tunnel strongly depend on the electron beam-loss distribution. It is thus important to estimate the beam loss exactly, which is the first step of shielding design. The measurement results of the electron beam-loss distributions during lasing (user time) using CTs are shown in Fig.4.1., including typical cases of the beam tuning operation and laser-seeding experiments (36). As shown in the figure, the beam losses occurred up to the first collimators (bunch compressor section) for all cases. During the user time, the electron beam loss is very low at the undulator section. On the other hand, the beam loss shows higher loss rates up to about 5 % in the undulator section during beam tuning and seeding experiments. Table 4-1 summarizes the measurement data and the estimated values for the shielding design. The design values were decided based on the analyses of the beam diagnoses and discussions between radiation safety physicists and accelerator physicists. These data show good agreement.

Source No.	Source Position	Beam loss Assumption（%）	Beam loss measurements (User time、%）
S1	At any points from gun to upstream collimator	15	~75
S2	Collimator(upstream)	60	
S3	Beam dump (50MeV）	(100)	
S4	Collimator(downstream)	4	~3
S5	At any point from down stream collimator to dump bending magnet	1	
S6	Dump bending magnet	1	
S7	Beam dump(250MeV)	19(100)	~22

Table 4.1. Electron beam-loss distribution at SCSS (design and measurements; source No.S1 through S7 are indicated in Fig. 4.1.)

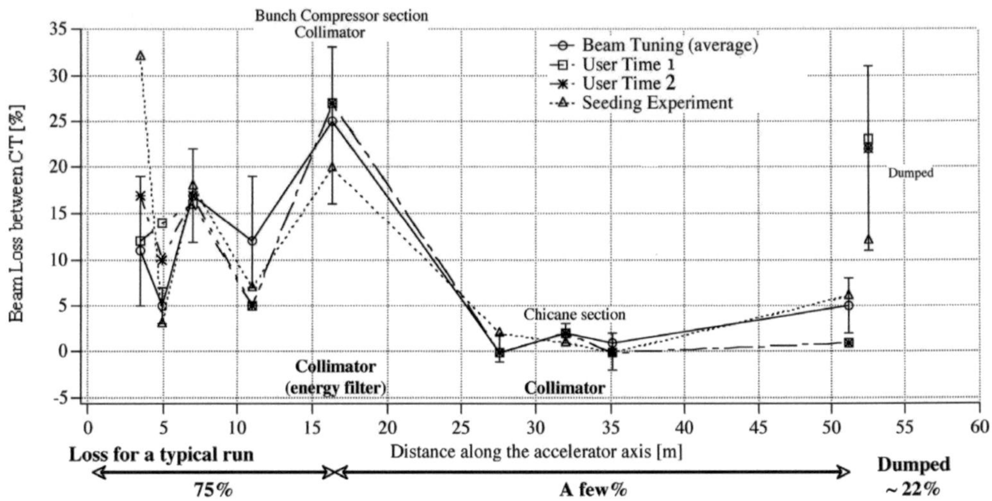

Fig. 4.2. Measurement results of the electron beam loss during three modes with 250 MeV operation.

Typical results of calculations using the analytical methods are indicated in Table 4.2. In these calculations, source reduction factors, f_1 and f_2 for 250 MeV electrons are set to be 0.09 and 0.22, respectively.

Estimate Point	Source Point	J (e/s)	Ee (GeV)	r (m)	Shield Thickness (cm)		Radiation	Dose (μSv/h)	Total dose (μSv/h)	Formula No.
E4	S4	7.5×10^9	0.25	3.2	O.C.	150	γ (X)	0.056	0.11	20
					Fe	8	N	0.053		22
E8	S6	1.9×10^9	0.25	3.2	O.C.	150	γ (X)	0.082	0.24	20
							N	0.020		22
	S7	1.9×10^{11}	0.25	3.2	O.C.	240*	γ (X)	0.016		20
					Fe	25	N	0.12		22
E10	S6	1.9×10^9	0.25	12	O.C.	280	γ (X)	0.19	0.19	15
							N	0.0026		17,18
	S7	1.9×10^{11}	0.25	10	O.C.	396*	γ (X)	9.0×10^{-5}		20
					Fe	25	N	6.9×10^{-4}		22

*Effective thickness

(Estimation points and source points indicate in Fig.4-1)

Table 4.2. Shielding design calculations using analytically empirical methods, for example.

SCSS prototype has two entrances as shown in Fig.4.1. Figure 4.3 shows the simulation results of the neutron dose distributions at two entrances due to electron injection into the dumps (A, electron energy of 50 MeV and source points S3; B, electron energy of 250 MeV and Source point S7) by using FULKA.

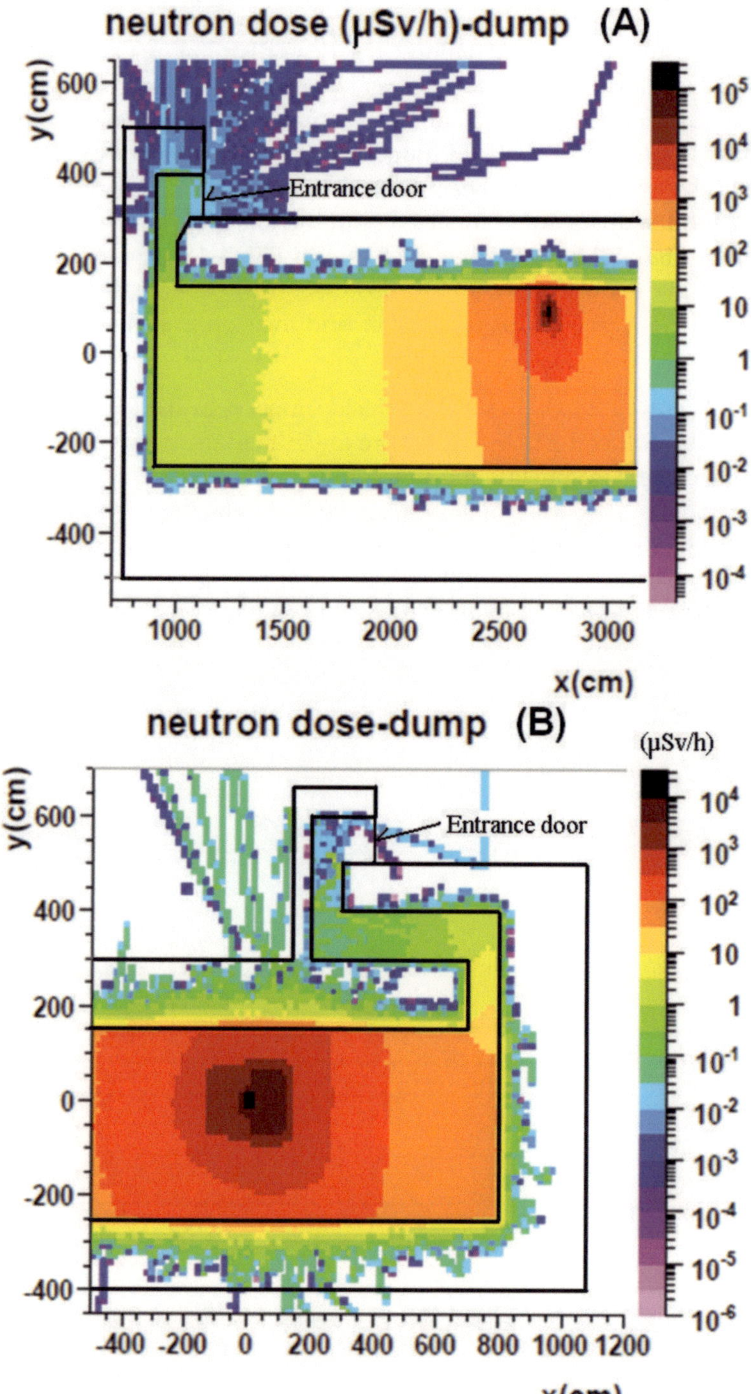

(A: 50 MeV dump(S3), B:250MeV dump(S7))

Fig. 4.3. Neutron dose distribution at the entrance of the shield tunnel of the SCSS prototype.

4.2 Dose measurements

To compare the results of the calculations, the dose distributions around the accelerator were measured in relation to the beam losses by using Gafchromic films and LiF thermo luminescent dosimeters. Andersson-Braun type rem counters and ion Chambers were used for measurements outside of the shield tunnel. Figure 4-4 shows, for example, the gamma-ray dose rate at the exit of the FEL beam from the tunnel (inner side of E12 as indicated in Fig.4-1). In Fig.4-4, the doses are normalized with the total electron charge measured at the first CT monitor. Figure 4-5 shows the dose rate in Fig. 4-4 as a function of the electron beam-loss rate at the undulator section. In this case, the doses are linearly proportional to the amount beam loss at the undulator section, and the slop is about 0.6 mGy/mC percent of the beam-loss rate.

The gamma and neutron doses outside the shield tunnel near the FEL beam transport line, as indicated by E10, E12 and E13 in Fig. 4.1 are lower than the detection level. At the SCSS, the maximum leakage dose outside the shield tunnel was about 2.0 μSv/h of the gamma rays (controlled area) for 50 MeV operation, and 0.5μSv/h of gamma rays and 0.15μSv/h of the neutrons (non-controlled area) for 250 MeV operation. The synchrotron radiation of the spontaneous emission light and FEL cannot penetrate the beam-transport pipe because of low energy. These leakage doses agree reasonably well with the calculations.

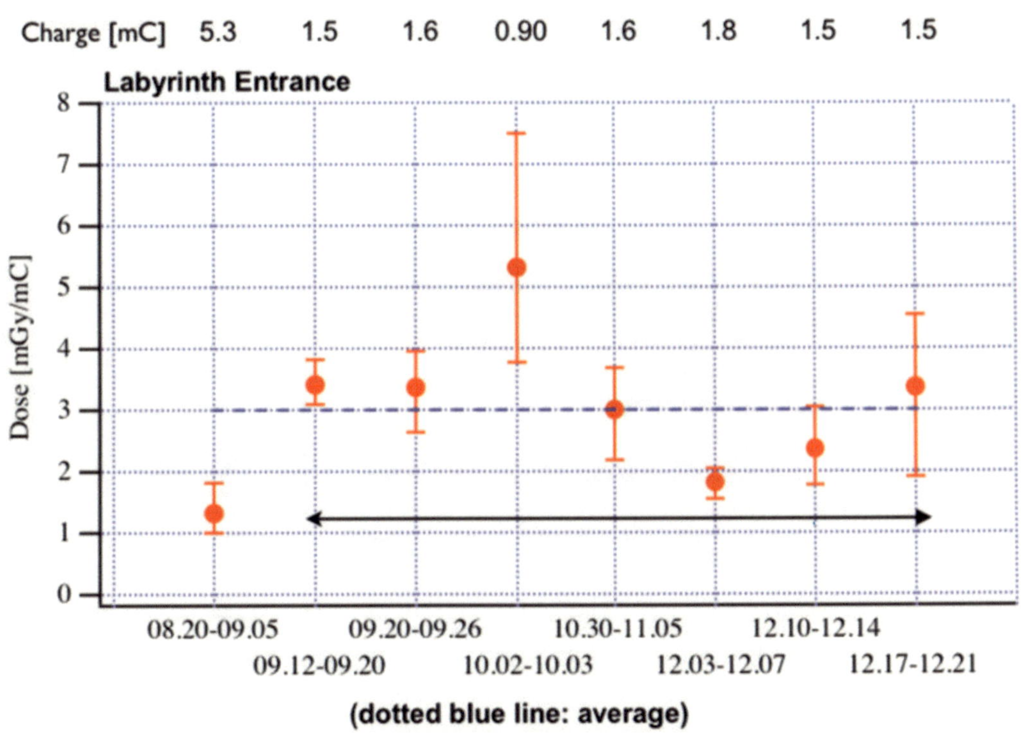

Fig. 4.4. Gamma-ray dose at the exit of the FEL beam from the tunnel (inner side of E12 as indicated Fig. 4.1.). The dose is normalized by the total electron charge at the first CT.

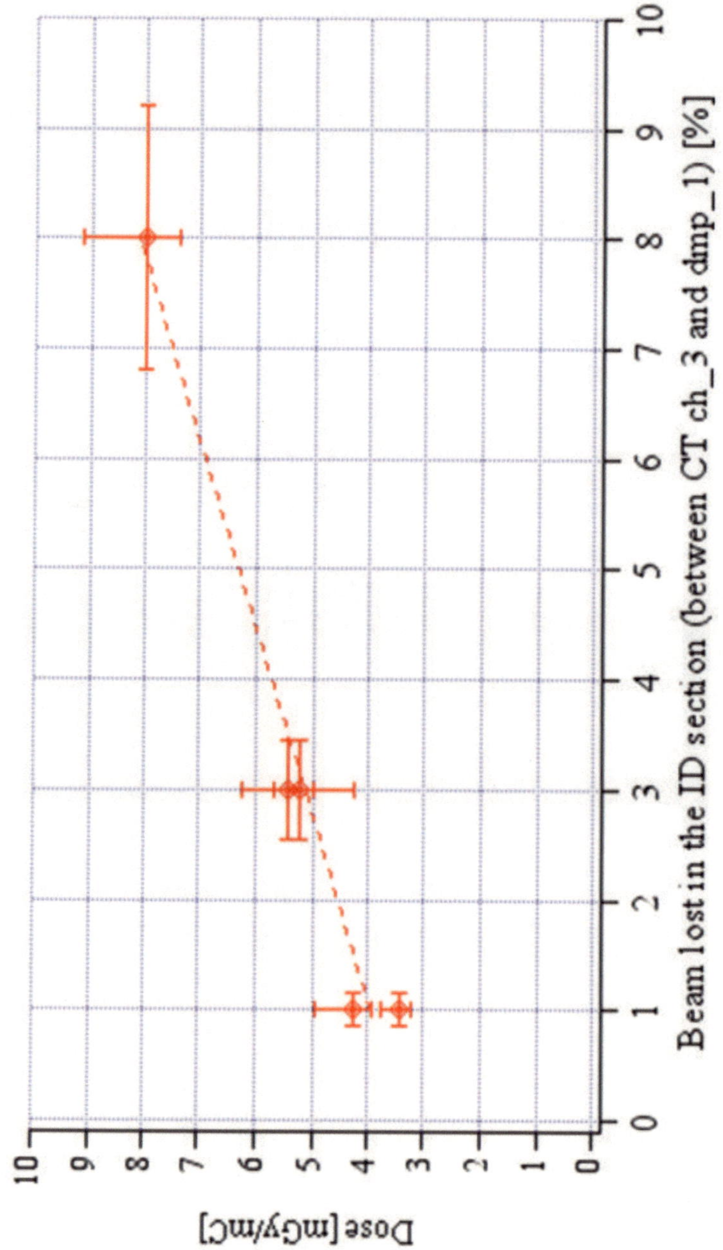

Fig. 4.5. Gamma-ray dose distribution at the FEL exit within the shield tunnel as a function of the ratio of the beam loss in the undulator section (doses are normalized by the total charge at the first CT)

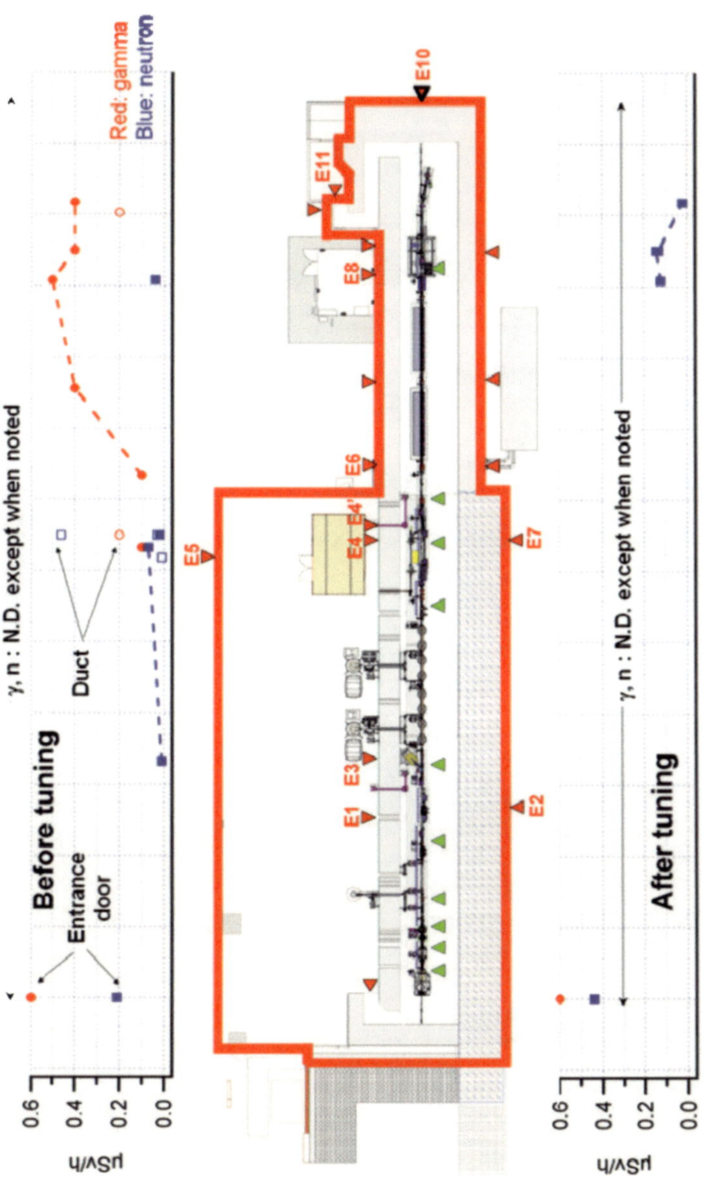

Fig. 4.6. Measurement results of the dose distribution outside the shield tunnel (250MeV operation). The red line indicates the boundary of the radiation controlled area.

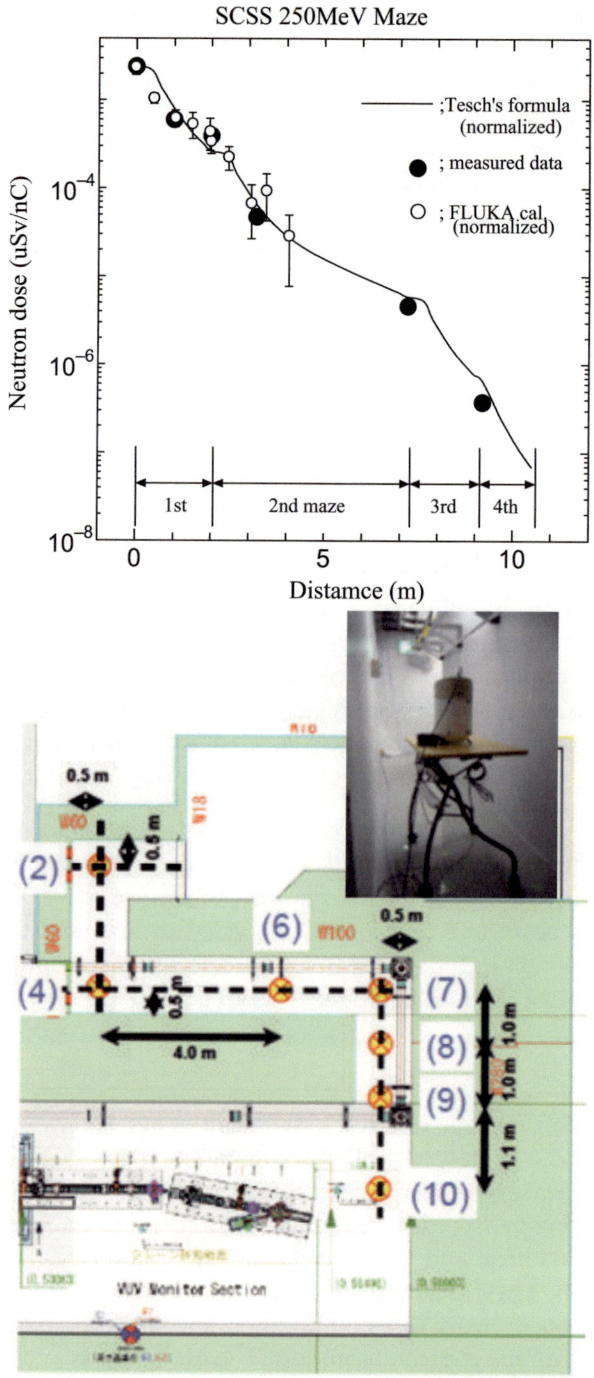

Fig. 4.7. (A) Neutron dose distribution within the maze in comparison with the calculation results of the Tesch' formula and FLUKA. (B) Neutron measurement points within the downstream maze including the photo of the Andersson Braun-type rem counter.

Figure 4.6 shows the typical dose distributions outside the shield tunnel before (beam tuning) and after (user time) the tuning operation. In Fig. 4.6, E1 through E11 are the measurement points, as shown in Fig. 4.1, and E1 through E5, except E2, are in the radiation controlled area, and others are at the boundary of the controlled area. In these cases, the characteristics of the dose distribution due to beam tuning are higher gamma doses near the undulators and 250 MeV dump (near E8) in comparison with that after the beam tuning. On the other hand, the neutron dose at the entrance door and near the 250 MeV dump are higher than that before the beam tuning. The higher neutron dose at the entrance door after the beam tuning seems to be due to harder tuning in comparison with before beam tuning at the first stage of acceleration. At E4 and E5 near the second (last) collimator, doses before beam tuning are higher than that after the beam tuning. These phenomena agree with the electrons accelerating to the last collimators without tuning and being scraped out by it. The gamma-ray dose at the upstream entrance door is due to scattered X-rays from the thermionic gun of 500 keV high voltage with over 1A current and the cut of the corner of the maze to carry on some instruments easily, as shown in Fig. 4.3(A).

The measured data of neutron doses within the downstream maze are shown in Fig.4.7(A), including the calculation results using Tesch's formula (29) and the FLUKA Monte Carlo code. Fig. 4.7(B) shows the measurement points in the Maze. In the figure, the neutron doses are normalized with the charge of electrons injected into the dump. Both calculations are normalized at the maze entrance in the shield tunnel (0 m in Fig.4.7, and (9) in Fig 4.7 (B)). Both calculations, Tesch's formula and FLUKA, agree well with the measurement results so that Tesch's formula can reproduce the photo-neutron dose distribution within the maze due to the 250 MeV electrons.

According to planning and expectation, the dose at the light beamline outside the shield tunnel (E12 in Fig. 4.1) was under the detection level because of a sufficient offset of 40 cm between the electron beam axis and the FEL light beam line, and the low energy of the spontaneous emission light including FEL. There are no significant induced activities, except for short life nuclei, at the machine components until now.

5. Summary

Radiation safety systems for Free Electron Laser facilities are overviewed. Practical cases are presented using the SCSS prototype facility in comparison with the shielding design and the measurement data of leakage doses. Three dose limits must be considered normally, one for radiological worker occupancy, including miss-steering condition, one for the boundary of the radiation controlled area, and the other for the site boundary. The shielding is design therefore required to be lower than the three criteria sufficiently, and the criterion of the site boundary is more sever in many cases. In addition, the shielding design and the radiation safety system depend on the electron beam loss scenario directly, so that intense scrutiny of the beam-loss scenario is necessary, and close communications with the radiation safety physicists and the accelerator physicists is important. Others are thermal and laser problems. Both problems will be become more significant with increasing the power of the free electron laser.

6. References

[1] ICRP publ. 60 " Recommendations of the International Commission on Radiological Protection, adopted by the commission on Nov. 1990". Annuals of ICRP, Permagon Press Oxford.(1991)

[2] ICRP publ. 103 " Recommendations of the International Commission on Radiological Protection,". Annuals of ICRP Vol.37, Permagon Press Oxford.(2007)

[3] ICRP publ. 74 " Conversion Coefficients for use in Radiological Protection against External Radiation", Annals of the ICRP Vol.26 (1996)

[4] S.Roesler & G.R.Stevenson " Deq99.f-A FLUKA user routine converting fluence into effective dose and ambient dose equivalent ", Technical Note CERN-SC-2006070-RP-TN EDMS No.09389 (2006)

[5] ICRU, "Quantities and Units in radiation protection dosimetry," Report 51 ICRU Publications (1997)

[6] J.Liu et al., "Comparison of design and practices for radiation safety among five synchrotron radiation facilities", Radiation Measurements 41 (2007)

[7] Y.Asano, "Characteristics of radiation safety for synchrotron radiation and X-ray free electron laser facilities", Radiation Dosimetry 146 No.1-3 (2011)

[8] H.A.Baldis, E.M.Campbell, and W.L.Kruer," Handbook of Plasma Physics", Elsevier Science Publisher, (1991)

[9] S.C.Wilks and W.L.Kruer, IEEE J.Quantum Electron 33 (1997)

[10] S.J.Gitmore et al., Fast ions and hot electrons in the laser-plasma interaction. Phy.Fluids 29, (1986)

[11] W.P.Swanson, "Radiological Safety Aspects of the operation of electron linear accelerators", IAEA technical Reports No.188 (1979)

[12] H.C.Dehne et al., Nuclear Instruments and Methods 116 p345 (1974)

[13] G.F.Knoll," Radiation Detection and Measurements,3rd edition", John Willy & Sons,Inc (New York ISBN 0-471-07336-5 (1999)

[14] H.Yasuda & T.Ishida," Time resolved photoluminescence from a phosphate Glass irradiated with heavy ions and gamma rays", Health Physics 84, No.3 (2003)

[15] A.Klett, A.Leushner, " A Pulsed neutron Monitor", ,IEEE 2007 Nuclear Science Symposium and Medical Imaging Conference, Honolulu USA (2007)

[16] F. d'Errico, "Radiation dosimetry and spectrometry with superheated emulsions", Nuclear Instr.& Methods B184 (2001)

[17] H.Hirayama, Y.Namito, A.F.Bielajew, S.J.Wilderman, and W.R.Nelson.," The EGS5 Code system", SLAC-R-730, KEK-Report 2005-8 (2005)

[18] F.Salvat, J.M.Fernandes-Varea,E.Acosta, and J.Sempau,"PENELOPE, A code system for Monte Carlo Simulation of Electron and Photon Transport", NEA/NSC/DOC (2001)19 ISBN:92-64-18475-9

[19] D.B.Pelowitz et al., "MCNPX 2.7.0 Extensions", LA-UR-11-02295 (2011)

[20] A.Ferrari, P.R.Sala, A.Fasso, and J.Ranft," FLUKA; a multi-particle transport code",CERN 2005-10(2005), INFN/TC_05/11, SLAC-R-773

[21] N.V.Mokhov & S.T.Striganov," MARS15 Overview", Fermilab-conf-07/008-AD (2007)

[22] S.Agostinelli et al., GEANT4 – a simulation tool kit", Nuclear Instruments & Methods A 506 250 (2003)

[23] K.Niita et al., " PHITS; Particle and Heavy Ion Transport code System V2.23 JAEA-Data/Code 2010-022 (2010)

[24] Y.Asano and N.Sasamoto,"Development of shielding design code for synchrotron radiation beamline", Radiation Physics & Chemistry 44(1/2) 133 (1994)

[25] T.Tanaka,"FEL simulation code for undulator performance emission" Int.Proceeding FEL2004 Trieste, (2004)

[26] Y.Hayashi,et al.," Estimation of photon dose generated by a short pulse high power laser", Radiation Protection Dosimetry 121 (2) 99 (2006)

[27] W.R.Nelson and T.M.Jenkins," The SHIELD11 Computer code", SLAC-Report 737 (2005)

[28] T.M.Jenkins,"Neutron and photon measurements through concrete from a 15 GeV electron beam on target-Comparison with models and calculations", Nucl.Instr.& Methods 159 265 (1979)

[29] H.Hirayama, and S.Ban," Neutron dose equivalent outside the lateral shielding of an electron linear accelerator oprating at 0.85GeV", Health Physics 56 6 (1989)

[30] A.Ferrari, M.pellicioni, and P.R.Sala, "Estimation of fluence rate and absorbed dose rate due to gas bremsstrahlung from electron storage rins", Nucl. Instrum. Methods B83 518 (1993)

[31] J.C.Liu, W.R. Nelson, and K.R.Kase, "Gas Bremsstrahlung and Associated Photo-neutron Shielding Calculations for Electron Storage Ring, SLAC-Pub. 6532 (1994)

[32] K.Tesch, "Attenuation of the PHOTON Dose in Labyrinths and Duct at Accelerators", Radiation Protection Dosimetry 20 169 (1987)

[33] K.Tesch " The Attenuation of the Neutron Dose Equivalent in a Labyrinth through an Accelerator Shield", Particle Accelerators 12 169 (1982)

[34] T.Shintake et al., " Status of SCSS; SPRING-8 Compact SASE Source Project", Proc. 8th European Particle Accelerator Conf. Paris 2002, European Organization for Nuclear Research (2002)

[35] Y.Asano, T.Itoga, and X.Marechal," Radiation Shielding Aspects of the SCSS prototype XFEL facility", Nuclear Technology 168 (2009)

[36] G.Lambert et al., "Injection of harmonics generated in gas in a Free-Electron Laser providing intense and coherent extreme-ultraviolet light", Nature Phys., 4 296 (2008)